J. Ian Collins
10 St. James Park
Beaconsfield,
Quebec H9W 5K9

OLD MAPS AND GLOBES

OLD MAPS AND GLOBES

with a list of cartographers, engravers, publishers
and printers concerned with printed maps
and globes from *c.* 1500 to *c.* 1850

RAYMOND LISTER

BELL & HYMAN
LONDON

Published by
BELL & HYMAN LIMITED
Denmark House
37–39 Queen Elizabeth Street
London SE1 2QB

First published in 1965 by
G. Bell & Sons Ltd
as *How to Identify Old Maps and Globes*

Reprinted 1965, 1970
Revised edition 1979

ISBN 0 7135 1146 X
Printed and bound in Great Britain at
The Camelot Press Ltd, Southampton

THIS BOOK
IS DEDICATED TO
ARTHUR K. ASTBURY
AS A TOKEN OF
A LONG AND HAPPY FRIENDSHIP

Acknowledgements

I AM grateful to all those who have helped and encouraged me in writing this book and in the somewhat exhausting task of compiling its *List of Cartographers, Engravers, Publishers and Printers*. In particular I should like to record my indebtedness to Miss Helen Wallis of the British Museum for much advice and many helpful suggestions, and to Mr. G. R. Crone, the Librarian of the Royal Geographical Society, who read the typescript of the chapters and bibliography.

As so often in the past, I have had much help from my friend and fellow author, Mr. A. K. Astbury, from the Librarians of the Royal Society of Arts and the London Library, and from Miss M. M. Sutherland of the University Typewriting Office, Cambridge.

I should also like to express my thanks to Mr. L. G. G. Ramsey, Editor of the *Connoisseur*, Mr. John Munday of the National Maritime Museum, Mr. John F. Maggs, Mr. P. G. M. Dickinson, and Dr. D. W. Dewhurst of the University of Cambridge Observatories.

I am grateful, too, as always, to my wife Pamela, for giving me a quiet haven in the midst of a busy life, without which I could not write.

Cambridge   R. L.

Contents

Contents

The Plates

which are arranged in approximately the same order
as the first references to them in the text

50. Detail from the map of Switzerland from Ptolemy's *Geographia* (1513)

51. Detail showing lettering from the map of Gt. Britain in Münster's edition of Ptolemy (Basle, 1540)

52. Detail from Emanuel Bowen's map of Huntingdonshire (1749)

53. Portion of 'Quarta Tabula Europae', showing western Germany, from Ptolemy's *Geographia*, published at Rome in 1478

54. Detail from map in Cassini's *Atlas Topographique des Environs de Paris* (Paris, 1768)

55. Detail from William Yates's map of Staffordshire (1775)

56. Detail, showing symbols for towns, etc., from Saxton's map of Norfolk (1579)

57. Plan of Westminster. From the 1723 edition of John Norden's *Description of Middlesex and Hartfordshire* engraved by Senex

58. Group of Terrestrial globes. Left to right:
    (i) Abbé Nollett (French), 1728: 27·9 cm diameter. Painted and lacquered pedestal
    (ii) Lapié-Langlois (French), 1815: 22·9 cm diameter. Stand supported by black and gold caryatides. The base has brass paw feet, and has a compass fitted
    (iii) Malby (English), 1845: 30·5 cm diameter
    (iv) Klinger (German), 1792: 30·5 cm diameter
    (v) N. Hill (English), *c.* 1750: 12·7 cm diameter. Fitted with an hour-glass
    (vi) Newton (English), 1836: 45·7 cm diameter. Fitted with a compass

    All have meridian circles or segments (vertical); all but no. iii have horizon circles (horizontal).

59. Horizon ring of terrestrial globe by Blaeu, 1602

CHAPTER ONE

An Outline of the History
of Maps and Charts

THE history of geography and cartography is an essential part of map identification, for it is from familiarity with such things as fashion-changes and development of scientific knowledge that maps, charts, plans and globes may often roughly be assigned to their correct periods. As we shall see, this is not an infallible rule, but at least it is a beginning. In the first place, therefore, I will broadly survey the history of the cartographer's craft, indicating as far as it is possible in the limited space available, the main points of its development and evolution.

The word 'map', which was first used, according to the *Oxford English Dictionary*, in 1527, is derived from the Latin *mappa mundi*, or world-map, derived in its turn from the Latin *mappa*, a napkin or towel, and *mundi*, of the world. The word 'cartography', the craft of map-making or the study of maps, is derived from two Greek words, χάριης, a leaf of papyrus or paper, plus γραφειν, to write.

Webster's Third New International Dictionary defines a map thus: 'A drawing or other representation that is usually made on a flat surface and that shows the whole or a part of an area (as of the surface of the earth or some other planet or of the moon) and indicates the nature and relative position and size according to a chosen scale or projection of selected features or details (as countries, cities, bodies of water, mountains, deserts).' The desire to make such 'drawings or other representations' as these is as old as the social history of mankind. Even primitive peoples have produced rough but tolerably accurate maps – still do, in fact, for the Eskimos and Bedouins make and use them.

Ancient Egypt, Babylon, Greece and Rome all made contributions to the development of cartography, although much of what was done in those remote times was vastly different from the elegant and colourful products of later times. The contribution of Ancient Egypt, for example, was in the nature of land-surveying rather than mapping proper, and was necessitated by the re-defining of boundaries after the inundations of the Nile, and for mining.

Although maps of a kind, on clay tablets, have survived from ancient Assyria, it was really the Greeks who made the first scientific maps – indeed, it was Herodotus who made the first written reference to a map. The Greeks made wide use of maps and itineraries, both in war and trade. It was in Greece that the theory that the earth is a sphere was first propounded – by followers of Pythagoras (it was usually, both in those days and later represented as a disc). The accuracy of Greek cosmography may be gauged by the fact that a Greek astronomer, Eratosthenes, worked out the circumference of the earth to within fifty miles of the correct figure.

In the first half of the second century A.D. lived and worked one of the most influential figures in the history of cartography. This was Claudius Ptolemy, an Egyptian astronomer and geographer of Alexandria, author of several works, but noted chiefly for his *Geography*, a work in eight books, illustrated by maps. It is to this follower of Hipparchus, the founder of scientific astronomy and discoverer of the precession of the equinoxes, that we owe most of our knowledge of ancient astronomy; in his geographical work, Ptolemy was a follower of Marinus of Tyre, and again it is due to him that we know of Marinus's work to-day.

Altogether Ptolemy's *Geography* contains twenty-six regional maps, and a *mappa mundi*. The geographical work attributed to this remarkable man remained for centuries the standard among Arab geographers, although he had little influence on western geography until the fifteenth century. There now exists no manuscript map based on Ptolemy earlier than the twelfth century, so it is not absolutely certain whether tradition has merely assigned the maps to him, on the presumption of his reputation as a geographer, or whether he was himself actually a map-maker. It is one of those questions over which scholars will argue for evermore; the important thing to us is that a certain tradition has grown up around Ptolemy's name and that certain maps drawn to certain projections have been traditionally ascribed to his invention.

In addition to the twenty-six regional maps and the *mappa mundi* just mentioned (known as the 'A' recension) there is another group of sixty-seven maps of smaller portions of the earth's surface (known as the 'B' recension). The first group was the one used in the fifteenth century Latin manuscript editions of the *Geography*. Apart from his *mappa mundi*, Ptolemy's maps are of rectangular conception, that is with meridians and parallels intersecting at right angles; the *mappa mundi* is, however, drawn on a simple conic projection as shown on Plate 2.[1]

Ptolemy's cartography, although remarkable for the times in which it was conceived, was far from accurate according to modern standards. For instance, he exaggerated the breadth of Europe and Asia and thereby distorted whole areas. He completely missed the inverted triangular shape of India, and made Ceylon (which he called Taprobana) a vast island, nearly as big as France and Spain added together. The Indian Ocean is displayed as a land-locked sea. But many of these errors were due to the fact that exploration was in its infancy. What many of the old cartographers did not know was filled in by guesswork – a tendency still apparent as recently as the

[1] Briefly, the term 'projection' indicates the reduction to a plane surface of the whole or part of the earth's spherical surface. There are various ways in which this may be accomplished, all of which involve the use of intersecting and co-ordinated lines. A simple explanation of the various methods is given in *A Key to Maps* by H. S. L. Winterbotham (London, 1947), Chapter IV, and a more detailed study may be found in *Map Projections*, by A. R. Hinks (Cambridge).

nineteenth century. Ptolemy's maps, although not entirely free from it, at least showed some awareness of actuality.

The Roman contribution (if such it may be called) to the development of cartography took such practical forms as road maps, although there were other manifestations of the science such as a famous map, *Orbis terrarum*, made by Augustus's son-in-law, M. Vipsanius Agrippa, which was referred to by Pliny in his *Natural History*. A form of map popular in later Roman times was the T-O map, so called from the fact that the three continents were symbolically shown by a T within an O (Fig. 1). In this

Fig. 1. The first printed T-O map from a work from Sr. Isidore of Seville, printed in 1482

the east was placed at the top of the figure, the outline of the O represented the boundary of the world, while the perpendicular leg of the T represented the centre-line of the Mediterranean and its horizontal arm the meridian stretching from the Nile to the Don. This type of map was popular, too, in mediaeval times, and its form was sometimes varied by having a Y instead of a T within the O, or a V within a square. These maps, however, are purely symbolic and of no practical geographical significance.

We must not overlook the Arab share in the development of cartography which was, however, scholastic rather than practical. Arabian maps, like the T-O maps (they were indeed often of this form), were symbolic and diagrammatic. The Arabs made far greater contributions to celestial than to terrestrial cartography (see Chapter Five).

Of mediaeval maps, an early and magnificent example is the remarkable circular *mappa mundi* in Hereford Cathedral, of *circa* A.D. 1300; it was probably made for Richard of Haldingham. It is thought to be a copy, near or distant, of a Roman original, but it has obviously been conditioned by Christian influence, for Jerusalem is shown at the world's centre. As in the T-O maps, the east is at its top. It is elaborately decorated with all kinds of drawings, including representations of the Last Judgement and Paradise. The British Isles are squashed into the perimeter and bear hardly any resemblance to their actual shape, and Scotland is apparently shown as an island. But mediaeval geography, like much that preceded it, was poetic rather than scientific.

More in keeping with maps as we know them to-day is the famous map of Great Britain by the Benedictine, Matthew Paris of St. Albans Abbey, made about 1250, and therefore somewhat earlier than the Hereford map, and now in the British Museum (MS. Cotton Claud. D.VI). It is on parchment and shows the rudiments of several comparatively modern conventions in map-making. The sea, for instance, is painted green, and the towns are represented by little buildings or groups of buildings. It is exaggerated and distorted, but on the whole it is a tolerable representation of this country. It shows features counted as important in mediaeval times, and from this point of view alone is of capital interest. Thus towns with monasteries and abbeys are given great prominence – something like nine-tenths of the towns marked on the map have such connexions. The towns are shown for the most part as little battle-mented buildings and in other cases as religious buildings (Plate 4), in some cases, like St. Albans, surmounted by crosses. Each town has a red line drawn around it, and it has been suggested that this represents the tiled roofs then taking the place of thatch in order to prevent fires from spreading. Roads are not shown, but rivers are given great prominence. It is thought that Matthew Paris deliberately distorted the shape of the whole so as to accentuate the route from Doncaster to Dover, which port was the usual point of embarkation for pilgrims bound for Rome and elsewhere.

Another important manuscript map of the British Isles is that known as the Gough map of *circa* 1360, now in the Bodleian Library at Oxford. It is even closer to modern conceptions than the Matthew Paris map. It measures 108 × 48·9 cm. Nobody knows who made it; its name is derived from the fact that it was mentioned by Richard Gough in his *British Topography*, published in 1780.[1] It shows roads, which are coloured red, and uses various symbols to indicate such things as towns, walled and unwalled, and monasteries.

By the thirteenth century attempts were being made by European cartographers to include Asia in world maps; this development may be seen, too, in a group of *mappa mundi* made by cartographers, mainly Majorcan, working in Catalonia. These cartographers were responsible for making the celebrated Catalan Atlas (before 1375). This is preserved in Paris, where it has been since it was presented to an envoy of the King of France by King Peter of Aragon. It is said to have been the work of a Jew of Palma, Abraham Cresques, 'master of *mappae mundi* and of compasses' to the King of Aragon.

The Catalan Atlas is a large *mappa mundi* on twelve leaves mounted on folding boards, like a screen; of these sheets, eight form the map itself, and the remaining four contain geographical information. The map is of elongated proportions, 69 centimetres high and 3·9 metres long, which means that extreme northern and southern regions are not included. It therefore abandons the circular form of earlier *mappae mundi*. Its main interest, however, lies in the fact that it is the first instance on a mediaeval map of Asia being represented in something like its real form, although the conception is still far from the actuality. It is, for instance, claimed that there are 7,548 'spice islands' off the coast of Cathay, and again the enormous island of Taprobana is shown as it appears on the map of Ptolemy.

Another celebrated Catalan *mappa mundi*, a circular one on parchment, and made *circa* 1450, is in the Biblioteca Estense at Modena, from which it is known as the Este map.

[1] It had first appeared as *Anecdotes of British Topography* in 1768.

The Catalan cartographers made a scientific approach to cartography. They correlated existing material and used only information that was supported by evidence. If some of the evidence proved to be faulty, the fault was not theirs, for they seem to have been well above the contemporary level in these matters.

The fifteenth century witnessed several interesting developments, some of which are illustrated by the *mappa mundi* of Fra Mauro, a monk from the island of Murano in the Venetian lagoon. Mauro's map was executed for the King of Portugal and finished in 1459; the monk had the assistance of a number of illuminators and a draughtsman and chart-maker named Andrea Bianco, who had worked on an earlier *mappa mundi*. A contemporary copy of it is preserved in the Biblioteca Marciana in Venice. Much of Fra Mauro's conception appears to be based on Marco Polo's journeys, for, although far from accurate, his delineation of China shows that he had some real knowledge of it. Incidentally, much of the detail in the Catalan Atlas appears also to have come from Marco Polo.

The name of Sumatra appears for the first time ever on Fra Mauro's map, and there are interesting details of Java and the trade in pepper and spices in that area. Here, too, for the first time, Japan appears on an occidental map. To the north of Java, Fra Mauro shows a small island which he calls Isola de Zimpangu. Japan was known in early times as Cipangu and the close resemblance between these two names makes it seem almost certain that this is what Mauro had attempted to represent. In addition Africa received more detailed treatment than in earlier maps and it is thought that Mauro got much of his information for this from members of the Abyssinian Coptic Church who had visited Venice. Yet another point of interest is that Jerusalem is at last abandoned as the world's centre, and with it was also finally abandoned the circular form of *mappa mundi*.

Side by side with maps of *terra firma* had been growing the production of sea-charts, which assumed an ever greater importance as the world explorations of the day ranged over wider and wider areas. At first mainly the work of Italians and Catalans,[1] these charts began to make their appearance in the second half of the thirteenth century and were from the first notable for their greater accuracy than terrestrial maps, this being largely due to the fact that they were made from data taken by direct observations, in which the newly invented mariner's compass was used as a basis. These sea-charts are often called portolans, but the name portolan or portolano is not strictly correct when used thus; originally it applied only to a harbour book or written sailing instructions, and not to an actual chart. But the name has willy-nilly come to be applied to a chart as well.

In England the portolan became known as a 'ruttier' or 'rutter of the sea' (from 'route'). Actual examples of portolans are of great rarity; probably, on the estimate of one authority, there have survived under a score for each century of the Middle Ages. This is a typical set of instructions from a fifteenth-century portolan:

From Sallo to Barcelona is 60 miles E.N.E. $\frac{1}{4}$ E. Barcelona is a city with a beach facing east and has a roadstead with a depth of 22 paces in front of the city. To the S.E. to S. from

[1] The Catalan Atlas, mentioned a few paragraphs back, was partly chart.

Barcelona is a low place called Lobregato.   On going out sail eastwards from the shore and watch for a castle which rises up out of a valley which leads to Sallo.

The inward mark for Barcelona is a high steep and solitary hill called Monserrat.   When you are N.E. of this, continue in the same direction and you will sight a low hill with a tower on it, which is called Mongich (Montguich).   Here is Barcelona.

The portolan is much older than the sea-chart, and is known to have been made as early as the tenth and eleventh centuries, with the growth of Italy's overseas trade, especially in the Levant.

To return to charts.   For the most part manuscript specimens are drawn on whole skins of parchment, some of them as large as 137·2 × 76·2 cm and more.   The most characteristic things about charts, however, are the straight intersecting lines drawn over their surfaces.   These lines, which are often drawn in alternating colours or in alternating continuous and dotted lines, are known as rhumb lines or loxodromes. Their purpose was to provide mariners with a rough indication by which they could plot the course they wished to navigate.   Thus the mariner would select the line nearest to his proposed course, and read off the directions he had to follow from the nearest of the compass roses, from the points of which the lines radiated (Plate 6). There were thirty-two lines radiating from each compass, representing the thirty-two winds of heaven.   The roses do not appear on the earliest examples, but are almost invariably present on later engraved charts.   The loxodromes were also used to show the bearing of one place to another, or, in other words, at what points of the compass they lie in relationship to one another.

Charts are normally, even in the earliest examples, provided with scales with their divisions subdivided into fifths, although, curiously, no unit of length is indicated. But it was not until the sixteenth century that they were provided with a scale of latitudes.

Of early manuscript charts may be mentioned the thirteenth-century Carte Pisane, probably of Genoese origin; the 1318 atlas of Petrus Vesconte; the 1327 chart of Perrinus Vesconte; and the chart of Angellino de Dalorto of about 1325.   They all cover comparable areas within the limits of the Mediterranean, the Black Sea, the Atlantic coast of Europe and part of the north African coast.

But we must return to Ptolemy, whose *Geography* received a great revival of interest in the fifteenth century.   A copy of his 'A' recension (see p. 16) was brought back from Constantinople in 1400 by Palla Strozzi, a Florentine, and translated partly by Manuel Chrysolorus and partly by Jacopo Angelus.   The translation was completed by 1406, and the maps redrawn with Latin legends by 1410 by Francesco di Lappaccino and Domenico de Boninsegni of Florence.   There were many subsequent manuscript versions of the *Geography*.   One, now in the National Library at Florence, and made by Henricus Martellus, shows the relief of the Alps represented by elaborately drawn shading in shades of brown and white, with green for summits.   Indeed, several of these manuscript versions of Ptolemy attempt to indicate relief by various forms of shading and colouring.   In lighter vein, a Florentine, Francesco Berlinghieri, made a rhyming version of the work (1482), although the maps that went with it were of high quality.

Ptolemy's *Geography* was first printed, without maps, in 1475, at Vicenza, and with maps in 1477, at Bologna; actually mis-dated 1462 on its title page, 1477 is usually accepted as the correct date. Another edition was printed in 1478 at Rome. The printed maps were, however, based on some dozen illuminated versions of the work, made by Dominus Nicholaus Germanus, a scholarly but shadowy figure who is known to have been in Florence and Ferrara from about the middle of the 1460's to about 1471. The method used for reproducing the maps in most editions was line-engraving, but various editions printed at Ulm, Venice, Strasbourg (see below). Lyons, St. Dié and Basle between 1482 and 1571 had woodcut maps. Another work bearing Ptolemy's name was published with woodcut maps at Cracow in 1512. It is of extreme rarity: *Introductio in Claudii Ptholemei cosmographiam* by Ioannes [Jan] Stobnicza.

The Bologna edition of the *Geography* of 1477 had its twenty-six maps drawn by the miniaturist Taddeo Crivelli (d. *circa* 1484), who had earlier (*circa* 1455) been engaged in painting the pictures for the magnificent Bible ('Bibbia Estense') of Duke Borso of Ferrara, and was now working at the court of Giovanni Bentivoglio at Bologna. They are drawn, as previously, on a conical projection, with climates and degrees of latitude and longitude shown in the margin. Although it is the first book to contain engraved maps it is not a good edition; its text is full of misprints, its maps incomplete and full of mistakes. It is thought to have been rushed through the press so as to appear before the Rome edition. It has even been stated that a workman was enticed away from the Rome printer and persuaded to reveal the methods used for printing it there.

The 1478 Rome edition, with its twenty-seven 'A' recension maps engraved on copper by Conrad Sweynheym, had its text edited by Domitus Calderinus. It is much superior to the Bologna edition. The maps are drawn on a rectangular projection, degrees of latitude and longitude are shown in the margins, and the length of the longest day is indicated. The lettering of the maps is in the 'Trajan column' style. Mountains are shown in elevation, not in plan, and look like groups of mole-hills. There was another edition in 1490, and yet another in 1507; in the last named, twenty-seven of the maps are reissues of the earlier editions, and there are six new ones of France, Italy, northern Europe, Palestine, Poland and Spain, making a total of thirty-three. A reissue was made in 1508.

The maps in the two early editions we have just been discussing are of the ancient world. The first edition of Ptolemy to contain maps of the modern world was the rhyming version of Francesco Berlinghieri of 1482, first mentioned. This in addition to the ancient maps, contains four modern ones, with names on them in their current form, and entitled respectively 'Gallia Novella' (France), 'Hispania Novella' (Spain), 'Novella Italia' (Italy) and 'Palestina Moderna' (Palestine), making a total of thirty-one. A reprint was issued in the sixteenth century. The projection is rectangular, and latitude and longitude are ignored. This is the only edition of Ptolemy printed with this type of projection, the one originally used in the earliest manuscript copies.

The next edition is interesting in that its maps are printed from woodcuts instead of from copper plates, and that it was the first edition of Ptolemy to be printed in Germany; the actual place of publication was Ulm. It was translated by J. Angelus, edited by Dominus Nicholaus Germanus. Some impressions were printed on vellum. It contained thirty-two maps on a trapezoidal projection, five of them modern; its world map was the first to show the discoveries being made by the great explorers of the time, and it is also the first map ever to be signed by an engraver: Johanne Schnitzer de Arnescheim. It is also the first edition of the *Geography* to contain a map showing Greenland. A later edition was published at Ulm in 1486, additions having been made to at least one of the maps (that of Germany).

An edition printed at Venice in 1511 and edited by Bernard Sylvanus had twenty-seven woodcut maps printed in red and black and with a world map drawn in a heart-shaped projection. One of the finest of all editions was published at Strasbourg in 1513. It, too, had woodcut maps, forty-seven in all. It had been commenced by Martin Waldseemüller, and was eventually published by Jacobus Eszler and Georgius Ubelin. Most of the blocks were re-used in Strasbourg editions of 1520, 1522, and 1525, and after that they were re-used in editions published at Lyons in 1535 and at St. Dié in 1541.

The great voyages of discovery in the fifteenth and sixteenth centuries added much to the progress of cartography; indeed, its development by this time had become more international in character. Whereas hitherto most of the original research had been made by Italian and Spanish geographers, those of other nations now began to make their contribution, and maps made or influenced by Dutch, English, French and Portuguese cartographers became more numerous. One particular and important advance resulted from the acquisition by the Portuguese of a large Javanese map which showed not only much of the New World and the world to the east of Africa as far as China, but included also details of Siam, Java and Brazil, and descriptions of Chinese navigation.

Many of these new discoveries were incorporated in the charts of the Portuguese cartographer, Diego Ribeiro who worked in Spain from about 1519 until his death in 1533. A chart he made in 1529, now preserved in Rome, depicts with considerable accuracy the whole of the world between the Arctic and Antarctic circles.

The science of navigation, too, became more refined and systematic, and by the opening years of the sixteenth century scales of latitude began for the first time to appear on sea-charts. Very few sea-charts from this period have survived, however, this being largely due to the fact that they were working documents which became worn out by continual use.

Yet many of the discoveries made at this time were slow to make their appearance on maps – the New World was not featured in a Ptolemy atlas until 1507, although it had been featured on separate charts a few years previously. At first cartographers were reluctant to alter their whole conception of the world map, and tried hard to fit in the new discoveries to existing outlines. However, attempts were at last made to combine the new discoveries with Ptolemy's conception, and finally truth prevailed with the firm establishment of the more scientific maps.

The first of such maps was engraved on copper by Francesco Roselli in 1506, after the design of Giovanni Matteo Contarini. The only surviving impression is in the British Library. It is drawn on a conical projection, looking down, as it were, with the North Pole as the focal point. It is possible that another such map was made of the southern hemisphere. It mainly follows Ptolemy for the delineation of the Middle and Far East, with corrections in the light of the Portuguese discoveries. It marked the beginning of the first stage of modern cartography, a stage that was completed with the publication in 1507 and 1516 respectively of Martin Waldseemüller's *mappa mundi* and *carta marina*.

Waldseemüller, the man who suggested the name of America for the New World, after Amerigo Vespucci, was a German, born about 1470 at Radolfzell, and educated at Freiburg. He settled in Lorraine at St. Dié, where he worked on an edition of Ptolemy (see p. 22). His world map was a woodcut made in twelve blocks in an edition of 1,000, and published in 1507. In its full title it declares its debt to both Ptolemy and the modern navigation of Amerigo Vespucci: *Universalis Cosmographia secundum Ptolomaei traditionem at Americi Vespucii aliorumque lustrationes.*

Like the Contarini map which we have just been discussing, only one impression has survived. The *carta marina* was also a woodcut and also was issued in twelve sheets. There are various reasons why such maps as those of Waldseemüller have survived only in small numbers or single impressions. Their sheets were usually stuck together on canvas or some other backing and used as wall maps. As such they were too large for glazing, and whether left hanging on the wall or rolled up and kept in cupboards, they became cracked, torn or simply rotted away. With few exceptions, the only surviving specimens are those bound up in book form, as atlases. At least a dozen maps of the sixteenth and early seventeenth centuries have survived in only single impressions. Some have disappeared altogether, like Waldseemüller's road map of Europe, which was printed in 1511.

The number of maps and charts published increased considerably during the sixteenth century, and there arose to meet this demand publishers who provided maps with all the latest information incorporated in them. Among such publishers, two of the greatest and certainly the most famous were Mercator and Ortelius, both Flemish.

Gerhard Mercator's real name was Gerhard Kramer; Mercator is its Latinised form. He was born in 1512 in Rupelmonde in East Flanders, but was of German descent. He studied at the University of Louvain. He was patronised by the Emperor Charles V, and in 1559 was appointed cosmographer to the Duke of Jülich and Cleves. Under the patronage of Charles V, he was in direct contact with the great Spanish and Portuguese explorers and map-makers, and so had access to all the latest geographical developments. He was a man of many parts: a practical engraver, a maker of scientific instruments and a land-surveyor, as well as much else. He was truly described as clever in mind and hand: '*ingenio dexter, dexter et ipse manu*'.

One of his most notable achievements was the introduction of the projection named after him and sometimes used to this day on maps and in atlases, despite what Lewis Carroll said of it in *The Hunting of the Snark*:

'What's the good of Mercator's North Poles and Equators,
Tropics, Zones and Meridian Lines?'
So the Bellman would cry: and the crew would reply
'They are merely conventional signs!'

Briefly, the special feature of Mercator's projection is that the parallels are at increasing intervals away from the Equator, proportional to the increasing distances between the meridians; for this the parallels were known as 'waxing latitudes'. It was first used by Mercator on his world map published in 1569, after he had already become famous as a cartographer, and it provided mariners with a basis for the type of chart they had long been wanting – one in which a line of constant bearing could be represented by a straight line. Only four impressions of this map and part of another have survived. It is made up of twenty-four sheets.

Mercator also introduced more refinement and accuracy into map-making than had obtained hitherto, and among alterations made by him were the reduction of the length of the Mediterranean Sea to nearer its true size, a more correctly proportioned representation of the land area of Europe between the Black Sea and the Baltic, and a more correct position for the Canary Islands, which had hitherto been misplaced by Ptolemy and his successors. These improvements had all been incorporated into his map of Europe, published at Duisburg in 1554, where Mercator was at the time a university lecturer. A second edition was published in 1572, and it contained yet more improvements. He published a map of the British Isles in 1564.

Some errors and misconceptions were not corrected for a long time. The island of Tierra del Fuego, off the southern tip of South America, was shown on Mercator's world map as part of the Antarctic continent. The shape of the whole of the American continent is distorted, and numerous small errors abound. But wherever he could improve on what had gone before, Mercator did so.

Mercator considered his world map as part of a larger plan to publish a collection of maps of various parts of the world. The first part of this was published in 1585 at Duisburg (when he was seventy-three). It consisted of fifty-one maps, and was published in three parts, each with a separate title page – namely, Belgium, France and Germany. The second part followed in 1589, and consisted of twenty-two maps of Greece, Italy and Slavonia. The third and final part was published in 1595, a year after Mercator's death, and consisted of thirty-six maps. The work was then published as a whole, with a general title page on which the word 'Atlas' was used for the first time for a book of maps: *Atlas sive cosmographicae meditationes de fabrica mundi et fabricati figura*. From 1595 to 1642 there were some forty-seven editions of Mercator's *Atlas*, with the text in various languages.

Mercator also made maps for Ptolemy's *Geography* which were, however, first published in Cologne without an accompanying text. Several editions were issued, the first one in 1578, the others, with text, between 1584 and 1750.

Artistically, Mercator's maps are in the best taste and are particularly remarkable for their fine plain italic lettering. He wrote a treatise on italic script which was published at Antwerp in 1540. On some of the maps hills and mountains are shown in elevation, as ranges of molehills, and towns are represented by little groups of buildings. Here and there he shows little groups of exotic animals, such as camels and elephants, drawn accurately and without exaggeration, while ships in full sail are shown on the seas, and an occasional marine monster may be seen. Some were issued plain, some hand-coloured. They are engraved on copper (Plate 1).

Mercator had several sons, of whom Arnold Mercator (1537-87) was a cartographer; he made town plans in addition to maps. Another son was named Rumold. He compiled the world map for the first collected edition of the *Atlas* and saw it through the press. Arnold's sons – Gerhard, John and Michael – were cartographers, and contributed in various ways to various editions of the *Atlas*. John became geographer to the Duke of Cleves.

A little earlier than Mercator was Abraham Ortelius, the Latinised form of his Flemish name, Wortels, sometimes spelled Ortel or Ortell. Like Mercator, he was of German descent; he was born at Antwerp in 1527. At the age of twenty, with the co-operation of his sister, he set up in business as a mapseller and colourer and met with considerable success. He became geographer to Philip II of Spain in 1575. He travelled widely in western Europe and came to England where he helped to persuade William Camden to produce his *Britannia* (see p. 39). He also knew Richard Hakluyt, the great English geographer. Ortelius was not a geographer in the same sense as Mercator, but was a scholarly craftsman. He died in 1598.

Ortelius issued a number of separate maps, including a world map in eight sheets in 1564, of which only one impression has survived. He also issued a map of Egypt during the following year, and two years after that a map of Asia printed on two sheets. His greatest and highly successful work, *Theatrum Orbis Terrarum*, was first published in 1570, and later editions printed at the press of Christopher Plantin. It contained in its original form seventy line-engraved maps on fifty-three sheets, including a world map, four continental maps and sixty-five regional maps, three of Africa, six of Asia and the remainder of Europe. It was an eclectic work, based on wide authority; one point of interest in this connexion is that authorities on which each map was based are mentioned. In the first edition the names of eighty-seven cartographers are mentioned; by 1603 this number had grown to 183. The maps are of uniform size and style. Between 1570 and 1612 there were well over forty folio editions published in several languages. It was undoubtedly the first modern atlas, although it was not called by that name. Like Mercator's *Atlas*, Ortelius's work was issued in both plain and coloured forms. Artistically, it is of high standard. From 1579 Ortelius added to his *Theatrum* a collection of 'ancient' maps entitled *Paregon Theatri*.

A great Dutch chart-maker of the sixteenth century was Lucas Waghenaer, author of the first atlas of engraved charts, which was published in 1583. It gave details of coastlines from Scandinavia to the south of the Iberian peninsula, but missed out Wales, Northern Scotland and Ireland. It was first published in Holland in 1583 as *Spieghel der zeevaert*; an English edition was published in 1588 as *The Mariner's Mirrour*. Features incorporated by these charts, and still used on Admiralty charts to-day, are silhouettes of coastlines as they appear from six to nine miles out at sea. For generations English seamen called any sea-atlas a 'waggoner', from the name of this chart-maker (see also p. 32).

Other great Netherlandish cartographers were Gerard de Jode, Jodocus and Henry Hondius, Peter Plancius, W. J. Blaeu and Jan Jansson.

Gerard de Jode (born at Nijmegen 1509; died at Antwerp 1591) published line-engraved maps, among them the world, Europe, Portugal and the Duchy of Brabant, formerly part of Lorraine. He also published in two parts in 1578 an atlas entitled *Speculum Orbis Terrarum*, containing respectively twenty-seven and thirty-eight maps, the latter completely devoted to the German Empire, the former of more general content.   It was engraved by Jan and Lucas van Doetecum and ran into several editions, one of which, issued after his death by his son Cornelis, contained eighty-three maps. Jode was a rival of Ortelius and some of his maps are superior to those of the younger man.

The two Hondius cartographers were father and son. Jodocus (his real name was Josse de Hondt – Hondius is the Latinised form) was born at Duffel in 1563 and died in 1611; Henry was born 1587 and died in 1638. Jodocus was an engraver as well as a geographer and mapseller, but Henry was a mapseller only.

When Jodocus was twenty-eight he emigrated to London where he remained for ten years, working as an engraver and typefounder and engraving the first English globes.   While he was in London he married his compatriot, Collette van den Keere, whose brother Pieter assisted him in engraving maps. He returned to Amsterdam with Pieter van den Keere about 1593-4, where, commencing work in 1605, he engraved several maps for the projected folio atlas of John Speed, to which we shall come later in this chapter.   It seems likely that Speed had seen and had been impressed by the work of Hondius when he was working in London. Among Jodocus's own publications was, in 1606, a new and enlarged edition of Mercator's *Atlas*, the copper plates of which he purchased in 1604.   By 1609 he had become the leading map publisher in Amsterdam.

Peter Plancius of Draneouter in West Flanders (1552-1622), a minister, theologian and former monk, is an important figure in the history of cartography and is said by some to have been second in his day only to Mercator.   So far as pure cartography is concerned his most important work was in the development of charts.   Plancius also was a prime mover of that characteristic Dutch cartographic publication, the world map printed on several sectional sheets intended for mounting as a wall-hanging. He issued one such map in 1592 in eighteen sheets; in it the world is depicted in hemispheres. Like so many other historic maps, only one impression has survived.

Plancius shared in the foundation of the Dutch East and West India Companies and his interest in the provision of good and accurate charts is therefore obvious.   He became recognised in his day as an expert on navigation, and in particular on the route to the East Indies.   For a time he advocated a route *via* the north-east passage, but later advocated the route round the Cape of Good Hope. He became official cartographer to the Dutch East India Company. Although he compiled some eighty maps he never published an atlas.

Willem Janszoon Blaeu (born at Alkmaar 1571; died at Amsterdam 1638) was the founder, in 1596, of the family firm of map-makers. He had his own shop at which his trade sign, a sundial, was displayed. The family was among the greatest map-makers ever known, and issued globes, atlases, maps and wall-maps. Blaeu, in 1633, succeeded Hessel Gerritsz as cartographer to the Republic. He was originally an instrument and globe-maker, and his first publications included charts, sailing instructions and declination tables. He invented a new form of printing press, referred to by Joseph Moxon, writing in *Mechanick Exercises* in 1683: 'The New-fashion'd Presses are used generally throughout all the Low-Countries.' His establishment was extensive; he had nine type-presses, six presses for copper-plate printing, a typefoundry, proof-reading rooms and rooms for all other departments in the processes of printing. He undertook binding and papermaking as well, and himself supervised every detail of the work.

Willem died in 1638, and was succeeded by his son Johann (1596-1673), who did much to standardise sea-charts. Another son, Cornelis (d. 1642), worked with Johann in the family firm. Their works include a world map of 1606, of which, once again, only one impression has survived, and *Theatrum Orbis Terrarum sive Atlas Novus* and *Atlas Maior,* both of which ran into several editions. The former included an atlas of the counties of England and Wales, based mainly on the maps of Speed and Saxton (see pp. 35 and 40) and it also included a one-volume national atlas of Scotland, with forty-nine maps, the first of its kind.

One of the greatest productions of the Blaeus was Johann's great world map printed on twenty sheets, and entitled *Nova totius Orbis Tabula* (1648), annotated and labelled in several languages. It displays Tasman's discoveries in Australia and New Zealand, in addition to showing considerable improvement in representation of other parts of the world's coasts. But there are other points that are not so accurate: for example, California is shown as an island. Previously it had been shown correctly as a peninsula, but it was now shown as an island owing to a misinterpretation of certain reports from Spanish explorers. Indeed, the form which California displays is a helpful guide for the dating of maps, for it is shown as an island between approximately 1622 and 1730.

After the death of Johann his workshop and much of the stock of engraved plates were destroyed in a fire, and so the Blaeus' work did not, like that of many cartographers, such as Mercator, outlive the family.

Jan Jansson (born at Arnhem 1596, died at Amsterdam 1664) was a rival of Blaeu, but his work is generally not so good, largely perhaps because he tried to publish much of it too quickly in attempts to forestall Blaeu's publications. Yet from a decorative point of view it is magnificent. He published an edition of Ptolemy's *Geography* in 1617, various separate maps and other atlases, including an atlas of England and Wales in 1646.

Other Netherlandish map publishers whose names may be briefly mentioned are

J. H. van Linschóten, publisher of *Itinerario* at Amsterdam in 1596; Leonhard Valk, who worked with Schenk; the Visscher family, who worked throughout the seventeenth and the first years of the eighteenth centuries. Also working in the seventeenth and eighteenth centuries were Allard, Danckertz and Ottens; and in the eighteenth century Covens and Mortier, Isaac Tirion and Pieter van der Aa. All of these people published important work; Ottens, for example, published a seven-volume folio atlas with 835 maps, but this was exceeded by van der Aa who issued a *Galerie Agréable du Monde* in sixty-six volumes and with some 3,000 plates. In sum, the late sixteenth and the seventeenth centuries in Dutch cartography represent one of the greatest periods in the craft's history. But by the eighteenth century Dutch maps had become decadent and imitative.

It is to France that we must now turn. Hitherto, French cartography had not been impressive. There were indeed some outstanding figures, such as Oronce Finé, whose Latin name was Orontius Finaeus, who in 1519 made a heart-shaped map of the world (published also in Basle) and another, of double heart-shape, in 1531. Yet another, again of single heart-shape, was engraved in 1536. Copies were later made in Italy. Other maps by Finé included a woodcut map of France on four leaves and a world map; no examples have survived. Another sixteenth-century French cartographer was Nicolas de Nicolay, who made a marine map for Medina's *Art of Navigation*. And there were Charles de l'Escluse (Carolus Clusius), Gilles Boileau de Bouillon, Jean Jolivet, André Thevet, Gabriel Symeone, Ferdinand de Lannoy and the Surhons, Jean and Jacques. The maps of all of these early French cartographers are of great rarity.

In 1594, at Tours, Maurice Bouguereau issued *Le Théâtre Français*, a national atlas, the first of its kind in France, containing fifteen line-engraved provincial maps of France and from one to three general maps, according to the copy. The atlas was reissued by Jean le Clerc, who had acquired the plates, as *Théâtre Géographique du Royaume de France*, seven times between 1617 and 1631, by which year its contents had grown to fifty-two maps. It was given yet another title by its final publisher, J. Boisseau, who issued it with a content of seventy-five maps in 1642 as *Théâtre des Gaules*. Another atlas was issued by Melchoir Tavernier in 1634, entitled, like one of the foregoing, *Théâtre Géographique du Royaume de France*. Another edition appeared in 1637. Some copies contain eighty maps, some ninety-five.

An important French sixteenth-century map was the large woodcut map of France by Françoise de la Guillotière, begun in 1596 and published about 1612-13. It was printed on nine sheets and a posthumous second edition was published by his widow in 1632.

The seventeenth century saw the rise of one of the great families of French cartographers, the Sansons, the first member of which (so far as maps are concerned), Nicolas, '*géographe ordinaire du roi*', was born in 1600 at Abbeville and died in 1667 at Paris. Sanson's work was carried out mainly by engravers from the Low Countries and Picardy, and thus must have been influenced, at least to some extent, by the Dutch school. Yet it is above all French and elegant in conception. He published about

300 maps, commencing in 1629 with a six-sheet map of ancient Gaul; he issued several atlases, of which his *Cartes générales de toutes les parties du monde* is best known. It was first published in 1658 and there were several editions. At first it had 113 maps, which number grew with each edition, but after his death another edition was issued by his sons Guillaume (d. 1703) and Adrien (d. 1708) with 102 maps. Between them the Sansons published at Paris 8vo atlases of the four continents, *L'Afrique* (1656 and 1660), *L'Amérique* (1656, 1657, 1662 and 1676), *L'Asie* (1652, 1653 and 1658) and *L'Europe en plusiers cartes nouvelles* (1648 and 1651).

The Sanson business was acquired from the sons of Nicolas by Charles Hubert Alexis Jaillot (born 1640 at Avignon, died at Paris 1712), who had started life as a sculptor and came to cartography through his marriage to the daughter of Nicholas Berey, a map colourer. He worked for a time with Guillaume and Adrien before taking over their plates, which he caused to be engraved on a larger scale, in addition to making maps and plans himself. Jaillot's maps are, from a decorative point of view, magnificent specimens of the art, and in the view of many have in this respect never been surpassed. He first issued an *Atlas Nouveau* in 1681, which ran through several editions. In 1695 it was reissued, enlarged, under the new title *Atlas François*, and thereafter ran into several more editions. Jaillot issued in 1693 another atlas with the splendidly baroque title of *Le Neptune Français ou Atlas Nouveau des Cartes Marines*. Jaillot's descendants continued to run the family printing house and to issue maps until 1780.

Among other French seventeenth-century cartographers there were Nicolas Tassin, who produced a number of maps and an atlas of French towns; Nicolas Sanson's son-in-law, Pierre du Val, publisher of a series of atlases, and Nicolas de Fer, who issued some splendidly decorative maps, many of them in atlases.

There was a ferment of scientific activity in France under Louis XIV. Maps and charts were in great demand in those great days of the French nation, and it was quickly realised that if they were to be of value against this scientific background, they must be based on precise scientific observations.

Those who were responsible for this more intensely scientific approach included Giovanni Domenico Cassini, a native of Perinaldo in Italy and later professor of astronomy at Bologna, and succeeding members of his family. Invited by Louis XIV to come to France, Giovanni Cassini became professor of astronomy at the Collège de France in 1671, and director of the Royal Observatory at Paris from 1669. Among new cartographic methods for which he was responsible was his use of the movements of the satellites of the planet Jupiter to determine longitude. It was due largely to Cassini's work that the new map of France was made, and a new survey of the country's coasts made by the Abbé Picard and Philippe de la Hire. One of the more obvious results of the survey was a perceptible recession of France's coastline as drawn on the map, which prompted Louis XIV sarcastically to remark that a disastrous war would not have lost him so much territory.

More refinements were used by another member of the Cassini family, Jacques, and his son, César François Cassini de Thury, who from 1733 onwards prepared a map based on a complete triangulation[1] of the whole country.   In 1747 this project received official support, only to be withdrawn almost immediately, and Cassini de Thury was forced to form a financial undertaking in order to see his work through.   When he died in 1784 he had completed the whole country except Brittany.   The work was finally completed in 1818, the State having assumed responsibility for it after the Revolution.

Cassini de Thury's completed map is remarkably modern in appearance, and at a glance its details might be taken as coming from an ordinary twentieth-century Ordnance Survey map, except that relief on it is but poorly indicated.   Symbols are extensively used to identify such things as churches, windmills, towns and so forth, and roads are clearly marked.   Yet it is not without decoration.

Indeed in the eighteenth century some of the great French artists of the day were employed in the decoration of maps – people, for instance, such as Boucher and Cochin. The result was the most scientific maps produced so far, with the added advantage that these monuments of scientific cartography were among the most beautiful and elegant ever seen. Yet one should qualify the claims for the accuracy of eighteenth-century French maps by stating that many of their makers engaged in a certain amount of guesswork rather than leave blank spaces.   In the words of Jonathan Swift:

> So geographers, in Afric-maps,
> With savage-pictures fill their gaps;
> And o'er unhabitable downs
> Place elephants for want of towns.

There were many other French cartographers in the eighteenth century in addition to the Cassinis.   There was, for instance, Guillaume Delisle, or de l'Isle (1675-1726), son of Claude Delisle (1644-1720), himself a teacher of geography and other subjects. Guillaume had studied astronomy under Cassini, and had been something of an infant prodigy, for, it is said, he could, before he was ten years old, draw maps of the ancient world.   In 1702 he was elected a member of the Académie; in 1718 he was made *Premier Géographe du Roi*.   He was patronised by Peter the Great of Russia.   As a cartographer, one of his chief influences on the craft was his honesty; for he was always prepared to leave a blank space where he lacked knowledge.   He issued about a hundred maps altogether, not of the highest standard as engravings, but showing, scientifically, a definite step forward.

Delisle's standards of accurate cartography were carried forward and added to by Jean Baptiste Bourguignon d'Anville (1697-1782), a scholar whose magnificent collection of cartographic material is now in the Bibliothèque Nationale in Paris. He did much to develop Asiatic cartography, and his works in this category included a survey of the provinces of China, carried out for the Society of Jesus.   From this he made a map of the Chinese Empire on sixty-six sheets, which was eventually published also in the Netherlands (1737) and England (1738-41), as part of J. B. du Halde's

---

[1] Triangulation is a method of measuring the earth's surface by resort to the geometric laws of triangles; thus, for example, a position may be determined by relating it to two other fixed points standing a known distance apart.

geographical description of China. D'Anville also issued maps of the continents of the world in two hemispheres.

D'Anville was succeeded by his son-in-law, Philippe Buache (1700-1773), who was chiefly notable for the advances he made in the representation of relief on maps.[1] One of the oldest methods of doing this was by the little conical 'molehills', already mentioned several times in this chapter (see also pp. 68-69, and Plate 7). Obviously, the representation of relief was a problem of considerable importance in the mapping of mountainous countries, and it was in fact a Swiss, Hans Konrad Gyger (1599-1674), who had earlier used a method at once aesthetically pleasing and practically effective: a system of shading which, however, made little differentiation for various levels, giving only relative reliefs. It was probably Gyger's method of shading that gave rise to the use of hachuring to denote relief. Hachuring is a system of lines drawn close together and running in the direction of the slope; the steeper the slope the heavier the hachuring (Plate 3).

Buache, however, used the method usually employed today: a line drawn through points of equal height (Plate 5), probably borrowed from an idea used in 1729 by the Dutch hydrographer, N. S. Cruquius, who used such a device to represent ocean depths. It is thought that the practice may have originated from a chart of magnetic variations published by Halley in 1701, on which he used 'isogonic' lines. In some present-day maps, both hachuring and contour lines are used on the same maps to denote relief (the Michelin maps are an example), and colouring is also combined with them to denote differing levels.

Buache issued several atlases, but in his own day was chiefly famed for his marine maps (Plate 6).

Other eighteenth-century French cartographers include Gilles and Didier Robert de Vaugondy (1688-1766 and 1723-86 respectively) who in 1757 issued an *Atlas Universel* of 108 highly decorative maps, based largely on Sanson maps; Roch-Joseph Julien (fl. 1751-76) whose *Atlas Géographique et Militaire de la France* was published in 1751, and who claimed to keep 4,000 maps in stock; Jacques Nicolas Bellin (1703-72) of the French Marine Office and G. L. Le Rouge (fl. 1741-79), each of whom issued a series of maps and atlases, Le Rouge being especially noted for plans of ports and fortifications. French supremacy in cartography dwindled at the end of the eighteenth century with the outbreak of the Revolution and for other reasons, and it was England that thereafter took the lead, an obvious development for a country whose maritime power was then reaching its height.

We have already noted early English manuscript maps, but it is to the sixteenth century that we must now turn to see the true foundations of English cartography so far as printed maps are concerned. In the early part of the sixteenth century local plans began to be made extensively in England, largely, it is thought, because of the requirements of people who had taken over monastic lands at the Dissolution in 1536 and the Suppression in 1539. Many of these – and town plans, too – were conceived as a combination of plan and elevation, a kind of bird's-eye view, a development, no doubt,

---

[1] For further consideration of this subject see Chapter Four.

partly from the symbols by which towns were represented on a map like that of Matthew Paris (see p. 17), and partly from illuminated miniatures. This method was used, too, on manuscript sea-charts, which were made in great numbers at this period because of Henry VIII's concern in developing the maritime power of England. Many of these charts were beautifully coloured and they often gave in a charming way an indication of the contemporary appearance of English villages.

One of the more important of the makers of this type of manuscript map and chart was Richard Popinjay, employed during the reign of Elizabeth I, from 1562 to 1587, in the south coast and Channel Islands ports. His maps are very detailed and show roads, dangerous rocks, safe anchorages and quays; they have scales, too, and moreover they have a restrained beauty.

Nevertheless, the first engraved map of the British Isles, drawn according to post-mediaeval geographical standards, was not produced in England at all, but in Rome in 1546, by George Lily under the influence, it is said, of a number of English Catholics living in exile, who co-operated for its realisation with local copper-engravers. (Formerly it was believed to have been compiled in England and then sent to Italy to be engraved.) It is drawn on a conical projection, and latitude and longitude are marked on its borders.

It took some time for the craft of copper-engraving to reach England, but it was eventually introduced here from the Low Countries in the second half of the sixteenth century by Thomas Gemini, a Belgian. Gemini had access to the copper-plate map, mentioned in the preceding paragraph, which had been printed in Rome and brought to England by George Lily after the accession of Mary I. Gemini altered it in various ways, including the insertion of his own name, and republished it in London. Like many another early map, only one impression remains: it is in the Bibliothèque Nationale in Paris.

Indeed, English map-making in the sixteenth century was in a large measure reliant upon Flemish engravers. The first English sea atlas, *The Mariner's Mirrour* by Lucas J. Waghenaer (published in 1588, soon after the defeat of the Spanish Armada), which contained forty-five charts, with bird's-eye views, and with ships in full sail and sea monsters, had three of them engraved by Hondius and nine by Theodor de Bry of Liège, both of whom were in England at the time. It was translated by Anthony Ashley. Incidentally, de Bry was a friend of Richard Hakluyt, the English geographer. Waghenaer's work had originally appeared in Holland under the title of *De Spieghel der Zeevaert*, with charts engraved by Jan van Doetecum (see also p. 26).

One of the finest copper-engravers working here in the sixteenth century was the Flemish Renold Elstrack (1571-1625?) a pupil of the English engraver William Rogers (b. *circa* 1545) who had himself learnt the art from the Wierix brothers at Antwerp. Among Elstrack's work were maps printed in Linschóten's *Itinerario*; in 1619 he engraved a map of the Mogul Empire, compiled by William Baffin. One of his most interesting maps, however, was an issue of the Italian Baptista Boazio's map of Ireland published in 1599. Boazio, a surveyor, is thought to have been brought back by the

Earl of Essex from his expedition to Ireland in 1599; his map of Ireland was the first of its kind to be produced up to that time, and had been first published by Hondius in 1591. It exists printed on silk. Elstrack attempted to perpetuate his name in this map by renaming one of the isles of Aran 'Elstrakes Isle'. It was on two plates, contained an abundance of fanciful detail and elaborate decoration and heraldry, and was dedicated to Queen Elizabeth, explaining that in it 'your Highness may distinctly see what Havens, Rockes, sands or Townes, in Ireland be'. Elstrack's name appears in one lower corner, its publisher's, 'Mr. Sudbury in Pope's Alley', in the other. It was later used in Ortelius's *Atlas* of 1602 and in subsequent editions (see also p. 25).

The earliest known map engraved by an Englishman is that of the Holy Land by Humfray Cole (1530?-91), a goldsmith at the Mint and a friend of Gemini. Cole was well known in his day as a maker of navigational and other scientific instruments. Such of these as have survived show him to have been an excellent workman. An astrolabe which he made (an instrument, that is, for the observation of the positions of the heavenly bodies) is in the British Museum; it once belonged to Henry Prince of Wales, the 'Incomparable Prince Panaretus', elder brother of Charles I. Cole's map of the Holy Land was engraved for the second edition of Archbishop Parker's 'Bishop's Bible' in 1572; his name is inscribed on it in a cartouche thus: 'Humfray Cole, goldsmith, a Englishman born in ye north and pertayning to ye Mint in the Tower, 1572.' Cole must have learnt the craft of engraving from Remigius Hogenberg, a Netherlandish engraver in the service of Archbishop Parker. Actually, the map itself is copied from one by Tilmann Stolz (or Stella) and engraved by another Hogenberg, Franciscus. It is engraved with the arms of Lord Burghley, has a scale showing English and Italian miles (said each to equal 1,000 paces), leagues and classical stadia, and various other decorations which include the rebus of the printer, Richard Jugge, a bird perched on a bush and chirping (showed by a lettered scroll issuing from its beak) 'IUGGE, IUGGE, IUGGE, IUGGE . . . '. The lettering is a clear italic. Towns are shown by little groups of buildings of various sizes, and various little scenes are engraved here and there. Hills are shown by groups and ranges of the 'molehill' type.

The convention of the bird's-eye view, mentioned in connexion with Popinjay's charts, was used in addition on certain printed maps and in particular on the town plans which were then making their appearance. Of the first of these to be published, Archbishop Parker was again the patron. It is a plan of Cambridge executed in 1574 by Richard Lyne, a pupil of Franciscus Hogenberg. It is a most attractive work, set within a beaded decorative border within which, in cartouches, are marked the four cardinal points. It has the virtue of showing the town in elevation, as it were, as well as in plan, thus combining the apparent with the systematic. In many ways it is like the perspective (if such it may be called) in Persian miniatures. A large decorative cartouche contains a brief description of the town; at the lower right hand side another contains a key to various features of the map, and other decorative features abound – swags of fruit, dragonflies, masks, arms and flowers; and the plan itself shows parterred and topiary gardens and herbaries, and horses and sheep grazing in peaceful summer meadows, richly irrigated by the many little streams with which Cambridge has from time immemorial been veined.

As on the Continent, cartographic science made considerable headway in the sixteenth century in England, progress that is marked by such publications as William Cuningham's *The Cosmographicall Glasse* (1559), the first English book on cosmography, Leonard Digges's *Pantometria* (1571), and others. Leonard Digges died about 1571, just before the publication of his book, which was undertaken by his son, Thomas. The date of his birth is unknown. He came of a family of gentlefolk, whose members had included sheriffs, judges, magistrates and members of Parliament. He studied at University College, Oxford, but took no degree, despite his devotion to the science of mathematics and land surveying. The full title of his most celebrated book was *A Geometricall Practise, named Pantometria, divided into Three Bookes, Longimetria, Planimetria, and Stereometria, containing Rules manifolde for Mensuration of all Lines, Superficies, and Solides*. Among important items featured in it is a description of a primitive theodolite (an instrument used in surveying to measure angles) and a long discussion of optics – a subject of vital importance to accurate cartography.

As in the case of Digges, we do not know the dates of William Cuningham's birth and death. We do know, however, that he was at Corpus Christi College, Cambridge, from which he took the degree of Bachelor of Medicine after studying the subject for seven years. He also studied at Heidelberg. He settled in London where he became, in 1563, lecturer at Surgeons' Hall. He wrote many books, but for our present purpose the most important was *The Cosmographicall Glasse* or, to give it its full title, *The Cosmographicall Glasse, conteinyng the pleasant Principles of Cosmographie, Geographie, Hydrographie, or Navigation*. Among other things, it contains a plan of Norwich, in which city Cuningham had resided for a time in the 1550's. Cuningham, although primarily a medical man, was a pioneer surveyor and cartographer, bringing to English cartography continental methods he had doubtless seen at Heidelberg, and using an astrolabe and compass, and a surveying chain to assist him in arriving at accurate measurements.

In the sixteenth century, too, appeared the first English terrestrial equivalents of portolans: highway tables giving details of thoroughfares in the country. They are in no sense maps, but tables of roads and distances, like this one:

Dover to London

From Dover to Caunterbury	xii myle
From Caunterbury to Sittyngborne	xii myle
From Sittyngborne to Rochestre	viii myle
From Rochestre to Gravisende	v myle
From Gravisende to Dertforde	v myle
From Derteforde to London	xii myle

The foregoing table is from a small book published in London and Canterbury several times between 1541 and 1561, and variously entitled *A cronicle of yeres*, *A Breviat Cronicle* or *A briefe Cronicle*.

It was against such a background as this that the early English map-makers began to practise their craft, and the first of these whom we are to consider are Christopher Saxton (*circa* 1542-*circa* 1610) and John Norden (1548-1626).

Saxton, the publisher of the first atlas of England and Wales and of the most complete survey of the country before the foundation of the Ordnance Survey in 1791, was a Yorkshireman, born at Dunningley near Leeds and may have been educated at Cambridge. After coming to London at an unknown date, he became attached to the household of Thomas Seckford, a leading lawyer of the times and Master of Requests at the Court of Wards, who became his patron and paid for the publication of his atlas. The maps, upon which all subsequent maps of the period were based, were produced between 1574 and 1579, and were undertaken on the Queen's authority and dedicated to her when they were published. Saxton had in 1577 obtained a ten-year privilege to make and market them. He was granted a coat of arms by Elizabeth – the only cartographer in our history to have received such an honour from the reigning monarch.

Saxton did not produce his maps without considerable difficulties, but he was nevertheless greatly assisted by an order from the Privy Council that he was 'to be assisted in all places where he shall come for the view of such places to describe certein counties in cartes, being thereunto appointed by her Majestie's bill under her signet'. Wales – a wild and remote place in those days – was the subject of a special order to its local authorities, who were instructed 'to see him conducted unto any towre, castle, highe place or hill, to view that countrey, and that he may be accompanied with ij or iij honest men, such as do best know the countrey, for the better accomplishment of that service; and that at his departure from any towne or place that he hath taken the view of, the said towne do set forth a horseman that can speke both Welshe and Englishe, to safe-conduct him to the next market-towne'. All of which shows that Saxton was not satisfied merely to copy precedent. He was a field-worker who went out and took his own details, and this in days when it was customary and all too easy to copy what had gone before.

Yet this is not the whole story, for Saxton must have had some foundation on which to base his new *Atlas*, and he doubtless drew upon existing topographical material. Nevertheless, he certainly worked quickly, for although, to take an example, he began his survey of Wales over half-way through 1576, all the maps resulting from it had been printed by the end of 1578. Saxton did not show any roads; they were in fact not shown on printed English maps until those of Norden (pp. 37-39).

In some ways Saxton is not consistent, despite the comparative accuracy of his work. The scales of his maps vary considerably from one another – Lincolnshire and Cornwall, for instance, are something under 4 miles to 1 inch, while Hertfordshire is 1¾ miles to 1 inch. He had, too, to contend with the practice of using 'customary miles' then in vogue. This practice meant that the unit of a mile varied from place to place; the old English mile was generally reckoned to be 2,140 yards (as against 1,760 yards to-day), but it varied by as much as four furlongs from place to place. Saxton seems to have varied his own scales and measurements accordingly.

Saxton's *Atlas* enjoyed a long popularity. It was issued several times in its original from between 1579 and about 1600, and subsequently, with various additions and alterations, until about 1770, by other publishers. Some of the subsequent editions, however, were bastardised affairs in which, to take an example, Saxton's maps were combined with Speed's ornamentation, plans or arms. These were issued in the seventeenth century by the London publisher, Philip Lea. They were reissued in a re-engraved format in 1607, 1610 and 1637 in Camden's *Britannia* (see p. 39). It was a beautiful production and had a sumptuous title page, with Queen Elizabeth depicted with all the oblique and direct flattery for which the age was noted. There were two versions of the original issue, showing, for example, differences in the Queen's dress; the first is of great rarity. The maps themselves are enclosed in decorative borders; titles are enclosed in highly decorative cartouches in Italian lettering rich with swashes. Heraldry is featured, and there are scales. Ships in full sail ride upon the seas, with here and there large fishes or dolphins. Towns are represented, according to size, by little buildings or groups of buildings, coloured red. Hills and mountains are shown by the conventional 'molehills' (some are curiously out of scale – Snowdon is shown three miles high), trees by little topiary growths of the 'Noah's Ark' type (Plate 7). On the whole, their hand-colouring is restrained but tasteful. An interesting point worth remembering in studying Saxton's maps is that in 1576 Seckford altered the motto on them from *Pestis patriae pigricies* to *Industria naturam ornat*.

The *Atlas* contained an index, and of this there were three separate and different editions. The first edition has the names of the counties arranged in a column, the numbers before them; the second has a narrower title, and the list is included in a line framework; the third is easily identifiable, as it is longer than either of the others and much more elaborate, and has coats of arms and tables of towns and other topographical features.

Saxton's maps were engraved on copper by several hands, among them Remigius Hogenberg (whose name we have already met in connexion with that of Humfray Cole; he probably engraved the *Atlas*'s frontispiece) and Leonard Terwoort, who between them engraved sheets dated 1575, containing a total of nine counties. Another of Saxton's engravers was Cornelius de Hooge (or Hogius in its Latinised form), a Dutchman who engraved the first two maps, dated 1574, later bringing his total up to five. These Flemish engravers brought the influence of the styles of the Low Countries to the maps they engraved; the cartouches, for instance, show the influence of Netherlandish engraving and goldsmithery, and many of the symbols are unmistakably Flemish. An English engraver who worked for Saxton was Augustine Ryther of Leeds, who proudly put the word *Anglus* with his name to distinguish himself from the numerous foreign engravers then working here. He kept a bookshop in Leadenhall Street, London. Two other English engravers, whose names appear on maps of 1577, were Francis Scatter and Nicholas Reynolds of London. Not every map in the *Atlas* was signed, but Saxton's name appears across the compass or on its scale on the majority of them.

Though much more expensive than hitherto, Saxton's *Atlas* is not beyond the reach of the collector of means, and single maps, considering their age, are often quite reasonably priced. It should be remembered that the maps were issued separately in their own day, as well as in bound atlases, and there are often differences between impressions taken at different dates and in the paper used. Indeed impressions were taken from the plates for many years after Saxton's death,[1] and for this reason great care should be taken in allocating a Saxton map to a definite date. The same may be said in the cases of the work of many other cartographers.

In 1583, again with Seckford as his patron, Saxton engraved on twenty plates a great wall-map of England and Wales. Only two impressions of the original edition are known: they are in the British Library and the Birmingham Public Library. Like his *Atlas*, however, this map was reissued many times by different publishers into whose hands the plates came, although, of course, alterations and additions were made to them. It was still being sold in 1795.

Next we come to John Norden, a friend of John Gerard, the herbalist, who once gave him some red-beet seeds which 'in his garden brought foorth many other beautifull colours'. He was born probably in Wiltshire or Somerset of a Middlesex father. He was a Master of Arts of Oxford, an attorney and lived at Fulham. Norden was as much a topographer as a cartographer and was the first Englishman to design a complete series of county histories, *Speculum Britanniae*, of which only the sections on Middlesex and Hertfordshire were actually published in his lifetime. The remainder was unknown for about 300 years after Norden's death. In the British Museum there is a manuscript by him, dedicated to Queen Elizabeth, entitled *A Chorographical Discription of the severall Shires and Islands, of Middlesex, Essex, Surrey, Sussex, Hamshire, Weighte, Garnesay, and Jarsay, performed by the traveyle and uiew of John Norden, 1595.* This was part of topographical work undertaken by Norden as the result of a Privy Council order from Hampton Court, dated 27 January 1593, addressed 'To all Lieuts, etc, of Counties', and requiring that 'the bearer, John Norden, gent.', should be 'authorised and appointed by her Majesty to travil through England and Wales to make more perfect descriptions, charts and maps'. He became surveyor of the Crown woods and forests in counties which included Berkshire, Devon, Surrey and the Duchy of Cornwall. He wrote much else besides the works I have just mentioned, concluding just before his death with a set of distance tables intended for use with Speed's county maps, and entitled *England, An intended Guyde for English Travailers* (1625) (Plate 8). These distance tables had attached to them a 'thumb-nail' map of the district, and a scale with open dividers above it. They were copied by Jacob van Langeren in various editions of his *A Direction for the English Traviller* between 1635 and 1643. They were once again used by J. Jenner in his *Book of Names* in various editions between 1657 and 1677. Finally, they were used by Thomas Cox in *Magna Britannia* in 1720. The idea is still in use to-day.

As a cartographer, Norden was an original craftsman, and as well as maps and plans to illustrate his own works, was responsible for some of the maps in the 1607 edition of

[1] See the Saxton paragraph in the *List of Cartographers* on p. 222.

Camden's *Britannia*, for separate maps used later by Speed in his *Theatre of Great Britain*, and for others besides. Especially fine and highly decorative, though very rare, are his larger maps of Hampshire, Surrey and Sussex, published between 1594 and 1596. The Sussex map was engraved by Christopher Schwytzer; Charles Whitwell, a pupil of Humfray Cole, engraved the one of Surrey. Each bears the arms of patrons of geography; Sussex those of William Sanderson, merchant; Surrey those of William Wand, Clerk to the Privy Council. The Hampshire map has not survived in its original state, but only as altered later by Peter Stent and John Overton, two publishers who had successfully acquired the copper plates.

Norden made several surveys of English counties, but those of them that were published (apart, as we have seen, from those of Middlesex and Hertfordshire) did not come out until after his death. That of Cornwall was published in 1728, that of Essex in 1840. *The Description of Cornwall* consists of upwards of a hundred pages. It contains 'a Map of the County and each Hundred; in which are contained the Names and Seats of the several Gentlemen then Inhabitants: as also, thirteen Views of the most remarkable Curiosities in the County'. An interesting feature is a grid system to help identify on the maps places mentioned in the text. The whole map is in each case divided into squares by horizontal and vertical lines. The vertical ones are lettered, the horizontal ones numbered; thus any square may be identified by quoting the number and letter at which it occurs, for example 24 i, 4 g, etc. (Plate 10). These had been used in European cartography since the end of the first quarter of the sixteenth century. The system also provided a ready-made scale, as the squares were of constant size (2 inches).

Town plans and views are an important section of Norden's work. One such, now lost, is mentioned in Leland's *De Rebus Brit. Collecteana* (1770): 'Mr. Norden designed a "View of London" in eight sheets, which was also engraved. At the bottom of which was the Representation of the Lord Mayor's Show, all on Horseback. . . . The View was taken by Norden from the Pitch of the Hill towards Dulwich College going to Camberwell from London, in which College, on the Stair Case, I had a sight of it. Mr. Secretary Pepys went afterwards to view it by my recommendation, and was very desirous to have purchased it. But since it was decayed and quite destroyed by means of the moistness of the Walls. This was made about the year 1604 or 1606 to best of my memory, and I have not met with any other of the like kind.' Town plans made by Norden include bird's-eye views of Higham Ferrers, Chichester, London, Westminster. Some of them were engraved by the Fleming, Pieter van den Keere (brother-in-law of Hondius; see p. 26) and they are among Norden's most attractive publications (Plate 57).

Norden also marked roads on his maps, a most useful innovation. Inset plans of towns were included, too. Probably also he was responsible for the introduction of the symbol of two crossed swords to indicate a place of battle, although he did not use this symbol exclusively, for soldiers are shown fighting at Barnet in his map of Hertfordshire. In any case his symbols are clear and concise (Plate 49). Details of agriculture and industries are also included on his maps.

Artistically, Norden's maps are comparable with those of Saxton. Conventional symbols, such as trees and hills, are treated in much the same way, and decorative cartouches and heraldry are used, though with a little more restraint. Bird's-eye view inset plans are a typical feature, and ships and marine monsters appear in places on the high seas. The lettering is clear italic, with swashes on some of the titling.

In many ways Norden was a pathetic figure. Several times he produced works in manuscript and tried hard to find patrons to pay for their publication – Lord Burghley and the Queen herself being his chief targets of appeal. He got little satisfaction and seems indeed, on the whole, to have been ignored. He spent nearly all he had in producing the two parts of the *Speculum* actually published, and in 1597 made two manuscript copies of his *Description of Hertfordshire*, which he dedicated to Burghley and the Queen, one to each. To the latter he wrote concerning the preparation of the *Speculum*: 'I was drawne unto them by honorable Counsellors and warranted by your royall fauor, I was promised sufficient allowance and in hope thereof onlie I proceeded . . . By attendaunce and . . . travaile in the business I have spent aboue a thousand markes and five years time . . . Onlie your Maiestie's princelie fauor is my hope without which I myself most miserablie perish, my familie in penurie and the work unperformed.' He does not appear to have had a reply from either of these great if reluctant patrons, so he published the Hertfordshire book himself, with a map in it engraved by William Kip.

Elizabethan map-makers with smaller output than that of these two great figures included Philip Symonson, who made a map of Kent on two sheets, engraved by Charles Whitwell; it is considered by many to be the finest specimen of Elizabethan cartography. Alas, the first edition no longer exists, and apparently there remains only one sheet, the eastern half, of the second; it shows main roads. Rare manuscript maps and plans by other Elizabethan cartographers survive, and especially there are a few engraved town plans. I have already mentioned Lyne's plan of Cambridge. In addition there were, among others, Hooker's plan of Exeter, of which there were two issues engraved in 1587 by Hogenberg – only one impression of each remains – and the five-volume work by Braun and Hogenberg *Civitas Orbis Terrarum* (1573; a sixth volume was added in 1618) which contained plans of many British cities and towns.

There was, too, Baptista Boazio, whose map of Ireland has already been mentioned (p. 33), who engraved in the Netherlands, a sea-chart of 'the famous West Indian Voyadge made by the English fleete (Sir Francis Drake Generall)'. It has been said that it was issued in the 1589 edition of the *Summarie and true Discourse* of this enterprise by Walter Bigges, but it is now thought to have been issued as a separate work. It was engraved by Renold Elstrack (see p. 32).

Before leaving the sixteenth century I must say something of William Camden (1551-1623), Clarenceux King of Arms, and his *Britannia*. Camden was an antiquary, historian and herald; he was educated at Christ's Hospital, St. Paul's School and Oxford, where he later founded a chair. The *Britannia* – an antiquarian investigation of Britain – was first published in 1586, and was followed by two more London editions during the next four years, as well as one published at Frankfurt. It was written in Latin and an

English translation did not appear until 1610. The last edition to be published in Camden's lifetime was that of 1607, and it was the first to contain a group of county maps, with their engraved surfaces measuring on the average 35·6 × 27·9 cm, drawn from those of Norden and Saxton and engraved by William Hole and William Kip. The maps appear also in the English translation of 1610 and its reissue in 1637.

Also worthy of remark is Michael Drayton (1563-1631), the poet whose *Poly-Olbion, or a Chorographicall Description of all the Tracts, Rivers, Mountaines, Forests, and other Parts of Great Britain* was published 1613-22, with eighteen and ten maps respectively, engraved by W. Hole. Geographically, they are of no great concern, but as decorative illustrations to the more romantic side of the age they are excellent. Nymphs, shepherds, nereids, civic figures, gods and goddesses all disport themselves in surprised English shires, personifying towns, rivers and other features (Plate 9). It is an England that Saxton or Norden would not have recognised, but which Edmund Spenser would have understood, a fitting accompaniment to Drayton's stately song:

> Of Albions glorious Ile the Wonders whilst I write,
> The sundry varying soyles, the pleasures infinite
> (Where heate kills not the cold, nor cold expels the heat,
> The calmes too mildly small, nor winds too roughly great,
> Nor night doth hinder day, nor day the night doth wrong,
> The Summer not too short, the Winter not too long)
> What helpe shall I invoke to ayde my Muse the while?

With the seventeenth century we come too to the *Theatre of the Empire of Great Britain* (1611), the work of yet another great Englishman, John Speed (1552-1629), who was born in Cheshire, the son of a merchant tailor, a trade Speed himself followed for a time; he became a Freeman of the Worshipful Company of Merchant Taylors. His interest in cartography began as a leisure-time occupation, and by 1598 he had made maps which he had presented to Queen Elizabeth and to his livery company. Among his cartographic activities were his copy in 1607 of Norden's map of Surrey for Camden's *Britannia* and his publication of a series during the next three years, of fifty-four royal folio *Maps of England and Wales*, some by Norden, some by Saxton, and most of them engraved by Jodocus Hondius. It was this series that, together with an accompanying description to each map, formed his *Theatre of Great Britaine*, published in 1611 by George Humble.

Speed himself was not a cartographer, and consequently made little if any contribution to this side of the subject. He said, indeed, 'I have put my sickle into other mens corne.' His maps are nevertheless notable for their wealth of antiquarian and historic detail and for their highly decorative appearance. They are rich with heraldry, decorative borders and cartouches. On each one a plan or view of the chief town is inset, with a decorative border. Figures are included, too: allegorical personages to bear swags and support cartouches, as well as 'real' people such as the four stately gentlemen in academic robes on the map of Cambridgeshire. The colouring, when present, is magnificent, although the sheets were issued uncoloured, and maps with contemporary colouring[1] are rare. The lettering, a mixture of italic and roman, is clear and decorative, with swashed letters on some of the larger legends. The map in the *Theatre* of 'The Kingdome of Scotland' is decoratively notable, and incorporated in the border are full-length portraits of James I, his Queen and his two sons. Inset is a map of the 'Yles of Orkney'. No less attractive is his general map of Ireland, with six figures in national costume. At least one impression of this is known printed on satin, and it is thought to be the first map produced thus, although silk had been used earlier (see p. 33).

Some of the maps include delightful views of famous buildings in the area covered by the particular map. The map of Surrey depicts Richmond and Nonsuch palaces, the latter with its parterred garden tipped up and drawn at right angles to the observer's eye, as if it were part of a Persian miniature. The map for Middlesex is of great interest in showing a view of Old St. Paul's Cathedral. On the Wiltshire sheet Stonehenge makes a stark feature with people of Speed's times perambulating among the trilithons, one couple followed by a little dog (Plates 11, 31 and 32). There are, too, in some, battle scenes, Roman coins and other antiquities.

The inset plans were taken from various other publications and incorporated into the main maps by the engraver, although Speed claimed that he was himself responsible for those of St. David's and Pembroke. Among the cartographers from whom his plans were taken were William Cuningham, William Mathew, John Norden, Christopher Schwytzer and William Smith.

There were many editions of Speed's *Theatre*, some of them now extremely rare, some showing interesting variations.[2]

Besides publishing maps, Speed wrote various other works, including such things as *Genealogies Recorded in Sacred Scripture* (*circa* 1611), which ran into thirty-three editions before 1640, and *A Cloud of Witnesses* (1616).

John Ogilby (1600–76), was a colourful character who started life as a dancing master. He injured himself, however, dancing in a court masque for the Duke of Buckingham and thereafter became attached to the household of the Earl of Strafford, Lord-Deputy of Ireland. In time he became Deputy-Master of the Revels in Ireland and even built a theatre in Dublin. When the Civil War broke out he lost everything and came back to England, being, however, shipwrecked on the way. Completely down and out, he made his way to Cambridge where a group of scholars, taking pity on him, taught him Latin, and without more ado he made translations of Virgil and Aesop, and proceeded to learn Greek. He took part in the arrangements for Charles II's coronation, and became Master of the Revels in Ireland in 1662, building himself another theatre in Dublin. Coming back to London he became a publisher, a trade he followed until his house and stock were destroyed in the Great Fire in 1666. But even this did not restrain the indefatigable Ogilby, and he obtained a commission, with his wife's grandson, to survey the city and plot out disputed property. After this he set up again as a printer and became King's Cosmographer and Geographic Printer. Among his publications was a series of atlases and maps; the most original are doubtless his series

---

[1] Cf. Chapter Three.   [2] For details, see Speed's entry in the *List of Cartographers*.

of road maps, of which his *Britannia* of 1675 contains the earliest strip road-maps of England. In fact, from this time onward the depicting of roads on maps became increasingly important, and they were added to those copper-plates of Saxton and Speed that were still in use. Ogilby himself was concerned with post roads, and it is interesting to note that it was because of the extension of the postal service beyond London in the seventeenth century that the statute mile of 1,760 yards became accepted all over the kingdom, although it was not finally and officially established until the passing of the Act of Uniformity of Measures in 1824.

Richard Blome (d. 1705), who according to Anthony à Wood was 'esteemed by the chiefest heralds a most impudent person, and the late industrious Garter (Sir W. D[ugdale]) hath told me he gets a livelihood by bold practices'. He published editions of Guillim's *Display of Heraldrie*, and various other works, including, in 1673, *Britannia: or a Geographical Description of the Kingdoms of England, Scotland, and Ireland, with the Isles and Territories thereto belonging*; *and there is added an Alphabetical Table of the names, titles, and seats of the Nobility and Gentry*; *illustrated with a Map of each county of England*. It was described, with some justification, by Bishop Nicolson as a 'most entire piece of theft out of Camden and Speed'. It does nevertheless contain an interesting map of London before the Great Fire, engraved by W. Hollar, and has an attractive border of the arms of the city and of some of the principal livery companies, some of them now extinct. It bears a dedication to Alderman the Hon. Sir Robert Vyner, with his arms.

Yet another map-maker of the period was Robert Morden (d. 1703) who first made maps and globes in London about 1668, entering into partnership about 1688 with Thomas Cockerill at the Atlas in Cornhill. His maps are not unattractive, and have pleasing cartouches and such decorative features as ships sailing on the seas. Geographically they are unoriginal and somewhat coarse. The old convention of 'molehill' mountains is maintained, forests are represented by 'Noah's ark' trees and towns are represented by the little church symbols which had been in vogue for so many years. Morden was a prolific publisher of maps and other works. His maps included *Pocket-Book Maps of all the Counties of England and Wales* (undated), various other county maps, some with roads shown, various European countries, a sea-atlas (1699) and a map of London, Westminster and Southwark (1700). His maps were used, too, for an edition of Camden's *Britannia*, published in 1695, the first collection of English county maps to use the prime meridian of London throughout.

Other eighteenth-century British map-makers were Emanuel Bowen, Thomas Kitchin and George Bickham. Bowen was an engraver of maps to George II and Louis XV; with Kitchin he published a series of attractive county maps between 1749 and 1780, which showed not only the geographical features in great detail, but also highly detailed historical notes in the free spaces, and views of principal towns. Cartouches of rococo design contain dedications, and others, with vignettes of the county's trades and manufactures, contain titles. George Bickham's 'maps' are really bird's-eye views of the counties, with the towns marked. They are, in a quaint way, very pleasing.

The British cartographers so far discussed are notable rather for their decorative virtues, or, in some cases, their comparative accuracy, so to speak, in reporting, rather than for their geographical originality. Indeed, with few exceptions, they followed geographical practices that had been used on the Continent for a considerable period before their time, and many of the maps, as we have seen, were the work of Flemish engravers. In the eighteenth century, particularly towards its end, British cartography took great strides forward, largely no doubt due to the expansion of the country's maritime influence, in particular in North America and India. There was, for example, John Mitchell's *Map of the British and French Dominions in North America*, which was published by Thomas Jefferys in 1755, and which showed the eastern part of North America from the Mississippi up as far as the southern part of Hudson's Bay. Another important milestone was the *Bengal Atlas* of Major James Rennell, published in London in 1779, and his *Map of Hindoustan*, published in 1782.

Thomas Jefferys (*circa* 1710-1771, the publisher of Mitchell's North American map, was a leading figure in the upsurge of British cartography. He was geographer to the Prince of Wales (afterwards George III) and was particularly notable for his charts of American coastal areas, based on actual soundings and measurements taken in the area by sailors. He was succeeded by William Faden, whose catalogue of 1822 lists over 350 items, including globes, plans and atlases, the latter made up, usually with the contents lists in manuscript, to suit customers' individual requirements. He published a North American atlas with thirty-four maps in 1777, and his published plans included those of the military operations of the War of Independence, and of New York and Philadelphia, and his maps those of Rhode Island, Carolina, Australasia and Africa. It was at Faden's workshop that the first sheets of the maps of the Ordnance Survey were engraved at the end of the eighteenth century, before that department possessed its own headquarters and staff.

John Cary (*circa* 1754-1835), who issued several atlases at the end of the eighteenth and beginning of the nineteenth centuries, was a member of the new class of map-maker, concentrating upon geographical excellence rather than on decoration, although aesthetically his maps are pleasing in their very plainness, and would for that reason be preferred by many collectors to their more elaborate predecessors. John Cary engraved his maps himself until about 1800, but thereafter devoted himself to the management of his concern, employing others to do engraving, colouring and other work. Through his firm, G. and J. Cary,[1] he issued a great variety of publications and other goods, from maps and plans to guides and road books, from celestial charts to geological maps and sections, from magic lanterns to orreries (see p. 80). His prices ranged from £10 10s. for a bound and fully coloured *General Atlas of the World* to £1 1s. for a four-sheet map of the world measuring a total of 178·8 × 96·5 cm, or for an extra 14s. it could be had mounted on rollers. By order of the Postmaster-General he supervised the survey of 9,000 miles of turnpike roads in Great Britain. Indeed, so many-sided was his work that he may be counted as one of the greatest influences in British

[1] The initials are those of his sons, George and John II; he is thought to have transferred the business to them, but to have remained an active partner; see also p. 83.

cartography since Saxton. The Cary plates were taken over by George Frederick Cruchley (fl. 1822-75) after the firm had closed down in about 1850. Cruchley's name is added to the maps thereafter, usually with acknowledgement to Cary. Cruchley's plates were acquired about 1876 by Messrs. Gall and Inglis. They were still being issued at the beginning of the twentieth century as cycling and motoring maps, but had been disfigured by many additions.

One of the greatest of the late eighteenth-century cartographers was Aaron Arrow-smith (1750-1823). He came to London from Durham in 1770 and was employed by Cary. According to some, he is supposed to have taken all the pedometer measurements and drawings for Cary's county maps; he certainly took the measurements for Cary's *Actual Survey of the Great Post Roads* (1784). He was independently established by 1790 in Long Acre, from which address he published his first map, now a rarity, *A Chart of the World upon Mercator's Projection, showing all the New Discoveries . . . with the Tracts of the most distinguished navigators since 1700.* This was followed by many other important maps and surveys, which enjoyed wide popularity abroad as well as in England. After his death in 1823, his sons Aaron and Samuel and his nephew John carried on his business, although his name continued to be used by them.

Arrowsmith's maps are plain, and apart from the title cartouches, without decoration. They were not intended primarily as decorative objects but for practical use, and their decorative qualities (which they most certainly possess) come from their Quakerish restraint, their splendid engraved lettering and their systematic and clear presentation of facts. They are large-scale maps and make splendid historical source material. Among his outstanding productions are his Pacific Ocean chart of 1798, in nine sheets (overall measurements 182·8 × 228·8 cm), his folio *Pilot from England to Canton* (1806; seven charts on twenty-three sheets), and maps of the United States (1796, 1802, 1815, 1819), Panama Harbour (1806), West Indies (1803, 1810), India in six sheets (1822), Africa in four sheets (1802, 1811), Asia (1801), Persia (1813) and Egypt (1802, 1807).

John Arrowsmith (1790-1873), nephew of the original Aaron, abandoned the large maps of his uncle, and in 1838 issued *The London Atlas of Universal Geography*, the best of its kind to be issued until that date. He knew many of the great Australian explorers and was able to use much of the information they had obtained and imparted to him in compiling his maps. Consequently, his maps of that part of the world are particularly fine.

Christopher (sometimes erroneously called 'Charles') Greenwood is decoratively among the most ambitious of the nineteenth-century cartographers. In 1834 he issued his *Atlas of the Counties of England*, remarkable for the excellence of its engraving. Each map is decorated with a view of one of the outstanding buildings in the county. He also issued separate county maps of 1-inch scale, such as Bedfordshire (four sheets, 1826), Cornwall (six sheets, 1827), Northamptonshire (four sheets, 1826), Yorkshire (nine sheets, 1817-18) and others.

There were two Greenwoods, Christopher and John. Christopher was born at Gisburn in Yorkshire in 1786 and died at Hackney in 1855. He first set up business as a surveyor at Wakefield, but set up his first London office in 1818 in Leicester Square.

In 1820 he entered into partnership with two George Pringles, father and son. By 1821 Greenwood's firm had become Greenwood and Co., county surveyors, with headquarters at 70 Queen Street, Cheapside, London. Christopher's younger brother, John, was a member of the company. Later the headquarters was moved to Lower Regent Street, but the firm's fortunes declined, and despite attempts by Christopher Greenwood to re-establish his fortunes, his cartographical activities had come to an end during the 1840's. John Greenwood had re-established himself as a surveyor in Gisburn by 1838.

Roughly contemporary with Greenwood, and one of his rivals, was A. Bryant who published twelve county maps and one of the East Riding of Yorkshire. His headquarters was at Great Ormond Street, London. Another of Greenwood's rivals was Thomas Hodgson of Lancaster with whom Greenwood carried on a public and highly sarcastic quarrel in the *Westmorland Advertiser*; but a more sensible attitude was advocated by a letter from a reader of that newspaper who declared, 'Instead of their ridiculous squabbles it is high time for them to attend to their proper business, or let subscribers manifest their disapprobation by withdrawing their names from the subscription.'

Greenwood sold several of his plates during his career. His 1818 map of Yorkshire was purchased in 1827 by the London map publishers, Henry Teesdale and C. Stocking, who published a lithographic copy of it without acknowledgement. Edward Ruff, a Fleet Street map-mounter, purchased Greenwood's map of Warwickshire and re-issued it, again without acknowledgement.

There were, too, the extremely attractive steel-engraved maps of Thomas Moule (1784-1851), published first in two volumes as *The English Counties Delineated* (1836-41, various editions) and later in *Barclay's Complete and Universal Dictionary* (various editions from 1842). The maps themselves are printed in considerable detail; roads are clearly shown. Heraldic charges, local views and allegorical scenes are all featured in abundance and the cartouches and lettering are conceived in the spirit of nineteenth-century romanticism. They have much charm (Plate 14).

Moule himself was an interesting character. He was not a cartographer in the generally accepted sense, but a scholar, a bookseller, a writer on antiquities and heraldry, and, for a period, an inspector of 'blind' or indecipherable letters at the G.P.O. Despite this he personally visited, in compiling his *English Counties*, every county except Devon and Cornwall, so at least he was conscientious. His other publications included *Antiquities in Westminster Abbey* (1825), *Heraldry of Fish, Notices of the Principal Families bearing Fish in their Arms* (1842), and he contributed commentaries to a number of picture books.

On the whole the development of Victorian cartography was along practical rather than decorative lines. It also became the province of governments and institutions rather than of individuals. In England its greatest manifestation was the Ordnance Survey, undertaken by the Board of Ordnance primarily for military purposes, first officially established in 1791, but which had not covered all of Great Britain until 1870. Many of these maps – Ordnance Survey and others – are very attractive and have by now achieved some sort of period charm.   They range from such miniatures as *A Hand-Atlas for Class Teaching* by Walter McLeod (1858), 15·2 × 9·8 cm and published at 3s. (Plate 15) to really large atlases published by such firms as Philip's and Bartholomew's, each of which is still supplying the public with maps and atlases.

In the nineteenth century the world map began to take on its proper shape.   True, not all areas were charted; in a lecture to the British Association in 1925, A. R. Hinks, a prominent geographer, mentioned that at that time Natal had still never been mapped, nor was this an isolated case.   But in the main the nineteenth century knew the shape of the world.   They knew Australia (it was still called New Holland by some) was an island; the Pacific had been accurately charted; China, Japan and the coastline of Antarctica had been resolved into their proper shapes; the world had, in fact, become less romantic, more familiar.   Perhaps that is one of the main reasons why collectors prefer maps issued before 1800 to later ones.

CHAPTER TWO

Celestial Maps and Charts

ALTHOUGH in comparison with terrestrial maps and charts celestial charts have been somewhat neglected, they do in fact form a most attractive branch of the subject, and moreover provide the collector with a category in which he can still find real rarities at something like reasonable prices.

Celestial charts existed in early times, for both Roman and mediaeval manuscript examples are known. One of the earliest printed specimens is a woodcut planisphere (i.e. the reduction of the sphere to a plane surface[1]) by Peter Apianus, dated 5 August 1536. As in the majority of examples made up to well into the nineteenth century, the constellations are represented by their actual figures as well as by names. This chart seems to have been issued separately, not as part of a book; but a number of books were issued at about this time with celestial charts as illustrations, notably those of Alessandro Piccolomini (1508-78), a Roman prelate whose works went through several editions and contained many woodcut celestial charts: some volumes had as many as forty-eight apiece. Piccolomini's collected works bore various titles, according to the edition – for example, *De la Sfero del Mondo e Dele Stelle Fisse* (1540) and *Della Sfera del Mondo* (1553); there were also editions in French. One characteristic of Piccolomini's charts was that he used Latin letters to distinguish individual stars, a method much used in the sixteenth and seventeenth centuries by French celestial globe- and chart-makers.

The first star-atlas proper was the work of Johann Bayer, a Bavarian astronomer, and was first published in 1603 at Ulm under the title of *Uranometria omnium asterismorvum, continens schemata nova methodo delineata*. Its line-engraved maps were the work of Alexander Mair. There were many editions, and it was reissued in various editions up to 1723. Bayer adopted the method of identifying the more important stars in constellations by giving them Greek and Roman letters, a method still used to-day, though it was slow at first to take root, particularly in England. Bayer's *Atlas* shows a total of 1,706 stars from the first to the sixth magnitudes.

Another German work, really a revised edition of Bayer, was *Cœlum Stellatum Christianum*, published in 1627 by Judas Schiller of Augsburg. Schiller may be said,

---

[1] The word 'planisphere' is now usually used to denote a plane polar projection of the celestial sphere with adjustable discs for showing the sky at any given date and time. An outstanding present-day example is Philip's *Planisphere*.

as the title implies, to have 'christianized' Bayer's astronomy. Andromeda becomes the Holy Sepulchre, Aries is strangely transformed into St. Peter, Hercules into the Three Magi, Taurus into St. Andrew, and so on. It is an idea that he may have got from the works of the English mediaeval scholar, the Venerable Bede, who is thought to have drawn a planisphere using such forms.

Yet another approach was shown by Erhard Weigel, Professor of Mathematics at the University of Jena, who made a *Cœlum Heraldicum*, in which the constellations are represented by the arms of European royalty. Weigel was also an architect, an inventor of scientific instruments and author of over a hundred learned books.

Ignatius Gaston Parides (1636-75), a French astronomer, mathematician, Jesuit and Professor of Rhetoric at the College of Louis the Great in Paris, made some excellent star-charts. His rare *Globi Cœlistis*, engraved on copper, went through three editions: 1673-4, published at Paris by Vallet; 1693; revised and published at Paris about 1700.

One of England's great astronomers, Edmund Halley (1656-1742), published in 1675 a first-class planisphere of the heavens, including a patriotic transfer of identity of the constellations of *Canes Venatici* to *Cor Caroli* (The Heart of Charles); it is said that a star in this constellation was unusually bright on the night before Charles II entered London at his Restoration, and so the constellation's name was changed by Halley. Halley himself was a celebrated mathematician and astronomer in his day, and his name is perpetuated for us by the comet whose period of orbit he discovered: Halley's comet. He became Astronomer Royal in succession to Flamsteed.

Among other seventeenth-century makers of celestial charts was the Frenchman Augustine Royer (fl. 1679-1700), whose work was based on that of Halley and Jacob Bartsch, and who was the first to name the famous constellation of *Crux Australis* (Southern Cross); the Dutchman Andreas Cellarius, whose work included *Harmonia Macrocosmica seu Atlas universalis et novus* (first published in Amsterdam, 1660; there were other editions up to 1708. It includes representations of the Ptolemaic and other pre-Copernican celestial systems); and Johannes Hevelius of Danzig (1611-87), author of the posthumous *Firmamenṭum Sobiescianum sive Uranographia* (1687-90). Many of the star-maps of this period indicated the magnitude of the stars by means of symbols of various sizes (cf. Plate 16).

An interesting, and now rare, little (18mo) celestial atlas was published in London some time between 1675 and 1690. This is John Seller's *Atlas Cœlestis*, the first pocket-size stellar atlas. Another atlas came a little later from the first Astronomer Royal, John Flamsteed (1646-1719); this also was called *Atlas Cœlestis* (1729 and subsequent editions). Its French edition, *Atlas Céleste*, was published posthumously in 1795. Flamsteed's celestial maps had figures drawn by Sir James Thornhill (1675-1734), famous as the painter of the inside of the dome in St. Paul's Cathedral, London. He used Bayer's method of designating stars by Greek and Roman letters, but supplemented them with numbers, a system which is still called 'Flamsteed's Numbers'.

John Bevis (1695-1771) published in 1750 a work largely based, with corrections, on Bayer's *Uranometria*, entitled, with a somewhat insular flourish, *Uranographia Britannica*, an insularity which is consciously emphasised by a frontispiece which shows

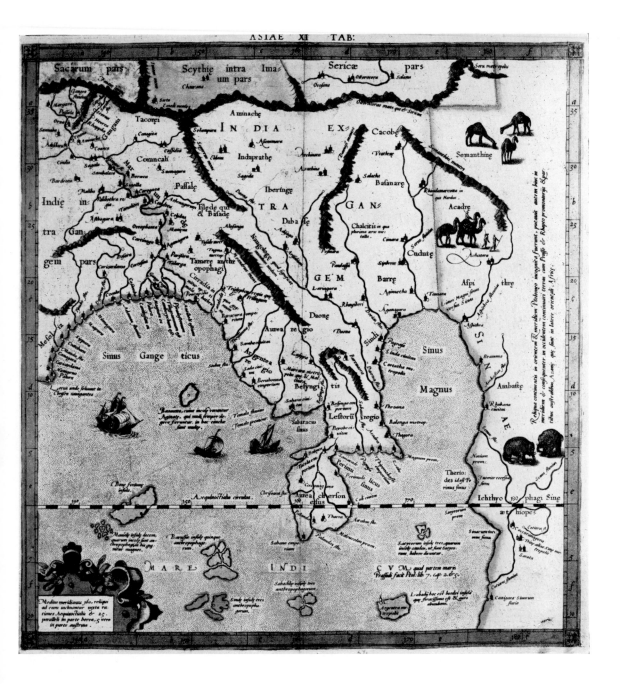

1. Mercator's Edition (Cologne, 1578) of Ptolemy. Detail: *Asiae XI Tabula.*   *British Library*

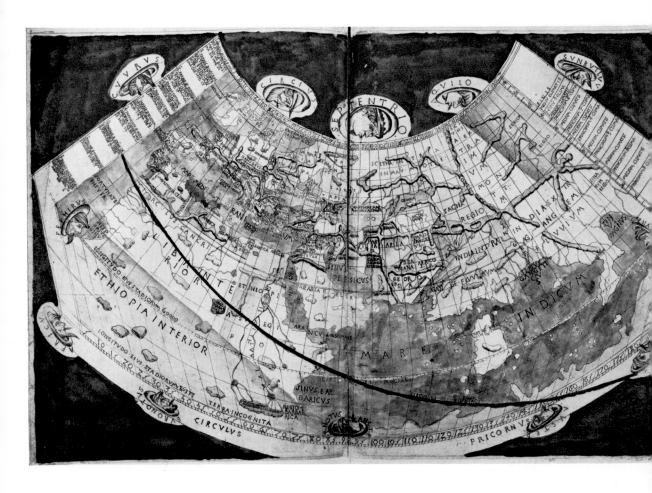

2. Map of the World, by Ptolemy, from *Cosmographia Bologna* (1477). The first engraved world–
atlas.
British Library

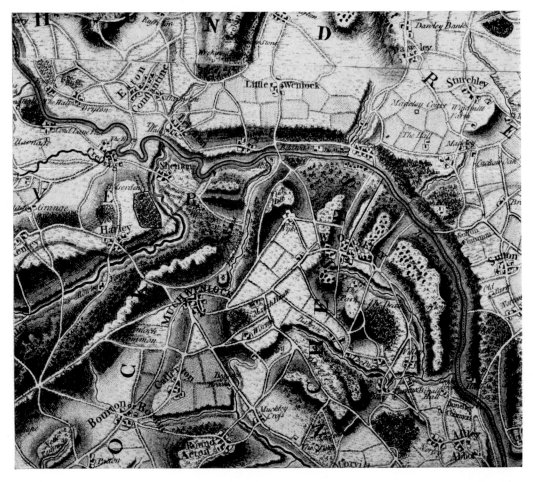

3. Detail from John Rocque's Map of Shropshire (1753). *British Library*

4. Detail from manuscript map of England *circa* 1250, by Matthew Paris (British Library, MS. Cotton Claud. D.VI).

British Library

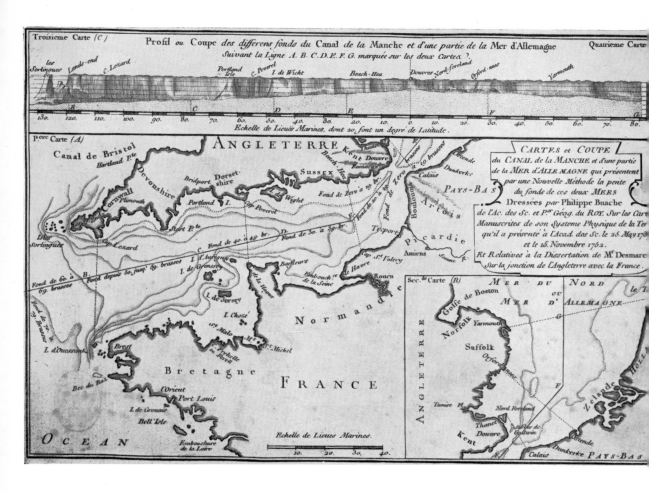

5. Philippe Buache: Map of the English Channel and German Sea, showing the use of isogonic lines (1752).

British Library

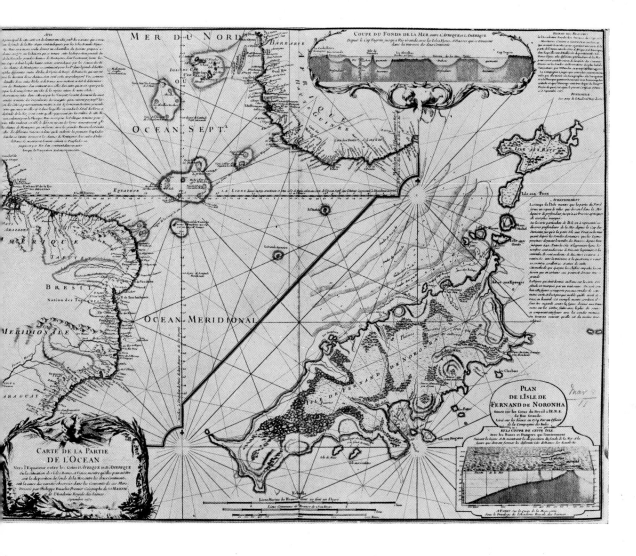

6.   Philippe Buache: Map of Part of the Ocean between the Coasts of Africa and America (1737).
British Library

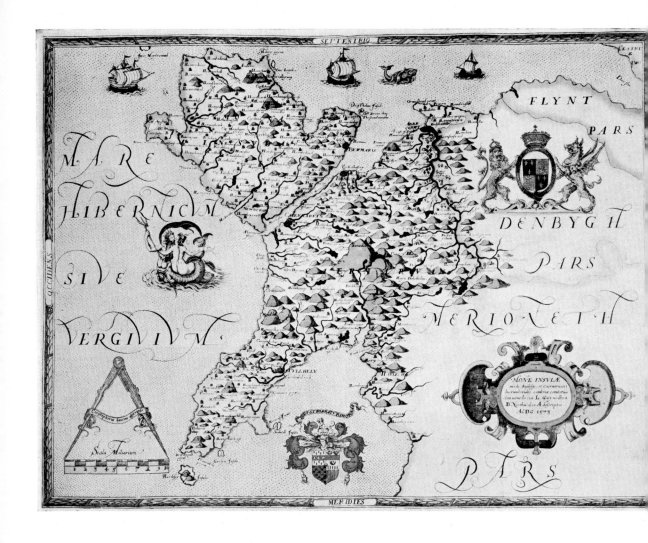

7.   Christopher Saxton: Map of Caernarvon and Anglesey.   *British Library*

8. John Norden: Distance Tables from *England: An intended Guyde* (1625). British Library

Somerſet.	Brightſtoll.	Bathe.	Welles.	Shepton.	Bruton.	Somerton.	Ilcheſter.	Glaſtonbury.	Bridgewater.	Taunton.	Charde.	Euell.	Wellington.	Willcombe.	Dunſter.	Croo-ke horne.	Froome.	Wincaunton.	Hunſpill.	Lamporte.	Miluerton.	Duluerton.	Mynehead.	Whatchet.	Pensforde.	Ilmiſter.
Ax-bridge.	12	16	8	11	16	14	17	9	13	18	23	21	24	23	28	23	18	20	10	15	23	32	30	22	11	21
Ilmiſter.	32	32	19	20	21	10	12	15	11	7	4	12	12	15	23	5	28	21	15	6	13	23	25	10	29	
Pensforde.	5	7	11	11	16	20	22	15	22	29	32	27	35	34	36	30	12	19	19	31	32	42	39	32		
Watchet.	34	36	25	27	31	22	25	22	10	12	20	27	12	7	5	24	36	33	14	19	11	11	7			
Mynehead.	46	43	31	34	38	28	30	29	18	17	25	34	15	11	3	27	42	38	20	24	10	11				
Duluerton.	43	46	34	36	40	30	31	41	20	16	22	33	11	3	8	28	45	41	23	24	11					
Miluerton.	34	38	25	27	30	19	21	22	11	6	14	14	3	13	18	25	31	14	15							
Lamporte.	27	26	14	15	15	5	6	8	8	11	12	11	15	17	22	8	22	16	12							
Hunſpill.	20	24	13	16	20	13	16	11	5	12	18	20	17	16	18	19	24	23								
Wincaunton.	24	22	12	9	3	12	10	13	23	25	23	12	30	33	27	19	10									
Froome.	16	8	11	8	18	18	18	15	25	30	31	20	36	37	40	27										
Crooke-horne.	33	22	20	20	19	10	8	15	15	12	6	8	16	20	27											
Dunſter.	38	42	30	31	35	26	27	26	16	15	23	33	13	13	8											
Wilſcombe.	36	40	38	29	31	22	23	23	13	8	15	25	5													
Wellington.	37	39	26	28	30	19	21	22	13	5	12	23														
Euell.	29	27	15	14	12	7	4	12	17	18	14															
Charde.	35	36	22	23	27	13	12	18	14	9																
Taunton.	31	33	20	22	24	15	15	16	7																	
Bridgewater.	24	27	14	16	20	11	14	13																		
Glaſtonbury.	19	18	4	6	10	6	8																			
Ilcheſter.	26	24	12	12	11	4																				
Somerton.	24	23	10	10	11																					
Bruton.	20	16	8	6																						
Shepton.	16	13	4																							
Welles.	15	14																								
Bathe.	9																									

The vſe of this Table.

THe Townes or places betweene which you deſire to know, the diſtance you may finde in the names of the Townes in the vpper part and in the ſide, and bring them in a ſquare as the lines will guide you : and in the ſquare you ſhall finde the figures which declare the diſtance of the miles .

And if you finde any place in the ſide which will not extend to make a ſquare with that aboue , then ſeeking that aboue which will not extend to make a ſquare , and ſee that in the vpper, and the other in the ſide, and it will ſhowe you the diſtance. It is familiar and eaſie.

Beare with defectes , the vſe is neceſſarie.

Inuented by IOHN NORDEN.

9. Section of Map from Michael Drayton's *Poly-Olbion*—Oxfordshire, Berkshire and Buckinghamshire. Engraved by William Hole.

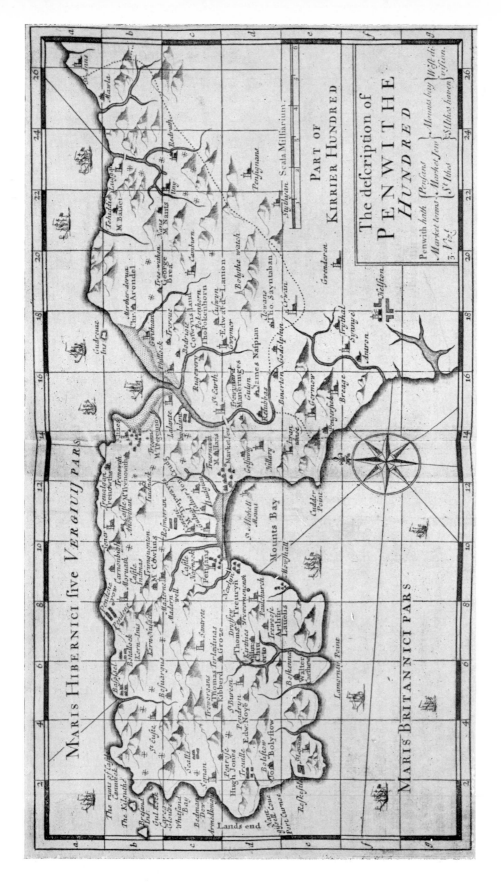

10. John Norden: Map from *The Description of Cornwall* (1728).

11. John Speed: Map of Wiltshire from *The Theatre of the Empire of Great Britain* (1611).   *British Library*

12. The Road from London to St. Davids. P. 32 of *Britannia Depicta or Ogilby Improved* (London, 1720). Engraved by Emanuel Bowen. 19·6 × 14·2 cm.

Author's Collection

13. The Road from London to Slough. P. 33 of *Britannia Depicta or Ogilby Improved* (London, 1720). Engraved by Emanuel Bowen. 19·6 × 14·2 cm.

Author's Collection

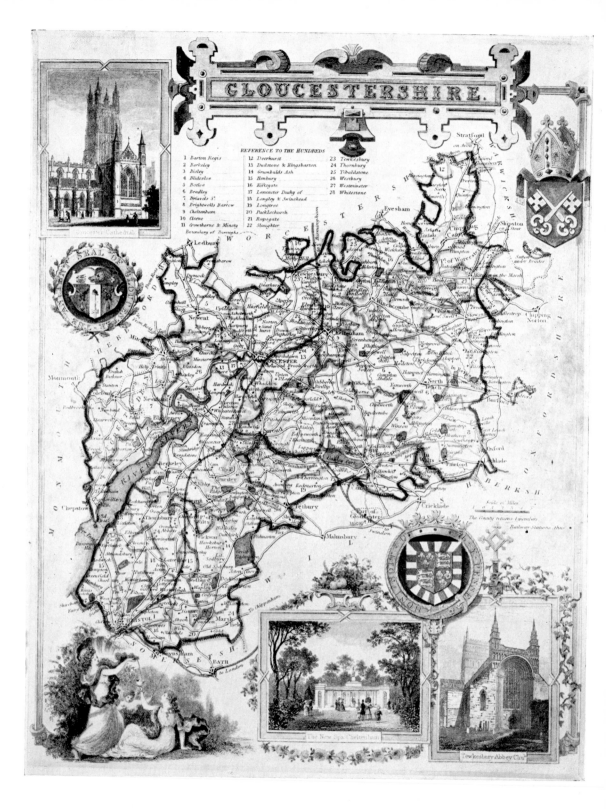

14.　Gloucestershire by T. Moule.　30·5 × 20·3 cm. (various edns. 1836–9).　*Author's Collection*

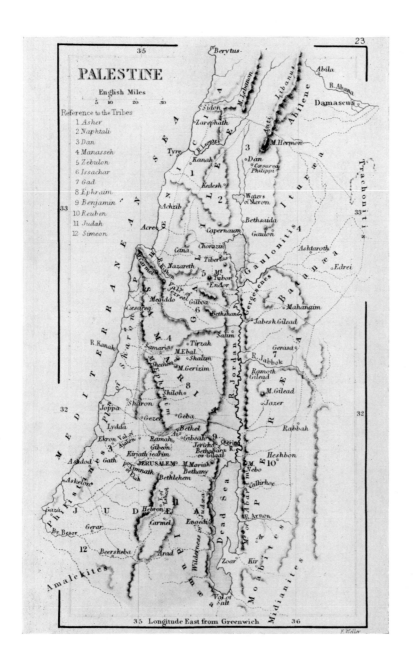

15. Palestine from *A Hand-Atlas for Class Teaching* by Walter
McLeod (London, 1858). Engraved by E. Weller.

Author's Collection

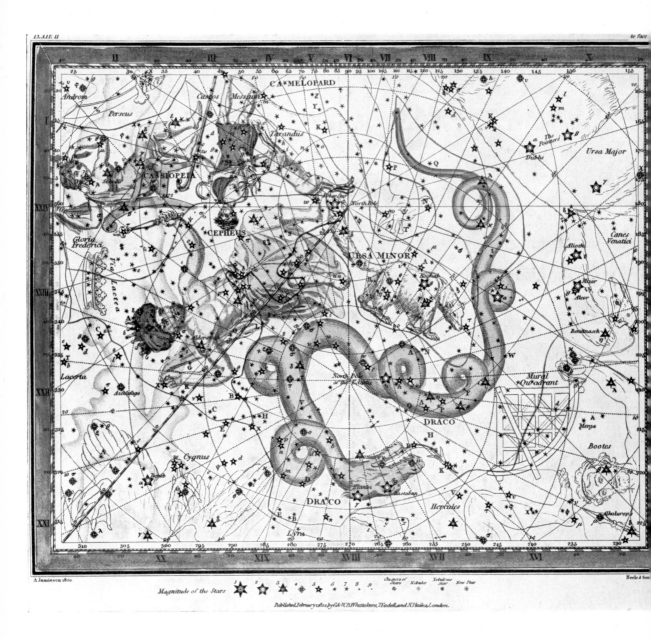

16. Alexander Jamieson: Map of Constellations from *A Celestial Atlas* (1822). The symbols for stars of various magnitudes are shown in the bottom margin. *Author's Collection*

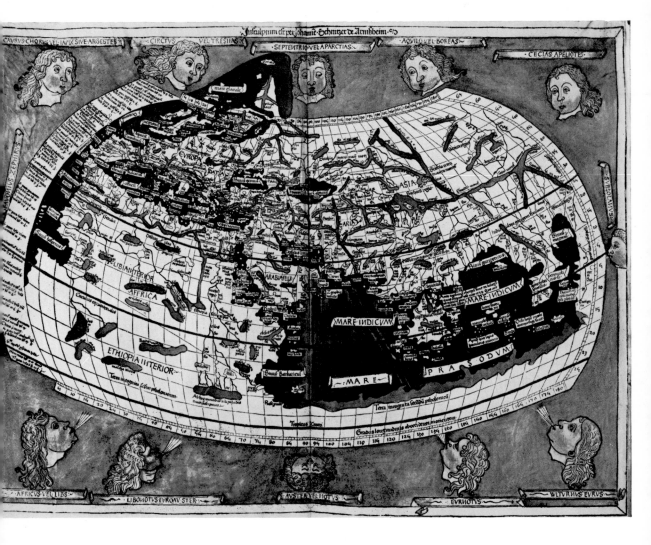

17.   World Map, from Ptolemy's *Geography*, published at Ulm in 1482.   *British Library*

18. Title cartouche from the map of the
New World in *Theatrum* of Ortelius
(Antwerp, 1590). *British Library*

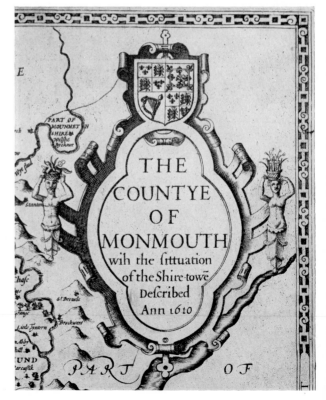

19. Title cartouche from John Speed's
map of Monmouth (1611 edn.).
British Library

the Hanoverian, Frederick Prince of Wales, dressed up as Caesar, accepting a copy of the atlas from the muse Urania. Bayer's system of lettering for indicating stars is followed, but any not shown in Bayer are marked by Gothic letters.

A German atlas by Johann Ehlert Bode (1747-1826) brought many of the constellations up to date and into the Age of Reason with a vengeance, with such appellations as *Globus Aerostaticus* (The Balloon, named after Montgolfier's balloon), *Tubes Herschelli Major* (Herschel's Great Telescope), *Machina Electrica* (The Electrical Machine) and *Officina Typographica* (The Printing Machine).

Incidentally, celestial charts may sometimes roughly be dated by such variations. The foregoing are obvious examples, but there are lesser-known constellations dropped from use from time to time. Such are the Pigeon and the Fly, obsolete since the early nineteenth century.[1]

During the eighteenth century much dissatisfaction was expressed at the method of showing constellations by figures (cf. Plate 16) and a method was devised of joining together the stars in constellations by means of connecting lines, a system still in use. There had been, too, some arbitrariness in the boundaries of each constellation, thus leaving some stars out of the scheme of constellations altogether. To overcome this, some astronomers and celestial cartographers introduced minor constellations of their own invention. During the nineteenth century this was all standardised.

The older system was, however, still used by some celestial cartographers such as Alexander Jamieson, whose *Celestial Atlas* (1822 – the maps are mostly dated 1820) contains some extremely attractive line-engraved charts (Plate 16) based largely on those of Bode. The engraving is beautifully executed and the charts are hand-coloured in some copies, plain in others – the former by far the scarcer.

As in terrestrial cartography the tendency in the nineteenth century was to get away from the picturesque and to concentrate upon accuracy of representation and to this end every modern technique was used, including photography. These developments are, however, outside the scope of this book.

For the sake of completeness, although they cannot be directly included as charts or maps, I should mention star catalogues, the first of which was compiled by a Greek about 160-120 B.C., and which have been made ever since. More closely related to celestial charts are solar and selenographic charts. Solar charts show such phenomena as sunspots, which were first recorded on diagrams forming part of a rare publication by John Fabricius, an assistant of Tycho Brahe, and published at Wittenberg in 1611. Another use for solar charts was to show eclipses; such charts appear in the several editions of Charles Leadbetter's *A Treatise of the Eclipses of the Sun and Moon* (1731, etc.), but they are more in the nature of diagrams and tables and hardly come within our scope. Yet the work is of considerable interest in that it gives horoscopes as well as scientific data, showing how, even at this comparatively late date, science had not entirely usurped more speculative subjects.

Selenographic charts perform the same service for the moon that solar charts perform for the sun, except that here there is more scope for the cartographer in that the moon's

[1] See also Chapter Five, p. 85.

surface is covered with prominent mountains, valleys and craters. Galileo first noted these features in 1610, and the first scientific map of the satellite was made in 1611 by Christopher Scheiner (b. 1675) a German. Many astronomers believed in those days that the moon was inhabited. Lunar maps are definitely a part of our subject and may be classed with stellar charts for their interest and aesthetic appeal. Many of the names used in those days to designate features on the lunar surface are still in use to-day. Among later important charts of the moon are those of Johann Heinrich von Mädler and Wilhelm Beer, two collaborators who between them charted and measured 7,735 craters, 1,095 mountains and 919 lunar formations.

There are, too, items of a highly specialised nature, such as maps of certain planets and charts showing the paths of comets. But these belong essentially to modern astronomy, and most of them fall outside the scope of all but purely scientific collections.

CHAPTER THREE

Methods of Map Production

THERE are three main printing methods by means of which maps have been produced: wood-cutting, copper-engraving and lithography. Mainly (although this is not an infallible rule) wood-engraving belongs to the earliest period of printed maps – the fifteenth and first half of the sixteenth centuries – lithography to the nineteenth century and copper-engraving to the whole of the period from the fifteenth century on.

At the outset one or two facts may be observed in connexion with the identification of maps by their means of production. For instance, lithography was not invented until 1796, so a lithographed map cannot belong to a date previous to that; indeed, it did not come into its fullest use until the second half of the nineteenth century, when it all but ousted copper-engraving. Again, wood-engraving was hardly used at all for separate maps after the middle of the sixteenth century and for maps in books after about the middle of the seventeenth century. But the collector should beware of applying these rules too rigidly.

The separate processes may be fairly easily identified if one is familiar with the methods used in executing them. We will consider each in turn.

WOODCUTTING or XYLOGRAPHY (Plates 17, 48, 50 and 51)

There are two ways of making wood-printing blocks: woodcutting and wood-engraving. In the former the wood is cut on the plank, those parts of the surface which are to show as black lines in the print being left standing, the remainder being cut away with knives, gouges and chisels (Fig. 2). Wood-engraving, a later invention, is carried out on the end grain of the wood. It is capable of much greater fineness than woodcutting. The lines (which in this case show white or colourless in the finished print) are cut by means of engraver's tools (Fig. 2). In other words, wood-cutting consists of black lines on a colourless background, wood-engraving of colour-less lines on a black background. Wood-engraving was brought to perfection by Thomas Bewick in the eighteenth and nineteenth centuries. But, in general, so far as map-making is concerned, the older form of woodcutting was used.

As will be seen from Fig. 3, printing from woodblock is a method of *relief* printing, in other words those parts of the block's surface from which the impression is taken are raised above the rest of the block. These are inked and applied under pressure to the paper or other material on which the impression is to appear. This method usually affects the paper and the ink. The pressure applied when the block is brought into

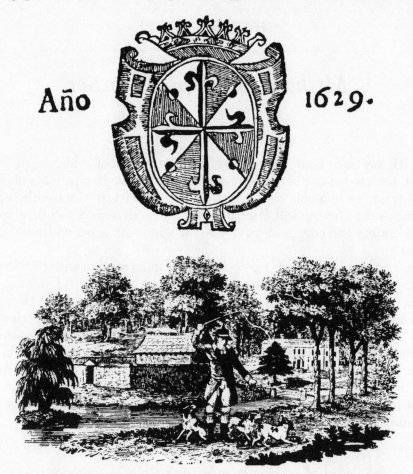

Fig. 2. Woodcutting *(top)* from the title page of *Historia del insigne* . . . by Mancano de Haro (Madrid, 1629)
Wood-engraving *(bottom)* by Thomas Bewick, from *The Chase* by William Somerville (London, 1814)

contact with the paper produces an indentation in the paper (Fig. 3); this is considered to be a defect by printers, but nevertheless it is usually present, although the passage of time may diminish its sharpness, particularly if the map has been bound up in a book or atlas, during which process the pressure applied in the binder's press will have more or less eradicated it. Paper distortion usually shows most clearly on the back of a map, but it may also be observed on the face; it is called the 'bite'.

The other effect, that on the ink, is called 'ink squeeze'. By the printing pressure, the ink covering the block's raised printing surfaces is sometimes forced out towards the edges of each area or line, making the colour of each more intense at its edges than at its centre.

Another characteristic sometimes present in maps printed from woodblocks usually needs a strong magnifying glass in order to be seen. It may be best described as a lack of continuity of a printed line or area. The paper on which old maps were printed is usually quite fibrous, with some portions raised to a microscopic degree above others, and if impressions are examined under a strong magnifying glass (e.g. ×12) it will frequently be seen that the lower portions are uninked, producing a somewhat mottled effect.

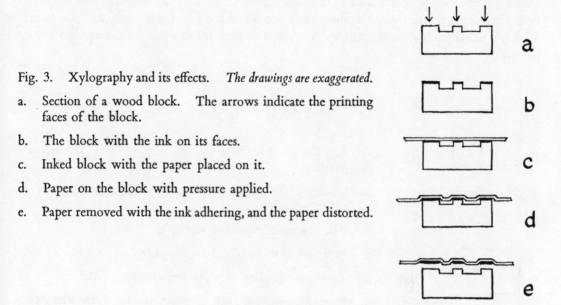

Fig. 3. Xylography and its effects. *The drawings are exaggerated.*

a. Section of a wood block. The arrows indicate the printing faces of the block.

b. The block with the ink on its faces.

c. Inked block with the paper placed on it.

d. Paper on the block with pressure applied.

e. Paper removed with the ink adhering, and the paper distorted.

Woodblocks were sometimes in use for many years, and in some cases they were brought up to date by having their lettering removed and replaced by movable type, usually of metal, through holes made in the blocks. A knowledge of type faces is useful in identifying such maps (see section on lettering pp. 67-68 of Chapter Four). Engraved lettering always has individuality and can vary from inscription to inscription and from letter to letter, whereas movable type is made to a standard, and such lettering will be the same in one place as in another. Moreover, it has a stiff appearance, as it cannot be curved or otherwise shaped into the map's outlines in the same way as engraved lettering.

Woodblock printing was a convenient method of map production from the printer's point of view, for it enabled him to produce his maps by more or less the same relief method as that used in printing the letterpress of his books, thus allowing him to print his illustrations from the same press.

Some of the most beautiful of woodcut maps are those produced by the Japanese *ukiyo-ye* artists.   A really poetic example is Hokusai's map of China, drawn by 'the old man mad about painting', as he called himself, in his eighty-first year in 1840-1; it is drawn in perspective, was engraved by Egawa Sentaro and published by Seiundo, probably at Edo.   For those interested in looking further into this aspect of maps, I have included one or two books in the first section of the *Bibliography* at the end of this book, under Hillier and Binyon.

COPPER-ENGRAVING, LINE-ENGRAVING *or* INTAGLIO-PRINTING
(Plates 12, 13, 32, 47, etc.)

This is basically the same method of printing as that used for printing banknotes, good-quality visiting cards and many postage stamps.   In it a metal plate is engraved or etched with the design,[1] the lines thus produced are filled with ink, and the paper, which is usually previously damped, is caused to pick up this ink by the application of pressure (Figs. 2 and 4).

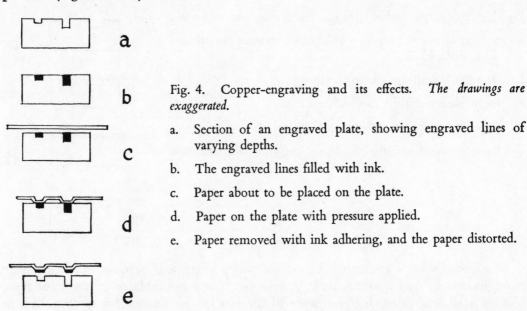

Fig. 4.   Copper-engraving and its effects.   *The drawings are exaggerated.*

a.   Section of an engraved plate, showing engraved lines of varying depths.

b.   The engraved lines filled with ink.

c.   Paper about to be placed on the plate.

d.   Paper on the plate with pressure applied.

e.   Paper removed with ink adhering, and the paper distorted.

The main characteristic of a map printed in this way is that the printed lines stand above the surface of the paper, as in a banknote.   This may usually be tested quite simply by lightly running one's finger over the surface, when the raised character of the lines can be felt quite distinctly.   There is, too, a slight distortion of the paper, similar to that described above under woodcutting, but in this case it is in the opposite

---

[1] It should, perhaps, be added that in this and other forms of engraving the engraver cuts the design in reverse.   Maps exist in which some versions have been printed in reverse, others correctly, as if the engraver had made an incorrect version then realised his mistake and corrected it. Certain Dutch maps of South America are examples.   Steel is sometimes used instead of copper for the plates, and sometimes the design is bitten in with acid instead of being engraved, in which case it is known as etching.

direction – that is, not *impressed* by the lines and areas of printing, but *pulled* by them (cf. Fig. 4). There is, too, usually a plate mark in the unprinted border, which is caused by the pressure of the metal plate during printing.

Also, generally speaking, the lines in copper-engraving are much finer than those in woodblock printing, the methods of shading, hatching, etc., are entirely different and the lettering is on the whole more graceful. By virtue of this greater fineness of line, copper-engraved maps were capable of greater accuracy in interpreting the geographer's data.

Conventional signs, such as those for towns (Plate 47), were usually impressed on the metal plate by a punch.

Often metal plates were kept in use for a very long time.[1] Even when they were worn down – and this usually happened only after 3,000 or 4,000 impressions had been taken – they could be re-engraved and their life further prolonged. Some of Saxton's plates were used for over 200 years, despite the fact that they had become geographically obsolete. In some cases attempts were made to bring the maps up to date, for engraving may be deleted by scraping down the lines and raising the surface by hammering the plate from behind. Often in such cases traces of the deleted impression remain. Such alterations make a fascinating study for the collector.

In both woodcut and copper-engraved maps tinting, when it was used, was applied by hand, in watercolour. More will be said of this later in the present chapter.

LITHOGRAPHY (Plate 16)

Literally, lithography means printing from stone, but other surfaces, such as zinc, are sometimes used. Unlike either of the processes just described, it is entirely planographic. Fundamentally, the method is based on the antagonism of water to grease. The stone is made chemically clean and is thus extremely sensitive to grease; the design is then drawn on its surface with lithographic ink or chalk. This done, those parts of the stone without drawn lines and areas are desensitised by the application of gum arabic and dilute nitric acid. The stone is moistened, and an inked roller applied over it; the ink will adhere to the drawn parts but not to the other areas.

Paper is then applied under pressure to the surface and a print obtained; innumerable prints may be taken in this way. By the use of separate stones for each colour, multi-coloured prints may be built up, and the advantages of the method applied to the printing of maps are therefore obvious. The colourful effects that may be obtained from lithography may be realised from the fact that such polychrome objects as cigar-box labels, book-jackets and postage stamps are frequently printed by it.

Needless to say, modern map production uses mechanical forms of the method just described, yet fundamentally it remains the same. In any case, we are not in this book dealing with maps produced after 1850, before which date the foregoing method was the one used.

[1] Under Speed's entry in *A List of Cartographers* in the present book, I have described his *Theatre of the Empire of Great Britain* through many of its editions in order to demonstrate the variations through which such plates can go during their lifetime.

It is essentially a nineteenth-century method. As I said at the beginning of this chapter it was not even invented until 1796, and it was some time after that before maps were first lithographed.

Maps printed by lithography may be recognised by the fact that their colours are often printed and not applied by hand with watercolour, and by the absolute flatness of the impressions. There are no upstanding lines, as in copper-engraving, and no distortion of the paper, as in both copper-engraving and woodblock printing. The impressions are in some cases not quite so sharp as in the other methods described, but the quality of line may be as fine as in copper-engraving or as bold as in woodblock printing.

It is possible for the same map to be printed in both copper-engraved and lithographed forms. This is done by taking an impression from the copper-plate and transferring it to the stone.   In the process it is comparatively easy to make adjustments and alterations to an impression, and this can provide much fascinating detective work for a collector; moreover, successive editions of the same map taken from successive lithographic transfers can provide even more such variations.

An excellent guide to the kind of thing to look for in maps of this kind is Harvey and Thorpe's *The Printed Maps of Warwickshire* (see *Bibliography*). Incidentally, it gives also variants in the editions of many copper-engraved maps.

☆   ☆   ☆

We must now consider the colouring or illumination of maps. As we have just seen, before the nineteenth century and the use of lithography, colouring by printing was almost unknown. Indeed, the only instance recorded of printed colour being used at all on maps before the nineteenth century was for lettering, as in the Ptolemy editions of 1511 and 1513, when some names were printed in red. Apart from that instance watercolour was used, certain conventions being followed which are largely still observed to-day: water areas in blue, countryside in green, mountains shaded in brown and red, and so on. Before engraved lines were used for indicating boundaries, painted coloured lines were used.

Colour was not, however, used invariably or even widely in the earliest days of map-printing. Many people felt that it added nothing to the chaste lines of engraving. Yet maps with contemporary colouring are to-day usually more highly esteemed than plain ones. From the sixteenth century onwards maps were sold either plain or coloured, and map colouring was a trade recognised by the painters' guilds. It was an offshoot of the art of the miniaturist and illuminator. Indeed Georg [Joris] Hoefnagel and Taddeo Crivelli (see *A List of Cartographers* and p. 21) were as celebrated as miniaturists as they were as cartographers.[1]

The type of wash applied to maps varied somewhat. For instance, in central European woodcut maps of the sixteenth century it is rather thickly applied in large areas. In many maps of the next century the colour was applied in light outline washes, and an added richness was obtained from the use of liquid gold.

[1] It is said that Jan van Eyck made a terrestrial globe for Philip the Good (Fazio, Bartolommeo, *De Viris Illustribus*.)

Some contemporary accounts survive of the methods used in colouring maps. Several chapters are devoted to it in William Salmon's *Polygraphice* (London, Fifth Edition, 1685). Another, by John Smith, blacksmith, tool- and clock-maker and Unitarian, was published in 1701 (there are several subsequent editions) in *The Art of Painting in Oyl*; another, by Robert Dossie, was published in 1764 in *The Handmaid to the Arts* (again there are several editions); yet another appeared in *The Complete Young Man's Companion*, published in 1807 at Manchester. Smith's is the most detailed of these, so far as the instructions are concerned (though Salmon gives more details of the actual colours). Smith commences by giving instructions for making a lye (wash) and a gum-water medium, and continues with a description of the colours used and instructions for making them. They are copper-green, myrrh (stone-colour), crimson, indigo, gamboge, red lead, blue bice, carmine, vermilion, umber and grass green. He then continues by giving instruction as to the method by which these colours are used and this is of such capital interest that I shall quote it in full.

On the Practice of Colouring Maps

The Colours being prepared as before is directed, you may proceed to Colour a Map in this manner, first take notice of the several Divisions in a Map which distinguish one Kingdom from another, or one County from another, which are known by certain Lines, or Rows of Pricks, or Points of several Sizes and Shapes agreeable to the Divisions they are to denote. As for instance, *Portugal* is distinguished from *Spain* by a row of large Points, or Pricks, and the Provinces of that Kingdom, or Shires, as we call them in *England*, are distinguished one from another by Lines of lesser Points or Pricks. Now if you were to colour the Kingdom of *Portugal* do this, first with a small Camel Hair Pencil in a Ducks Quill, colour over all the Hills within the Large prick Line that divides it from *Spain* with the *Tincture of Myrrh* very thin; then if there be any Woods, dab every Tree with the point of a very fine Pencil dipt in Grass Green, made of *Copper Green* tempered up with *Gum-Boge*, but in dipping your Pencils into any Colour, stroke it against the sides of the Pot or Glass in which you put it, that the Colour may not drop from it and spoil your Work; then with another Pencil dipt in *Read-Lead*, tempered thinly with Gum-Water, let the *Principal Cities and Towns* be done over that the Eye may more readily perceive them. Lastly, with a Ducks Quill Pencil dip in some Colour, as *Copper Green*, and trace out the Bounds of one of the Provinces, keeping the outmost Edge of the Pencil close to the Pricks, and be careful to lay your Colours all alike, and not thicker in one place than in another, and when 'tis almost dry, take another clean Pencil of the same Size, and dip it in Water, stroaking the Water out well, and therewith rub upon the inside of the coloured Line, till it take away most of the Colour on the edge, and make it grow faint and lose it self by degrees, and continue so to do till you have gone quite round; then take Yellow made of *Gum-Boge*, and go round the inside of the Pricks that divide the next Province, sweetning over the innermost Side of it, when almost dry, with a Pencil dipt in Water, as you did before, do over the next to that with the *Crimson Tincture* made with *Cochinele*, or thin *Carmine*, and the next to that do round with *Red-Lead*, and the next to that with *Grass Green*, and the next to that with any of the former Colours that will so agree with the Work, that two joining Provinces may not be coloured with the same Colour, for then you could not distinguish them so well by sight.

And in this Work of dividing, observe, That when your Boundary Lines pass through

Woods already coloured, or Hills; observe then, I say, to miss the Colour of those Woods and Hills in your drawing a Colour round the Province, and be careful also not to draw any Colour over the Cities or Towns that are painted Red, for that spoils the Beauty of it.

And when you have coloured over or divided all the Counties, then colour *the Sea-shoar*, and all *Lakes of Water*, if there be any, with thin *Indico*, working of that side of the Colour which is from the Land faint, with a wet Pencil as before was taught, and if there be any Ships, colour the Water shaded at the bottom with the same *Indico*, painting the Hull of the Ship with *Umber*, the Sails with Tincture of *Myrrh*, and the Flags with *Vermillion* or *Blue Bice*; and if they are represented as firing their Guns, let the fire be done with *Red Lead*, and the Smoak with very thin *Bice*, and as for the Margent or *square stick of Degrees, as the Gravers term it*, which goes round the Map, let that be coloured either with Yellow or Red-Lead, or Crimson, none but those three Colours serving well for this purpose.

As for the Compartment or Title, which consists generally of some neat Device to set the Map off, and make it appear more beautiful, it may be coloured according to the Nature of it. As for instance, *Crowns* or any thing representing Gold with Yellow, shadowed in the darkest parts of the Graving with Orpiment, the Hair of Men or Women with Tincture of *Myrrh*, or if Black, with half Water half common Ink, or with burnt *Umber*; the Flesh of Women or Boys with a very little of the Tincture of *Cochinele*, in a large Quantity of Water, and Garments either with thin Green shadowed with thicker, and with the Tincture of *Cochinele* made thin with Water, and shaded with the same Colour thicker, and thin *Bice*, and shadowed with a thicker mixture of the same, or with *Vermillion* shaded with *Carmine*. In general observe, That the Colour must be laid on the lightest part of all Garments, very thin and deeper in the Shades, for then the more beautiful it will appear; the thick of the same Colour being the most natural Shade for most Colours, except *Yellow* and *Blue*, for *Blue* sometimes requires to be shaded in the darkest Places with a Black, or at least with a thick *Indico*; and *Yellow* requires *Red-Lead* or *Crimson*, and sometimes it appears very pleasing when shaded with *Green*.

If you are to paint Clouds, do them sometimes with Tincture of *Myrrh*; and in some Cases, with a very thin Crimson, and for Variety, you may do some with thin Ivory Black, ground very fine, and tempered up with much Gum-Water. Smoak is best represented with very thin *Blue Bice*, and if you are to colour any Representation of Sea Waves, do it with *Indico*.

If you are to colour any Representations of Land, do the lightest parts over with very thin Yellow that represents a Straw Colour, shading it in some places with Orpment; and in others let a light Green be laid, and shade it with a deeper Green: Rocks must be done with Tincture of *Myrrh*, or of *Soote*, and the Trees some with *Copper Green*, some with *Grass Green*, and some with thin burnt *Umber* and *Gum-Boge* mixt; Houses may be done with *Red-Lead*, and the Tiles with *Vermillion*, or with *Bice* to represent Blue Slat, Castles may be done with Tincture of *Myrrh* in some parts, in others, with thin *Red-Lead*, and the Spires and Pinacles with *Blue*.

But when all is said that can be said, the only way to colour Maps well, is by a Pattern done by some good Workman, of which the *Dutch* are esteemed the best; three or four such Maps colourest [*sic*] by a good Artist, is sufficient to guide a Man in the right doing of his Work: But if he cannot obtain this, he may by a few Tryals grow a good Artist in a short time; for this is only attained to by Practice, and if a Man does spoil half a score Maps in order to get the knack of colouring well at last, there's no Man that is ingenious will grumble at it, or grudge at the Charge.

The hardest thing in this Art is, to know rightly how to make and prepare the Colours which here is taught faithfully: And if your Paper be good and bear the Colours well, without suffering them to sink into it, all that are here mentioned will lie fair and pleasant to the Eye, and 'tis the Fairness of the Colours that is most esteemed in this Art of Map-Painting: But if the Paper be not good and strong, no Art can make the Colours lie well; therefore in buying *Maps*, chuse those that are Printed on the strongest or thickest Paper: For they colour best, provided the Paper be well sized, and indeed it will be found, *when we have taken the greatest Care we can*, that Colours will lie fairer, and look more bright and pleasant on some Paper, than on other sorts, tho' they seem to be as strong.

☆ ☆ ☆

Another important factor in the identification of maps is the watermarks of the paper on which they are printed. There are literally thousands of such marks and their variations, and to attempt any sort of complete list would require several large folio volumes. I have therefore thought it best, in order to give some minimum indication of the problems involved, to reproduce the excellent paper on the subject by the late Edward Heawood, which is printed in the *Appendix*. Even this covers only a part of the subject and for those who wish to look into it more deeply, I have enumerated in a special section of the *Bibliography* some standard works.

A word of caution to the collector should be added here. The identification and dating of old maps is a difficult and complicated business, and the evidence should all be very carefully sifted before a conclusion is reached. As I have already said, blocks and plates were often used over and over again before being destroyed, and an impression might, judged only by its printed surface, belong to any year before that. One should look for evidence from the impression itself: was the plate altered or 'edited' during its lifetime, and if so does the impression show such alterations or modifications? If it does, then it must belong to a date after that at which the alteration was made. Does the impression show signs of wear on the plate? If it does, it is almost certain to be a late impression. Having ascertained such points as these, the paper should be examined for its watermark and that in turn looked up in one of the watermark lists (see *Bibliography*). As will be seen from remarks in the *Appendix*, even this is no infallible guide, but it may quite often enable us to date a map within ten years.

Geographical evidence on the map itself should be used for dating with the greatest caution. Map plates were used over such long periods, often with no amendments at all, not even of the date, that all one can say, for example, in the case of a map which shows a discovery made in 1750 is that it cannot belong to a date previous to that, which is after all not much.

☆ ☆ ☆

Finally, since this chapter deals primarily with methods of map production we will briefly consider some oddities of cartography. Among the most attractive of these are maps printed on silk or satin. Two of these have been already mentioned in Chapter One (pp. 33 and 41) – namely, Baptista Boazio's rare map of Ireland, one impression of which is known printed thus, and John Speed's map of Ireland printed on satin; again but one impression is known in this state. Maps of China were, in the Han

Dynasty (200 B.C. – A.D. 200), executed on both silk and wood; but these of course were manuscript examples.

Miniature maps have sometimes been made. John Speed's *Prospect of the Most Famous Parts of the World* was issued in miniature editions in 1646, 1668, 1675. Miniature copies of Saxton's maps, wrongly attributed to Speed, and reduced and engraved by Pieter van den Keere were issued as *England, Wales and Scotland described and abridged* by George Humble in several distinct editions, all dated 1627, but varying in the settings of the text. Abraham Ortelius issued a pocket-size *Epitome Theatri Orbis Terrarum* in 1601. Yet another which might be mentioned, although it is well outside the period covered by this book, is the *Atlas of the British Empire*, produced for Queen Mary's Dolls' House by Edward Stanford Ltd., *circa* 1928. It contains twelve coloured double-page maps and measures only 4·4 × 3·4 cm.[1]

Several packs of playing cards with maps on them have been issued at different periods from Tudor times onwards. Such are *The 52 Countries* [sic] *of England and Wales, described in a Pack of Cards* by Robert Morden (1676) and *Recreative Pastime by Card-play* by W. Redmayne.[2] It was convenient to the makers of these cards that there were then fifty-two counties in England and Wales. Of the two packs Morden's is the more attractive. The map in each case occupies the centre of the card, in a panel at the bottom of which the county's length, breadth and circumference in miles are given, together with the name of the county town, its distance from London and its latitude. In a panel at the top are the details of the suit and the card's value. In each suit the King and Queen are shown as Charles II and his queen, Catherine of Braganza. The knaves are shown as young men, different in each suit.

Geographical games with maps include *Middleton's New Geographical Game of a Tour through England and Wales*, published in 1829.

Tapestry maps also are known. Brilliant specimens were made on William Sheldon's looms at Weston and Barchester in Warwickshire in the late sixteenth and early seventeenth centuries. The counties represented on them include Berkshire, Gloucestershire, Oxfordshire, Warwickshire and Worcestershire, and London and district are included too. Four of these tapestries were purchased by Horace Walpole and presented by him to Lady Harcourt who displayed them in a specially built room at Nuneham.

[1] For details, see *Catalogue of the Library of Miniature Books collected by Percy Edwin Spielmann*, London, 1961.
[2] See also p. 81 for cards issued by Joseph Moxon (1627-1700).

CHAPTER FOUR

Decoration and Conventional Signs

AS in all other forms of visual art – and it is obvious from what has already been said that they are that in addition to being scientific documents – maps reflect the styles of the ages in which they were made. Here, just as much as in pictures, one may discern fashions and styles developing and giving expression to themselves over the ages. Thus in the cartouches and lettering of the maps of Christopher Saxton and John Speed we have all the pageantry and colourful courtliness of the Elizabethan Age; in the maps of de Hooge and the Blaeus, the very quintessence of the baroque; in Bowen's we have the rococo, in those of Delisle eighteenth-century elegance. Moule, Langley, and Pigot are as essentially a part of nineteenth-century romanticism as Pugin's Scarisbrick Hall, or Maclise's illustrations to Moore's *Irish Songs*.

A knowledge of this aspect of map-making is as essential to the collector as its more technical side, and it can be of considerable help to him in arriving at identifications. At least, in an unsigned map, it can help him to assign it to a period of origin. As we have already observed several times (and the point cannot be laboured too much), plates and blocks were frequently in use for many years after they were engraved, so some caution must be observed, yet style and ornament can at least provide a starting-point in assigning a map to a period, a date, a cartographer.

We will deal with various types of ornament separately and under headings, with suitable illustrations in each case. In general, however, it will be observed that the earliest maps are but sparingly decorated, that decoration reached high peaks in the sixteenth and seventeenth centuries, that it was more sparingly used as maps became more scientific in design in the eighteenth century, although the cartouches were still elaborate, that decoration tended to return with many maps in the nineteenth century.

CARTOUCHES

These were used on maps to contain either the title, a key, the maker's name, the dedication, the map's scale or for some other such purpose or purposes.

In the earliest woodcut maps they were mostly very restrained, in most cases being little more than simple rectangles, although some were plainly decorated (Plate 17). Real elaboration came in with the magnificent copper-engraved maps of the sixteenth century, although earlier examples of maps produced by this method had quite simple cartouches. Saxton and Speed (Plates 19, 21, etc.) had some of the most elaborate

61

cartouches ever made, proliferations of strapwork,[1] volutes and heraldry. In some cases there are several cartouches on one map, joined together by interconnecting straps. Some contain views, with legends on flying scrolls, some contain pictures of local antiquities and topographical features. Nor was the elaboration seen only on English features. Ortelius's Map of the New World from his *Theatrum* (Antwerp, 1590) has a cartouche containing its title and decorated with masks, sphinxes and swags of fruit as well as entablatures, straps, volutes, palms and ribbons (Plate 18). Other maps by Ortelius contain equally elaborate cartouches and one can at random pick out the sprawling nude figures of men and women, vases of flowers that would be worthy of a painting by Van Huysum, and birds both mythological and merely exotic.

These cartouches blossom out into architectural schemes of vast pomposity on the title pages, as may be seen, for instance, in Waghenaer's *The Mariner's Mirrour* with contemporary navigational instruments also incorporated, and, most magnificent of all, in Saxton's *Atlas*. In this, Queen Elizabeth, as if in a portrayal by Nicholas Hilliard, sits beneath an architectural canopy that could have come direct from a setting for a masque, while below is a double cartouche decorated with fruit and musical instruments. In many ways it resembles late sixteenth-century Dutch and Flemish cartouches, which showed much Renaissance detail, taken by engravers from pattern books of the period. It was probably engraved by Remigius Hogenberg.

On the whole seventeenth-century cartouches were less elaborate than those of the sixteenth century, although here and there elaborate examples may be seen. Those illustrated in Plates 22 and 24 are typical. More naturalistic detail than hitherto is in evidence and often the cartouche itself is composed of naturalistic elements. Explorers and natives are shown trading, and the title appears on a tent of sail tied to a group of trees (Plate 23). The tendency is definitely towards the baroque and in some cases the baroque spirit breaks through entirely in all its magnificent abandon, as in the map of the Territory of Frankfurt in Blaeu's *Novus Atlas* of 1640, where the map itself is made into virtually an outsize cartouche by being surrounded with coats of arms and allegorical figures.

On Blaeu's maps many of the cartouches have figures on or supporting them, symbolic of the trades or inhabitants of the area or country portrayed. Some seventeenth-century cartouches are comparatively plain, as in the notable case of the maps of Robert Morden (Plate 36).

The tendency for greater naturalism continued in the eighteenth century, although rococo was introduced about 1720 to good effect. This is particularly noticeable in the maps of Emanuel Bowen, whose cartouches often have a pretty Chippendale appearance (Plates 12 and 13). Many were more elaborate, like the cartouche on the map of Barbados by W. Mayo (1722) in which the rocaille effect is completed by a pair of dolphins blowing water through their nostrils into a shell pool.

Other eighteenth-century fashions, such as chinoiserie and romantic ruins, are to be seen on many maps. Imaginative scenes, too, are often incorporated; a common motif was that of a native riding on the back of a crocodile; it was used by Homann,

---

[1] Strapwork originated from designs of interwoven and interlaced straps of leather with their ends curling.

Moll, Senex, Seutter and others. In de la Rochette's map of the Cape of Good Hope the title, author's and engraver's names and the date appear on a forbidding-looking cliff in front of which are native huts and trees, while on the plain spreading in front of all of this two elephants are hunted by horsemen and a third lies already dead.

Many eighteenth-century cartouches were completely plain and in some cases, notably the maps in J. Cary's *New and Correct English Atlas* of 1787, had become a plain tablet bearing the map's title (Plate 25).

In the nineteenth century this plainness predominated, decoration being usually in the form of vignettes and views. But, on the other hand, some tended to return to an elaboration almost comparable to that of Saxton, except that it was coarser and heavier; in short, typical of Victorian revivalism. Obvious examples are the maps of T. Moule (Plate 14).

In coloured specimens of early maps, the cartouches are sometimes left uncoloured, although in special copies for important clients or patrons, they were elaborately coloured and illuminated with liquid gold.

BORDERS

The borders around the edges of maps are closely related to their cartouches, and in a more limited manner go through much the same stylistic changes and developments. They were, however, used far less than cartouches and indeed are comparatively rare, at least so far as decorative specimens are concerned.

On some early maps they are in the form of completely plain frame-lines. In others, as in the Ptolemy *Geography* printed at Ulm in 1482, they are put to practical use in incorporating degrees of latitude (or more precisely, clima or zones of equal daylight) and longitude (Plate 17), a device still in use.

More elaborate borders are to be seen in maps of the sixteenth century. Some, such as the world map in Peter Apianus's *Cosmographia* (Antwerp, 1545) form pictures of mythology and allegory on which the map is, so to speak, laid down. Others, as in the *Theatrum* of Ortelius (Antwerp, 1590), are extended into corner ornaments of elaborate Renaissance patterns. Saxton's maps provide border patterns which are probably derived from picture-frames or architectural cornices (Plates 33, etc.), and Speed's show conventional patterns which sometimes, as in his magnificent map of Cambridgeshire, blossom into a riot of heraldry (Plate 28). Similarly, Blaeu's borders, as in the map of America from his *Novus Atlas* (Amsterdam, 1635), sometimes contain whole series of views, figures and plans. Later the tendency was almost overwhelmingly to use the border for showing degrees of latitude and longitude, although some contained decoration, too, right into the nineteenth century.

A selection of border patterns is in Plates 26, 27, 29-34, 37, 38 and 40.

SCALES AND COMPASS ROSES

These are of practical value, but often are decorative too. Usually on the scales a pair of dividers is extended, and in some cases this idea is elaborated. In the 'Scale of Miles', for instance, on John Speed's map of Cornwall, the dividers are held by an allegorical boy. In some of Saxton's maps he incorporates his signature into the scale design. Some cartographers use the sensible idea of a scale incorporating the standards of several nationalities. Theodore de Bry's chart of southern England from the Isle of Wight to Dover has its scale shown in English, Spanish and Dutch leagues, the whole framed in a decorative cartouche surmounted by a mask. In Ortelius's map of Russia, Muscovy and Tartary, from his *Theatrum Orbis Terrarum* (Antwerp, 1570) the scale is shown in Russian, English and Spanish miles; in this case, apart from one or two meagre scrolls at the ends of the tablet, it is plain. On some maps the scales are completely unornamented, but more have cartouches as elaborate as those used for the titles.

Compass roses are frequently beautiful objects in themselves. In some cases, where loxodromes are used, several may appear on one map, as in Plancius's map of France from the 1606 English edition of Ortelius's *Theatre of the Whole World*. Many ingenious variations are worked out for the design of compass roses by various cartographers. Gastaldi, for instance, uses a spire to represent the nothern pointer and a cross to indicate the eastern pointer, presumably because the Holy Land lies to the east. Baptista Doetecum used the latter device too. One of the commoner designs for the northern pointer is the fleur-de-lys. The number of points varies, but is always based on the number four; many have thirty-two, others sixteen or eight. Most have all the pointers, except that of the north within a containing circle, but some, like Speed's map of Anglo-Saxon Britain, have all the points outside the circle like a bursting star or *feu de joie* (Plate 39).

Here and there roses are represented by semicircles. One example is the map of Guiana in Dudley's *Dell' Arcano del Mare* (Florence, 1646-7); another is the map of Middlesex from Speed's *Theatre* (London, 1611-12), in which the scale, surmounted by dividers and suspended from strapwork, has a semicircular rose at its base.

Scales and compass roses are not present on all maps. For specimens, see Plates 7, 25, 26, 41, 42, 44, 45, etc.

FIGURES, MONSTERS AND SHIPS

It was a long time before cartographers ceased to illustrate ships and whales on the high seas, natives on the land, and every kind of phantasmagoria everywhere. In the words of Othello:

> . . . the Cannibals that each other eat,
> The Anthropophagi, and men whose heads
> Do grow beneath their shoulders.

Ships were depicted on maps from early times. One interesting woodcut example occurs on Sebastian Münster's map of the New World in his *Cosmographiae Universalis* (Basle, 1540), where Magellan's ship is shown in the Pacific Ocean, and the Portuguese and Spanish standards are shown on flag-poles stuck in the Atlantic, the former off Africa, the latter off Hispaniola (Haiti). It seems that ships were used as conventional signs to denote stretches of ocean, for they occur on maps right up to the beginning of

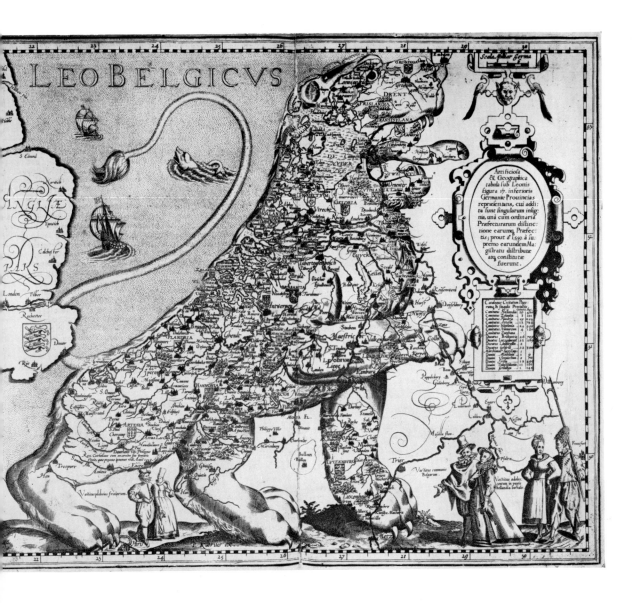

20. Pieter van den Keere: Map of the Seventeen Provinces of the United Netherlands in *Germania Inferior* (Amsterdam, 1617).

British Library

ANGLIA

homini numero, reruniq fere
omniu copijs abundans, sub mi-
tissimo Elizabethæ, serenissima
et doctissimæ Reginæ, imperio,
placidissima pace annos iam
viginti florentissima.

An. Dni
1579

OCEANVS

21.   Title cartouche from Saxton's Map of England and Wales, from his Atlas of 1579
British Library

22. Title cartouche from the map of Valois in *Novus Atlas* by W. J. Blaeu
(Amsterdam, 1635). *British Library*

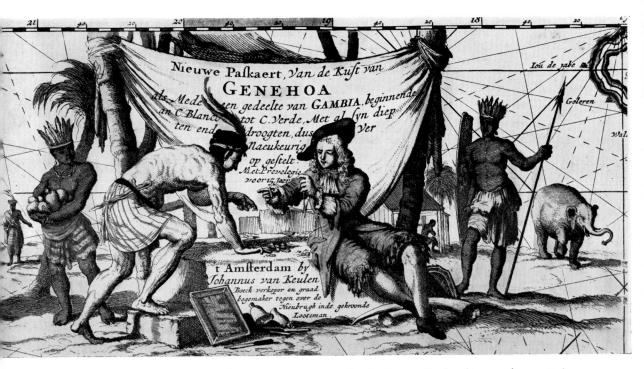

23. Cartouche of the Coast of Guinea in *De Zee-Atlas* by J. van Keulen (Amsterdam, 1681).
By permission of the Trustees of the National Maritime Museum

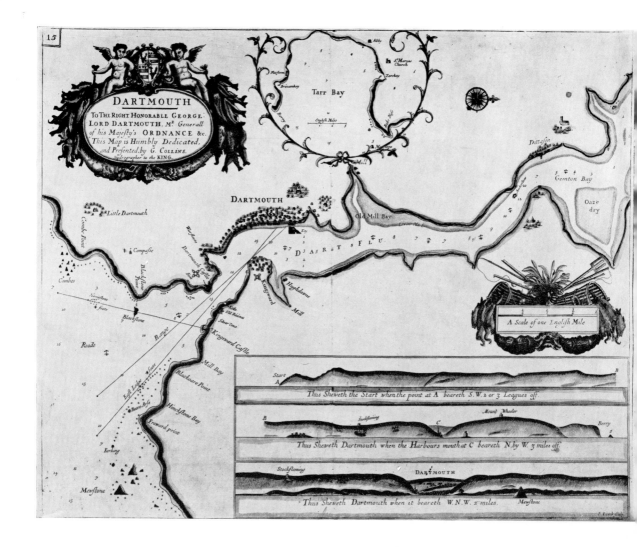

24. Captain Greenville Collins: Chart of Dartmouth with inset of Torbay, in *Great Britain's Coasting Pilot* (London, 1693). *British Library*

25. Title cartouche and compass rose from the map of Hertfordshire in *New and Correct English Atlas* by J. Cary (1787). *British Library*

26. Portion of border ornament from the map of Gloucestershire from *The Theatre of the Empire of Great Britain* by John Speed (1611). *British Library*

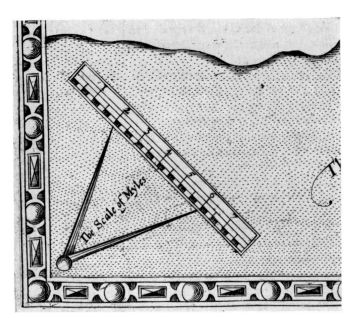

27. Portion of border ornament from the map of Devon from *The Theatre of the Empire of Great Britain* by John Speed 1611). *British Library*

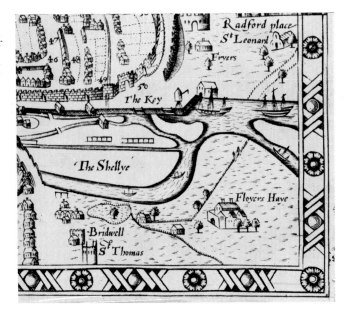

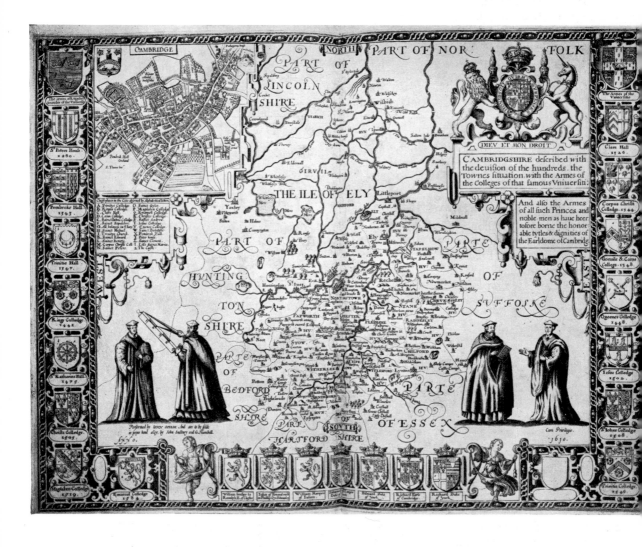

28. John Speed: Map of Cambridgeshire from *The Theatre of the Empire of Great Britain* (1611).
British Library

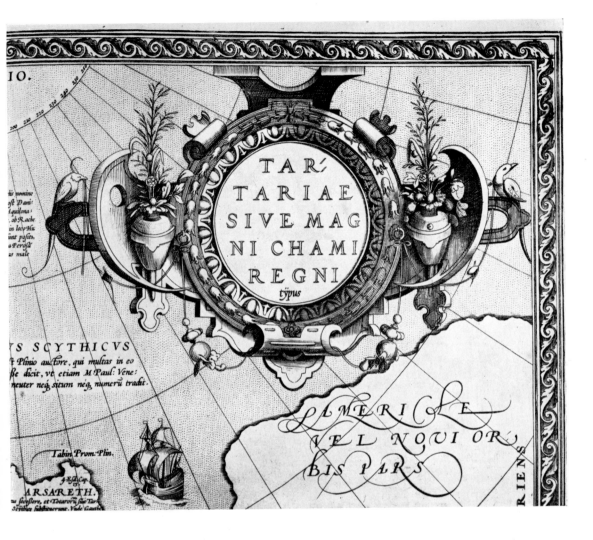

29. Cartouche and portion of border ornament from the map of Tartary in Ortelius's *Theatrum*
(Antwerp, 1570).   *British Library*

The gulfe

THE CHANELL

THE SEA COASTES
of England, from the Sor:
linges by the landes end to
Plymouth with the hauens
and harbrowghes

30. Cartouche, portion of border ornament, monster and ship from the map of the coasts of
England from Scilly to Plymouth in L. J. Waghenaer's *The Mariner's Mirrour* (London, 1588).
British Library

31.   Portion of border ornament and inset view of Old St. Paul's Cathedral,
London, from the map of Middlesex in Speed's *Theatre* (1611).

British Library

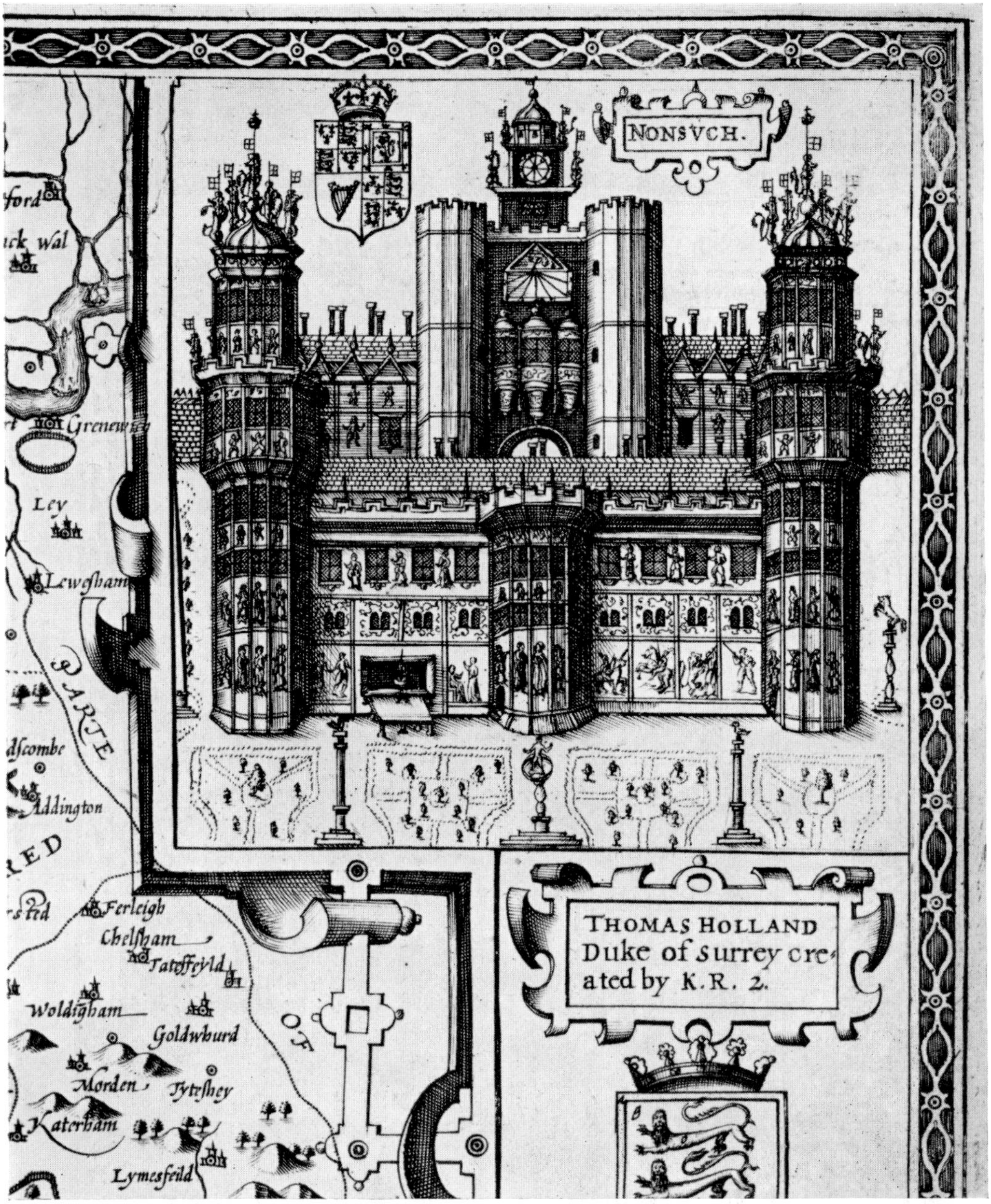
NONSVCH.
THOMAS HOLLAND
Duke of Surrey cre-
ated by K.R. 2.
Grenewich
Ley
Lewesham
Addington
Ferleigh
Chelsham
Tatesfeyld
Woldigham
Goldwhurd
Morden
Tytesley
Katerham
Lymesfeild

32. Portion of border ornament and inset view of Nonsuch from the map of Surrey in Speed's *Theatre* (1611). *British Library*

33. Portion of border ornament from the map of Dorset in Christopher Saxton's
Atlas (1579). *British Library*

34. Portion of border ornament, ship and boats from the map of Lincolnshire and
Nottinghamshire in Saxton's *Atlas* (1579 edn.). *British Library*

35.   Frederick de Wit: Map of Candia (Crete) (Amsterdam, *c.* 1680).   *British Library*

36.   Title cartouche from the map of Cumberland by Robert Morden in Camden's *Britannia* (1695).   *British Library*

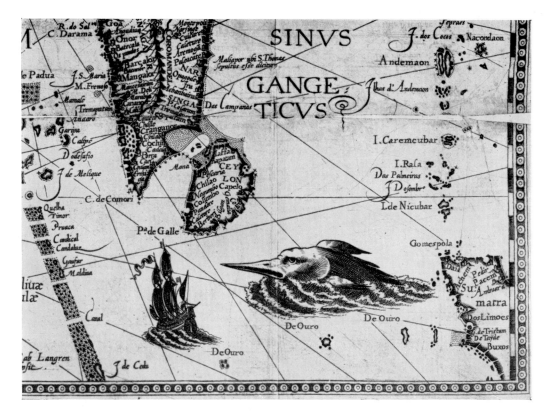

37. Portion of border ornament, monster and ship from the map of the Indian Ocean in Linschóten's *Itinerario* (Amsterdam, 1596). *British Library*

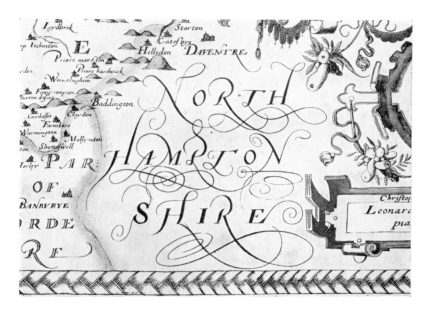

38. Detail, showing portion of border ornament and lettering from Saxton's map of Warwickshire and Leicestershire (1576). *British Library*

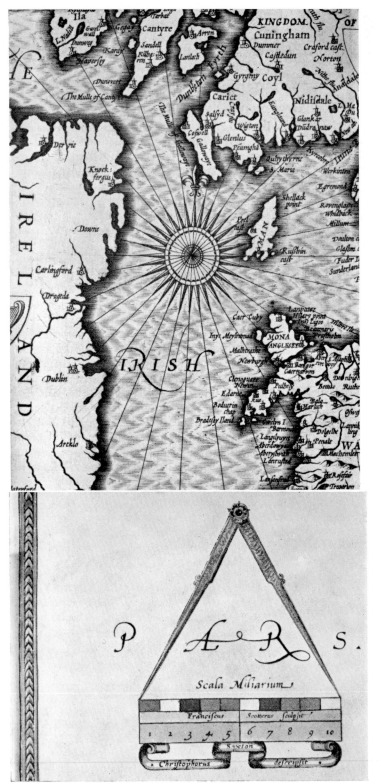

39. Compass rose from the map of Anglo-Saxon England in Speed's *Theatre* (1611).
British Library

40. Detail, showing dividers, scale and border ornament from Saxton's map of Norfolk (1579). *British Library*

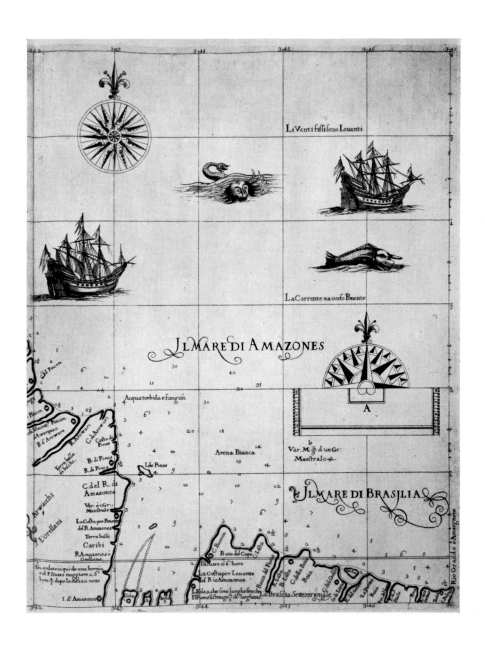

41. Portion of the map of Guiana, from Dudley's *Dell'Arcano del Mare*
(Book VI), published at Florence, 1646–7. *British Library*

51

50

40

5 30

42. Detail showing monster and ships from Plate XVII of Saxton's large map of England and
Wales, 1583: Cornwall and the Scilly Isles.   *British Library*

the nineteenth century, each succeeding period showing contemporary ships, and it must be admitted that they add enormously to the decorative effect of a map, whether they be isolated, with only whales and walruses sharing the ocean with them, as in John Seller's map of the northern coasts of Europe from *The English Pilot* (London. 1671), or sailing in dignified squadrons as in the map of Dunkirk and Mardyck in Schenk's *Atlas Contractus*.   A knowledge of the history and design of sailing ships can be of great value in assigning a map to its period of origin (Plates 30, 34, 41, 42, 43, 48, etc.).

Monsters generally become less fanciful as time passes, and the great sea-monsters (doubtless meant for whales, but in fact nothing like them) in the maps of such cartographers as de Bry, van Langren, Saxton, Hondius and others give way in time to naturalistic whales, flying fishes and other animals, used decoratively rather than as statements of fact.   Some animals were more or less naturalistically represented from the beginning; camels and elephants, for example, which could even then be sometimes seen in noblemen's zoos, thus providing living models (Plates 1, 23, 37, 42, 43, etc.).

While on the subject of monsters, we may mention Pieter van den Keere's map of the United Netherlands (Amsterdam, 1617), in which the provinces are represented by an enormous lion stretching up the coast from France to Friesland (Plate 20).   Incidentally, the ocean on this map contains a whale and two sailing ships, and figures in national dress are shown along its base.

A map of northern regions by Olaus Magnus (printed from woodblocks in 1539; engraved on copper and published in 1572) is celebrated for its representations of kayaks, whales and other features of the sea.

In figures we may include such allegorical personages as the winds represented by blowing heads on the world map from the 1482 (Ulm) edition of Ptolemy's *Geography*. These beautiful woodcut heads are absolutely typical of their period and are a striking instance of how the art of a period can be used to provide evidence of a map's date (Plate 17).   In the Strasbourg 1522 edition of the same work, there is a woodcut map of India, on which is represented among other scenes a devil burning a woman, probably an allusion to suttee.

The allegorical maps in Michael Drayton's *Poly-Olbion* have been mentioned elsewhere (see p. 40).   They are unique in their prolific use of mythological figures (Plate 9).

Cannibals, tritons, warriors, gods, goddesses, heroes, natives of every kind and hue, these are among the vast assembly of actual and mythological figures represented in maps from the earliest times, right into the nineteenth century.   Like ships and monsters, they, too, for similar reasons, can be used for dating and identifying maps.

Figures are also briefly mentioned under *Cartouches* on pp. 61-63.

HERALDRY

Heraldry forms a colourful part of the decoration of maps throughout the period with which we are dealing.   Sometimes it is the arms of the cartographer's patron or patrons, sometimes arms connected with the country, district or town represented, sometimes it is included simply in order to flatter members of the nobility and gentry. In some cases it breaks out into a whole display of heraldry, as in John Speed's map of Cambridgeshire (Plate 28), which in addition to the royal achievement, contains twenty-six shields, all but one contained in an elaborate border design. But even that does not represent the watershed of heraldic decoration on maps; one of Saxton's maps of England contains in its borders eighty-five coats of arms!

One use to which heraldry was sometimes put was to indicate places. An example of this usage is to be found on Pieter van den Keere's map of the United Netherlands (Plate 20) mentioned under *Monsters* above, in which important towns and provinces are indicated by their arms.

Cartographic heraldry may be used in map identification and dating; a knowledge of the periodic changes of arms can be so applied. To take an obvious example, the quarterings and supporters of the royal achievement of Elizabeth were different from those of James I; Saxton's maps, which were engraved in Elizabeth's reign, contain arms with the Tudor quarterings, whereas Speed's, which were engraved during the reign of James I, contain the Stuart quarterings (Plates 19 and 21).

But as in many other things connected with maps, caution should be used. It does not mean, again to use one of the foregoing examples, that because a map contains the arms of Elizabeth, that it was printed in her reign; an impression could be made at any time during the plate's life. All that one can say on the basis of that evidence alone is that the map could not have been printed before those arms came into use. But that in itself provides at least a beginning.

VIGNETTE VIEWS

Vignette views and scenes are yet another decorative embellishment on old maps, which help to give them their charm. The earliest printed maps sometimes have them, and they appear here and there right up to the mid-nineteenth century. An early example is Sebastian Münster's edition of Ptolemy, published at Basle in 1540, which has on its woodcut map of Africa Minor a lively view of the shipwreck of St. Paul. But the hey-day of vignette views came in with the general use of copper-engraving. John Norden's work affords some attractive examples, such as his bird's-eye view of Chichester on his map of Sussex (engraved by Christopher Schwytzer); Speed's map of Surrey in his *Theatre* (1611-12) contains two delightful vignettes, one of Richmond, the other of that strange and romantic pile, Nonsuch, while his map of Middlesex from the same work shows us Westminster Abbey and Old St. Paul's (Plates 31 and 32), and elsewhere we are regaled with views of Stonehenge (Plate 11). St. Paul's shipwreck is represented again on the map of Malta in Ortelius's *Epitome Theatri Orbis Terrarum* (1601).

Indeed, vignettes appear right through the history of maps almost until our own day. Ratzer's plan of New York (engraved by Kitchin in 1776) contains a view of the city, unrecognisable as the congested skyscrapered townscape of to-day. Another famous vignette view of New York, *circa* 1662, is on Visscher's map of New Belgium, New England and Virginia. Mayo's map of Barbados (engraved by John Senex in 1722) gives us a view of the island's landscape, with palms, houses, windmills (probably used for pumping), while a European, assisted by two Negro slaves, surveys the landscape. The nineteenth-century vignettes are full of romanticism, scenes that existed in that form only in the imagination of their contemporaries. Langley's map of Kent from his *New County Atlas* (*circa* 1818) contains views of the shipyard at Chatham and a hopfield, with Kent's magnificent prancing-horse shield propped against a convenient tree. But it is T. Moule's work, already mentioned several times, that illustrates the zenith of this aspect of nineteenth-century map-making. His map of Gloucestershire contains vignette views of Gloucester Cathedral, Tewkesbury Abbey and the New Spa, Cheltenham, in addition to three coats of arms and a pretty little pastoral scene of garlanded maidens drinking from a fountain, obviously an allegory of the Spa (Plate 14).

LETTERING

The lettering on maps, whether it be engraved (as in the majority of cases) or printed from movable type, follows, generally speaking, the typographical and calligraphic fashions of its period. Early manuscript maps and some early printed woodcut maps have wholly or partial gothic lettering – 'Black letter' as printers call it (Plates 50, 51). But from about 1470, with the rise and influence of the great Italian printers and writing masters, humanistic lettering (Malvolio's 'sweet roman hand') was used almost exclusively, either roman or italic (Plates 21, 30, etc.), each of which assumed dozens of different forms. It must, however, be said that cartographic lettering is in the overwhelming number of cases absolutely clear. The purpose of lettering on maps is to indicate clearly their features, and extravagance in style was therefore nearly always suppressed. In some titling, however, more decoration was allowed, as in the magnificent swashed lettering in Saxton's cartouches (Plate 38) which holds its own amid the strap-work and volutes. But to indicate towns and other small features plain lettering predominates.

Another point is that the lettering on copper-engraved maps is freer, less stiff than on woodcut maps. In the technique of the copper-engraver the graver is used almost as freely as a pen, but in woodcutting the letter has to be laboriously carved out of the wood, three-dimensionally, with a knife. In both methods, the lettering is, of course, cut in reverse.

Except in the earliest specimens, on which roman capitals alone are used, lettering on maps is graded according to the importance of the subjects indicated: roman capitals for countries, oceans and continents, minuscules for towns and cities, italic for forests, moors and so on.

To summarise the main styles of engraved lettering in use at various periods, the reader is referred to Plates 12, 18, 21, 29, 33, 36 and 56. Of these, Plates 18, 21 and 29 represent those in use in the sixteenth century. Plate 18 is based upon movable type; Plates 21 and 33, show italic based on manuscript models; they are both taken from Saxton's maps. Plate 29 is based upon ordinary contemporary handwriting. During the seventeenth century lettering grew closer to movable type as in Plate 36, although for special purposes a form of lettering resembling handwriting was used (Plate 12).

From the end of the seventeenth century lettering resembled movable type even more closely, and became in many cases practically indistinguishable from it. A study of contemporary type-faces illustrates this. Titles become a vehicle for all kinds of fancy lettering – even gothic was revived under the Germanic influence that came in with the Hanoverians – yet it is rarely without dignity and clarity.

CONVENTIONAL SIGNS

As a general rule, lettering is used on maps only where symbols and conventional signs cannot be used. Mercator aimed at an ideal map on which no legends need appear, but of course such an ideal is impossible to realise. Nevertheless, to indicate a wide variety of features on maps, certain symbols have been employed for long periods with great success. Among the features so depicted are buildings, towns, rivers, mountains, parkland, woods and the sea.

So far as mountains are concerned, they were first often represented by great slabs of rock (Plate 53), a convention that persisted until 1500. Then this gave way to. a shaded representation (Plate 50) giving the effect of downland rather than of mountains, but in some cases giving an effect of true relief, though this was probably unintentional. Up to the early seventeenth century mountains were, particularly on English maps, frequently represented by little hummocks, sugarloaves or molehills. Thus an inconsistent element was introduced into maps, which are essentially *plans*: the mountains are shown in elevation, as bird's-eye views; moreover, without any real attempt being made to render them in any kind of scale or proportion to the rest of the map, or even to one another (cf. p. 36). As to their gradients, no attempt is made to depict them individually; the hummocks are indeed no more than symbols (Plates 1, 56, etc.). Yet there are certain conventions in their representation, such as that they are always shaded on their east and south-east sides as if lit by the sun from the west, though this is probably due to the fact that it would be the natural way for a right-handed draughtsman to shade them.

The representation, in the form we know to-day, of mountains and other elevations in plan, was first established, though somewhat crudely, from about 1680 (cf. p. 31), by lines of vertical shading, which was varied to represent steep slopes by being made heavy and close, to represent foothills and gentle slopes by being made more widely spaced and fading out (Plate 15). A similar convention was used in the eighteenth century to represent conjectural mountains in unknown regions and is known as 'hairy caterpillars' (Plate 55). Yet side by side with these developments the molehill convention was still used by some cartographers.

The Ordnance Survey introduced many improvements, such as spot-heights, but the true and accurate representation of mountains and contours is a modern refinement. Yet it should not be forgotten that the contour line, as used on present-day Ordnance Survey maps, was, as we have already seen (p. 31), first used in 1729[1] by a

---

[1] The device of connecting lines was first used in 1701 on Halley's map of the Atlantic, but in this case the lines are of equal magnetic variation, not of heights or depths.

Dutch hydrographer, N. S. Cruquius, to indicate points of equal depth on sea-charts; the idea of using the device to indicate terrestrial levels does not appear to have occurred to anybody until modern times.

The 'bird's-eye view' convention used in the representation of mountains was also used in other conventional signs: woods and forests, for instance, which are shown as little groups of separate trees, shaded, like the mountains, on the east (Plate 56). Sometimes they are naturalistic, sometimes 'Noah's Ark' shapes, as in Drayton's *Poly-Olbion* (Plate 9). This remained the standard practice until the eighteenth century. But from *circa* 1750 to *circa* 1820 it gave way to a more impressionistic rendering by clumps of woodland (Plate 54).

From the late sixteenth century on the Continent, and from the early eighteenth century in England, agricultural land was shown in the form of ploughed furrows (Plate 3). And other countryside features were depicted, such as hedges, windmills and parkland, the latter being shown by palisades or fences (Plates 49, 52 and 57). The proliferation of such symbols may be judged by this extract from John Ogilby's preface to his road atlas *Britannia* (Vol. 1):

> The *Road* . . . is express'd by double Black Lines if included by Hedges, or Prick'd Lines if open . . . *Capital Towns* are describ'd Ichnographically[1] . . . but the *Lesser Towns* and *Villages*, with the *Mansion Houses, Castles, Churches, Mills, Beacons, Woods, &c.* Scenographically, or in Prospect. *Bridges* are usually noted with a Circular Line like an Arch . . . *Rivers* are *Decypher'd* by a treble wav'd Line or more, and the lesser *Rills* or *Brooks* by a single or double Line, according to their Eminency.

So much for the country. Towns were, in the early days of map-making, shown as groups of buildings with towers and red or blue roofs with flags flying, and surrounded by walls, the whole dominated by the city castle. This remained the standard symbol up to about 1520, after which the church instead of the castle dominated the symbol, because city castles largely fell into disuse after that time. Such symbols were used up until nearly the end of the eighteenth century, and when they were coloured, red was the dominant shade – the colour of the roofs on early manuscript maps. And we must not forget Drayton's representation of towns by allegorical figures, splendid in their nudity, but this is a type of symbol rarely met with (Plate 9).

There were hundreds of variations of the representation of towns by buildings throughout this long period, a few of which may be seen on Plates 55-57.

In some maps from the last decade of the sixteenth century onwards, a rough profile of the town was sometimes shown (Plate 13). Villages were indicated by churches or towers. In many cases the features threw eastward shadows (Plate 49). During the seventeenth century, true plans came into use (Plate 35).

After 1800 it became the standard practice to use dots and circles of various types to indicate towns and other communities, according to size.[2] And soon the Ordnance

---

[1] ichnographically = in plan.

[2] They had been used before 1800. Norden, for example, sometimes used the device. Indeed, his work is particularly important in the use of symbols and conventional signs.

Survey introduced true symbols – a cross for a church and so on – where actual plans were impracticable.   It should be remembered that in the sixteenth century 'buildings' symbols sometimes had a dot or circle incorporated in them (Plate 47).

Finally, sea-shore and sea, the early representations of which show dangerous waves, whirlpools, shipwrecks, monsters and tritons.   The sea was indeed in those days a forbidding element, and the cartographers drove home the point with emphasis. Until nearly the middle of the sixteenth century the sea was represented by swirling and billowing lines, tempestuousness being the keynote.   But this changed from the 1540's onward, when all water was shown stippled or dotted (Plates 34–35). The cartographers of the Low Countries, many of whom came to England as refugees, depicted water by a moiré, 'watered silk' pattern, and this was adopted by English cartographers such as Speed (Plate 43).   But from about 1630, the sea itself was left blank, except for ships and monsters, although it was usually made blue when colouring was used.

The sea-coast itself was sometimes hatched, a method which, although used earlier, was used with great prominence from the last decade of the eighteenth century onwards (Plates 21, 48, etc.).   Another convention was the use of lines running around the coasts, in appearance somewhat like contour lines (Plate 24).   These were in use from *circa* 1730.   In some sea-charts the coast was shown in elevation or silhouette and anchorages and soundings were marked just as in present-day maps (Plate 46).

When conventional signs are grouped together on a map in a cartouche or tablet, the grouping is called a 'characteristic sheet'.   The first map on which such a key appears was the map of Franconia (*Das Francken Landt Chorographia Francia*, Ingolstadt, 1533). The 'Lily' map of the British Isles of 1546 (see p. 32 *note*), bears in some of its editions keys to the symbols used, but the first map definitely made and engraved in England to bear such a key was Norden's map of Middlesex (see p. 37 and Plate 49).

CHAPTER FIVE

Terrestrial and Celestial Globes and Armillary Spheres

TO adapt Walter Pater's remark on music, we may say that the globe represents the condition to which all map projections aspire – the representation of the earth in its true shape. The idea of the earth as a globe is not a modern discovery for we hear of both celestial and terrestrial globes in Greek antiquity – from the Pythagoreans, from Aristotle and from others – although there was an alternative idea which saw the world as a disc, surrounded by an ocean. Some believed that it was a cylinder moving within a hollow sphere. Even in the Middle Ages, when so much of classical learning had been lost or forgotten, there was an awareness of the value of globes. But it was with the dawn of modern times at the end of the fifteenth century, and the world discoveries that were made from that time onward, that their true importance became obvious.

Joseph Moxon (see p. 81) defined globes thus:

> A *Globe* (according to the Mathematical Definition) is a perfect and exact round Body, contained under one Surface.
>
> Of this Form (as hath been proved) consists the *Heavens* and the *Earth*: And therefore the Ancients with much Pains, Study and Industry, endeavouring to imitate as well the imaginary as the real appearances of them both, have Invented two *Globes*: the one to represent the *Heavens*, with all the *Constellations, fixed stars, Circles* and *Lines* proper thereunto, which *Globe* is called the *Celestial Globe* and the other with all the *Sea Coast, Havens, Rivers, Lakes, Cities, Towns, Hills, Capes, Seas, Sands,* &c. as also the *Rumbs, Meridians, Parallels,* and other *Lines* that serve to facilitate the Demonstration of all manner of Questions to be performed upon the same: And this *Globe* is called the *Terrestrial Globe.*

What is probably the oldest extant celestial globe is in the Royal Museum at Naples: 'Atlante Farnesiano' as it is traditionally known. It is supported by an Atlas figure and the whole is constructed of marble. It is thought to date from *circa* 300 B.C. The constellations are represented on its surface by actual figures.[1]

Although no armillary sphere survives from such early times, there is plenty of evidence that these too were in use in classical antiquity. An armillary sphere is one

---

[1] With celestial globes the observer is imagined to be in space outside the apparent sphere of the heavens.

constructed of rings, usually of brass (though other materials, such as alabaster, are known) representing the apparent paths of certain celestial bodies in the apparent sphere of the firmament, and pivoted, within a horizon and meridian on its polar axis (Fig. 5). Another device, the astrolabe, represents on a plane surface the apparent movement of the stars, in the celestial sphere, to the earth; the word was applied later to a graduated circle used for taking altitudes at sea. Methods of making astrolabes and celestial spheres were described in great detail by Ptolemy in his *Megale Syntaxis* or *Megiste Syntaxis* or, to give it its Arab name, *Almagest*. Globes, too, are used symbolically on certain Roman coins. In Byzantine times globes were constructed by Leontius Mechanicus, a scholar.

Fig. 5.   Armillary Sphere.   Pasteboard armillae, wooden base.   By Jean Fortin (1780)

The early Persians guarded the science of astronomy as a treasured secret, on account, it is said, of a prophecy that the Christians would overthrow their empire by means of knowledge derived from it. When their country was conquered by the Arabs the details of Arabian astronomy were enshrined for posterity in Persian literature. The mediaeval Arabs contributed to the development of globes just as they did to that of plane map-making. Their deep interest in astronomy from early times led them to develop both celestial and armillary spheres and other forms of astronomical instruments which by the thirteenth century had already reached a very advanced state. Arab globes were usually made of engraved brass, although silver ones also are known. Some have picturesque inscriptions; one, for example, in the museum of the Royal Asiatic Society of London (it is probably Persian), is inscribed in Khufic characters,

'Made by the most humble in the supreme god, Mohammed ben Helal, the astronomer of Mosul, in the year of the Hejira 674' (A.D. 1275).

The influence of Arab astronomers on the science is the reason why so many stars have Arab names, like Algol, Aldebaran, Alpheratz, Altair, Dubhe and Fomalhaut.

In the Middle Ages, in Christendom, the idea of a spherical earth was rejected by most scholars, although a few, like the Venerable Bede, adhered to it. The nearest that the average scholar came to the idea, was to regard the earth as a disc. But the true theory was never entirely dismissed and we read of many manifestations of it; for instance, the Hohenstaufen Frederick II is said to have commissioned from an Arab geographer a celestial globe of gold with the stars represented by pearls, and there is a fourteenth-century French miniature which shows the earth as a sphere, surrounded by stars and clouds. Alfonso X of Castile caused an extensive astronomical treatise to be made, in which is given interesting details of the material from which globes were made in those days: gold, silver, copper, brass, iron, lead, tin (or two or more of these metals combined), clay, stone, wood, cloth, leather, or parchment in layers. Nevertheless, the author is at pains to state that few of these materials are really suitable. It is doubtful whether they were all used.

With the world discoveries of the fifteenth century, as we have seen in Chapter One, there came a great improvement in the standards of cartography, in which globes shared. The earliest surviving modern terrestrial globe was constructed by Martin Behaim (*circa* 1459-1506), a native of Nuremberg, and a member of a local patrician family. Behaim's parents intended him to follow a commercial career, and he was educated with this idea in view. He claimed to have studied under Regiomontanus (Johann Müller), the astronomer, and it is probable that it was this relationship that first aroused in the lad an interest in cosmography and mathematics.[1] From 1476 to 1484 Behaim was in the Netherlands, where he had in the first place been sent to develop his knowledge of commerce, although he returned to his native city on occasions, and in fact was there once sentenced (in 1483) to a week's imprisonment for dancing at a Jew's wedding during Lent.

In 1484 he went to Lisbon, to further his commercial interests, but became involved there in the (to him) more interesting matters of cosmography and mathematics, and in about 1484 or 1485 he was appointed a member of a *Junta dos mathematicos* instituted by King John II to agree upon rules to be observed for determining latitude from meridian altitudes of the sun. Moreover, he took part in one or two voyages of exploration down the African coast. Later he was knighted by the King. In 1490 he returned to Nuremberg, when at the suggestion of one of its members, the City Council commissioned him to make a globe showing recent Portuguese and other discoveries.

The globe itself is a colourful object, a fine example of the illuminator's art. Its

[1] It should be stated that there is some doubt about the extent of Behaim's mathematical knowledge, his supposed study under Regiomontanus and his share in the construction of the globe. Since, however, his name is so closely associated with it, I have quoted the traditional story. See 'Martin Behaim, Navigator and Cosmographer; figment of imagination or historical personage?' by G. R. Crone, Vol. II of *Das Actas* of Congresso Internacional de História dos Descobrimentos. (Lisbon, 1961.)

20-inch diameter shell was constructed on a spherical matrix of loam, on which were pasted several layers of paper or vellum. When it was dry, the paper sphere thus made was cut round its equator and removed from the loam core. The two halves were glued together again, having been reinforced inside with wooden rings. The outside surface was coated and the gores, or segments, were glued into place, with, as was usual, discs for the polar regions.

Sheets of gores which have never been cut out, may sometimes be found. Two kinds of several known forms are illustrated in Fig. 6. Of these the former is the

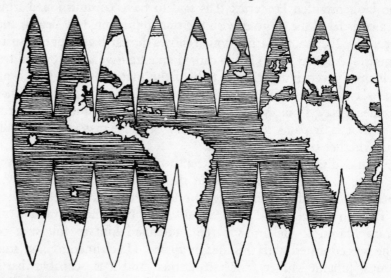

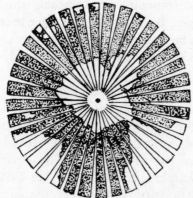

Fig. 6.

ABOVE: Rough sketch illustrating the principle of gore-maps for covering globes. Based on a 16th-century example. The original has 12 gores. On this sketch the 3 gores containing India, etc., are missing.

BELOW: Rough sketch showing arrangement of globe gores of Antonius Florianus (1555). Southern hemisphere.

more usual; its gores or biangles vary in number from approximately twelve to thirty-six, with twelve and twenty-four as the commonest.

But to return to Behaim's globe. It is now inferred that his source was a large map by Henricus Martellus made *circa* 1490; an example is in Yale University. It appears to be one of the maps which Columbus used for working out his project of a westward route to Asia. The details were painted on the globe by Georg Glockendon, an illuminator. It took him fifteen weeks to complete it. The meridian and horizon (see p. 84) were made of iron by a blacksmith, and the whole instrument was mounted on a wooden stand. This, too, has now been replaced by one of iron.

Geographically, Behaim's globe has much in common with world maps of the time, as described in Chapter One, and indeed it may be taken as a general rule in identifying and dating globes that they follow the contemporary characteristics of plane maps. It is rich in heraldry, colour, gold and silver, the bright blue of the sea dominating everything else. The ecliptic is marked with the successive zodiacal signs, each contained in a little ultramarine and gold medallion. National flags mark ownership of land areas, sea monsters and ships are to be seen on the high seas, and in some places the inhabitants may be seen: a turbaned prince in Tartary, naked aborigines in Africa.

The method of construction used in making Behaim's globe was not unique; during the late part of the fifteenth and throughout the sixteenth centuries there were made metal globes and ivory globes, with details engraved on their surfaces, wooden globes with painted surfaces or with manuscript gores on parchment or vellum pasted on to them; some, too, were constructed of wood covered with papier mâché; some were covered with metal plates or wrapped in cord, some made of a combination of plaster and papier mâché. Within this period came the first globes bearing gores printed on paper; they were printed sometimes from woodblocks and sometimes from copper plates. The shape of the gores was carefully worked out to cover globes of a given size and thus they could be manufactured quickly and in quantity. Present-day globes are still made like this, unless they are made of plastic.

Meridian and horizon circles were nearly always of bronze or brass, although when precious metals have been used in the construction of the globes themselves, these, too, are of the same materials.

Map-makers who were producing printed globes in the late fifteenth and early sixteenth centuries included Waldseemüller (see p. 23), whose gores were among the first to show America. They were engraved in 1509 and are twelve centimetres long. It is thought that a complete Waldseemüller globe may have been issued with a tract printed at Strasbourg, issued in 1507, which states 'how you shall understand the globe and the description of the whole world . . .'. Apparently, in those days, descriptive books or pamphlets were frequently issued with globes, for other such examples are known.

The Lenox globe in the New York Public Library is an example of a metal globe of the period. It is an engraved copper ball, 127 millimetres in diameter, in two hemispheres joined together at the equator. The engraving is good. It has no stand or rings. From the drawing of the maps, it has been assumed to date from 1510.

One of the greatest of sixteenth century globe makers was Johann Schöner of Nuremberg (1477-1547), astrologer, astronomer, geographer, physician and writer of forty-six books on these subjects. He was born at Carlstadt in Franconia, and received his education at Erfurt, where he studied with the idea of entering the Catholic priesthood. His superiors, however, objected to his interest in mathematics and other scientific subjects, and he abandoned this career, joined the Protestant Church and became professor of mathematics at Nuremberg University. He made his first globe in 1515 at Bamberg; two specimens of it survive. It measures twenty-seven centimetres in diameter, is on a wooden base, and has brass rings. An interesting point about this globe is that it indicates a strait of water between what it calls 'America' (corresponding to what is now known as South America) and the large land-area he calls 'Basilie Regio' (now known to be the relatively small island of Tierra del Fuego). This strait was actually discovered by Magellan during his circumnagivation of the earth and in 1520 was named after him. Schöner shows also a north-west passage, but it is now generally thought that both were indications of his wishful thinking rather than of any remarkable knowledge of the area. Schöner's name for North America is Parias.

Globes were made of precious metals in the sixteenth century too. One, known as the Nancy globe from the fact that it is in the Lorraine Museum at Nancy, is made of silver. It dates from about 1530, is sixteen centimetres in diameter and is supported by a statuette of Atlas; land areas are gilded, sea areas enamelled blue. It is surmounted by a small armillary sphere. Other metal globes were made of copper or iron with their lines of engraving undercut and filled with gold and silver wire, an old technique known as damascening and widely used by armourers. Another use of precious metals on globes is illustrated in a portrait in the National Portrait Gallery in London: Queen Elizabeth is portrayed wearing earrings in the form of armillary spheres.

A German example, in Dresden, consists of a hollow terrestrial sphere which, upon being opened, discloses within a celestial sphere. It is constructed of brass throughout. Globe goblets form an interesting extension of metal globe-making. Usually of silver, often gilded, their upper hemispheres may be removed, leaving the lower hemispheres as drinking cups; variations on this form occur, of course, in different specimens. The stem is sometimes a statuette of Atlas, Pan or some other pagan deity (Fig. 7).

Caspar Vopel (1511-64) was a map- and globe-maker of Cologne, nine of whose globes and armillary spheres have survived to the present time. One of these, dated 1541, is in the National Museum at Washington. It is eleven and a half centimetres in diameter, with a small globe at its centre, enclosed in the outer armillary sphere of eleven metal rings, which are engraved with scales and symbols. Another specimen of his work, a celestial sphere covered with engraved gores, is at Cologne.

Mercator (see pp. 23-25) constructed globes, complete examples of which survive to-day. Stars are marked on his terrestrial globes, as a guide to travellers at night. Some twelve sheets of his engraved terrestrial gores are known, some of them on globes, some still in sheet form. They were designed to cover a sphere of forty-one centimetres diameter. Loxodromes and compass roses are outstanding features on them and so is the lettering: roman capitals for continents, empires and oceans and italic for kingdoms, provinces and rivers, with a variant for races. Instructions are included to enable the traveller to find his position at night by means of the stars. On one of the gores the map is inscribed 'Edebat Gerardus Mercator Rupelmundanus cum privilegio Ces Maiestatis as an sex Lovanii an 1541'.[1]

---

[1] 'Published by Gerard Mercator of Rupelmunde under the patent of His Imperial Majesty for six years at Louvain in the year 1541.'

In 1551 Mercator published engraved gores for a celestial globe, based upon Ptolemy's forty-eight constellations, with additions, and with a slightly astrological bias.

Mercator's method of construction was to form a rough sphere of narrow strips of wood on which was pasted a cloth covering, and then to cover the whole with a thin coating of plaster. To give it its finished shape the sphere was coated to a thickness of about six millimetres with a mixture of plaster, sawdust, wood-fibre and glue. The gores were pasted on to this. Mercator also made wooden pocket globes. But he also

Fig. 7. Globe-goblet, supported by figure of Pan. Gold. Last quarter of 16th-century

worked in unusual materials; he is supposed, for instance, to have made for the Emperor Charles V a celestial globe of glass, with engraved figures and with a wooden terrestrial globe at its centre.

A French cartographer whose globe gores were certainly influenced by Mercator was François de Mongenet of Franche Comté. He published both celestial and terrestrial globes measuring about eighty-five centimetres in diameter. The first edition, published in 1552, was printed from woodblocks; a second edition, undated, but thought to be about 1561, was published in Italy, having been printed from engraved copper plates.   In the 1552 edition the unmounted gores are outlined by a black border and on the terrestrial sheet, a zodiacal band is printed below them for cutting out and mounting on the globes.

The Venetian Antonius Florianus published a line-engraved gore map in 1555 with a different construction from those discussed so far.   The globe was divided into northern and southern hemispheres, each divided into thirty-six radiating gores.   It is commoner than most early gore maps.

Italian globe-making in the sixteenth century was of high quality, and embraced the whole range of the art – globes made from printed gores, metal globes, armillary spheres, every imaginable form, in fact.   One particularly pleasing example, made by a member of the Volpaja family of Florence, noted instrument-makers, was an armillary sphere of gilded brass enclosing a crystal sphere.   Like many other spheres, it contains a compartment for housing a magnetic needle (see p. 85).   Other important sixteenth-century globe-makers were Giovanni Maria Barocci, Emanuele Filiberto, Francesco Basso, Ignazio Danti, Hieronymo Boncompagni and finally Mario Cartaro, who made globes of solid wood of about sixteen centimetres diameter, covered with engraved gores.

Johannes Praetorius (1537-1616), a German globe-maker, made some magnificent metal globes, beautifully engraved and set in stands decorated with all the ornamental flamboyance for which the age is renowned.   Human figures, lions' masks and paws, scrolls, musical and scientific instruments, all combine to evoke the atmosphere of a period when to be scientific was to have aesthetic sense as well.

Towards the end of the sixteenth century there was a general improvement in the art of globe-making, and there are extant a number of really beautiful examples that rank as works of art of a high order.   Outstanding among them is a celestial globe in the Metropolitan Museum of Art at New York, made by Gerhard Emmoser of Vienna in 1573.   It is made of silvered bronze and is supported by a spirited statuette of Pegasus in the same material.   Nor is this all; the globe contains a clockwork mechanism that causes it to rotate on its equatorial axis once in twenty-four hours. Globes like this are essentially for millionaires or museums (although one does occasionally see moderately priced late eighteenth- and nineteenth-century globes with clockwork mechanisms); they do, however, represent a notable development in the history of the art of globe-making.   The same may be said of the globes of George Roll and Johannes Reinhold, Antonio Santucci, Jost Bürgi, Peter Apianus, and those of the important figure, Emery Molyneux, who in 1592 made the first English globes.   There were, too, globe-makers who made only one or two globes, but which were of such outstanding quality or interest that they take their place in the story of the craft's development beside more prolific makers.   Such was Conrad Dasypodius (1532-1600), maker of the famous Strasbourg clock, which as part of its system has terrestrial and celestial globes which revolve mechanically once every twenty-four hours; such, again, was the great astronomer, Tycho Brahe (1546-1601), whose great brass globe was 182·8 cm in diameter, and who from his island observatory of Uranienburg exerted a strong influence on the many young cartographers and globe-makers who visited him, including Blaeu (see p. 27).

At the end of the sixteenth century there flourished in the Netherlands one of the greatest of all families of globe-makers, the Van Langrens, who were of Danish extraction. There was the father, Jacobus Florentius (his classical name) and three sons, Arnoldus Florentius, Henricus Florentius and Michaele Florentius. Arnold became globe-maker to the archdukes in the Spanish Netherlands, and later Royal Cosmographer and Pensioner. The van Langren globes were built on a foundation of wood or papier mâché, which was given a coating of plaster on which engraved gores were mounted. They were decorative too: the terrestrial globes featured sea monsters, animals, vegetation, human figures and ships carrying their national ensigns. The geographical features show the improvement of knowledge which had by this time developed, and there are loxodromes.

Jodocus Hondius and his son we have already dealt with as makers of maps (see p. 26); they were also prolific makers of terrestrial and celestial globes. Hondius globes vary somewhat in their basic construction. Some are of wood, some of papier mâché covered with varnished plaster. They are covered with engraved gores and are sometimes coloured by hand. Dedication cartouches with armorial bearings are sometimes featured and sometimes such things as compass directions are given in Dutch instead of in Latin. In general their geographical and decorative features are the same as those on their plane maps. Diameters vary from twenty-one to sixty centimetres. Jodocus Hondius engraved, among others, the gores for the terrestrial and celestial globes constructed by Emery Molyneux under the patronage of William Sanderson, who also was a patron of the great English cartographer, Norden (see pp. 37-39). Examples are in the Library of the Middle Temple in London.

Willem Blaeu (see p. 27) made mathematical instruments and globes in addition to maps. It is thought almost certain that he made instruments for Tycho Brahe; in fact, some of his globes state in their inscriptions that he was a pupil of Brahe.[1] Geographically, his terrestrial globes were based upon the maps of Mercator; their sea areas are covered with loxodromes. His globes were constructed by pasting hand-coloured engraved gores on to a prepared sphere. In at least one case the hollow sphere is constructed of metal and covered with plaster; others are made of papier mâché and plaster. His first terrestrial globes were made in 1599, his first celestial globes in 1603.[2] These were quite small (thirty-four centimetres in diameter); although he produced still smaller ones after this, he also made some really large globes, measuring up to seventy-six centimetres in diameter.

Geographically, Blaeu's terrestrial globes are a faithful reflection of contemporary knowledge. Artistically they are pleasing, with sea monsters, ships and other such

---

[1] There are in the Blaeu atlas well-known representations of scenes at Tycho Brahe's observatory at Uranienburg.

[2] Blaeu's early globes of 1599, 1602, 1603, 1606 were not issued, it appears, until after 1617. All his terrestrial globes show the discoveries made by Schouten and Le Maire on their voyage round the world, 1615-17, the chief one depicted being that of Cape Horn. This is an illustration of the dates on globes being considerably earlier than the date of issue.

features, and our attention is drawn to such information as 'Patagone regio ubi incolae gigantes' (The Patagonian district, in which place dwell giants). Dedications are enlivened with heraldry. The celestial globes likewise incorporate recent discoveries; stars down to the sixth magnitude are represented, each magnitude with its own symbol. Nebulae are also given a special symbol.

Other globe-makers of the first half of the seventeenth century were Peter Plancius (see pp. 26–27), Jan Jansson (see p. 27), Isaac Habrecht (1544–1620) of Strasbourg, Adam Heroldt (fl. 1649), German maker of armillary spheres, Giuseppe de Rossi (fl. 1615), who reproduced Hondius globes in Italy, and Matthäus Greuter (1556–1638), a native of Strasbourg who worked in Italy.

Greuter appears to have been one of the most prolific globe-makers of his age. All of his globes are fifty centimetres in diameter; they are made of engraved gores pasted on to the usual spherical base. Ships, sea monsters and loxodromes are all there. Names are given in many languages; sometimes the language native to a region is used to identify town and other features in that region; the New World names are given in Spanish or Portuguese; the Old World and the seas are given in Latin; elsewhere there are other languages, including English, Dutch and French. His celestial globes are based on those of Blaeu and on the discoveries of Tycho Brahe, a fact he mentions in his dedications; names on his celestial globes are given in Arabic and Latin.

As the seventeenth century wore on, the tendency was for globes to become more elaborate and extravagant in size, as well as in conception. An example in the Museum of Natural History at Frederiksborg Castle in Denmark is a combination of an armillary sphere with an orrery inside it (an orrery is a machine representing the movements of the solar system); it was made by Andreas Busch in 1657. It is 120 centimetres in diameter, but was dwarfed by one presented in 1713 by Duke Frederick of Holstein-Gottorp to Peter the Great of Russia; this measured 335·5 cm in diameter. A terrestrial map was on its outer surface; the inside was in effect a small and rough version of the popular planetarium in Marylebone Rd., London, for one could enter it by means of a door and from a platform observe the heavens, which were shown on the globe's inner surface. A mechanical device set it in motion and the stars could be seen rising and setting. About twelve people at a time could be accommodated in it. An even bigger globe was described by a German globe-maker, Erhard Weigel (1625–99), to which he gives the high-sounding name of *Pancosmo, o Mondo Universale*; this piece of Brobdingnagian cosmographical equipment measured 976 cm in diameter and was supported by a couple of 244 cm-high statues of Athene and Herakles. It would be a truism to say that such an object is beyond the scope of the ordinary collector.

A 457·5-cm diameter globe, which people could get inside, was made for Louis XIV at Versailles by P. Vincenzo Maria Coronelli (1650–1718), a Venetian member of the Franciscan Order, who was a prolific globe-maker and cartographer. He founded a learned geographical society, the Accademia Cosmografo degli Argonauti, and was given the title of Cosmographer to the Most Serene Republic, a pension of 400 florins a year and a copyright privilege extending to twenty-five years. His globes, mostly of

43. Detail, showing ship and monster, from Speed's map of Cornwall
(1611). *British Library*

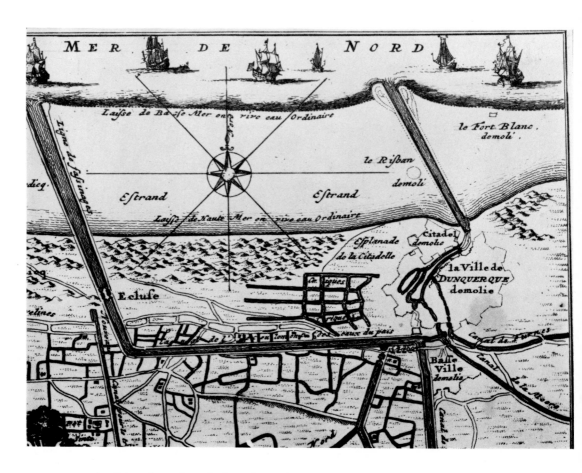

44. Detail of the plan of Dunkirk and Mardyck in Schenk's *Atlas Contractus* (The Hague, after 1713), published by Anna Beek. *British Library*

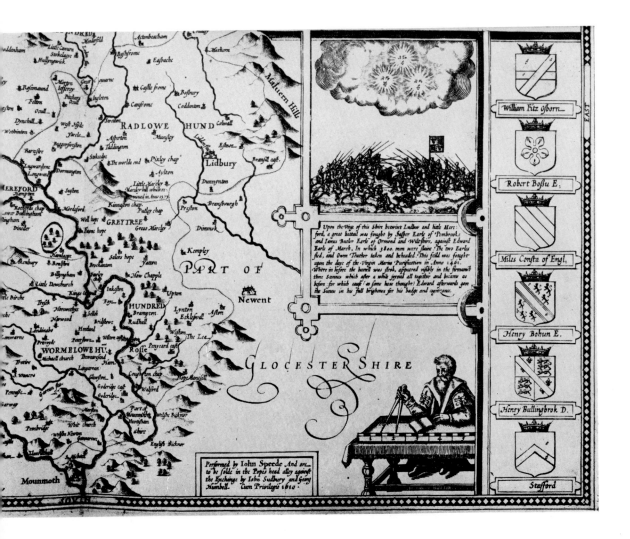

45. Detail from Speed's map of Herefordshire (1611).   *British Library*

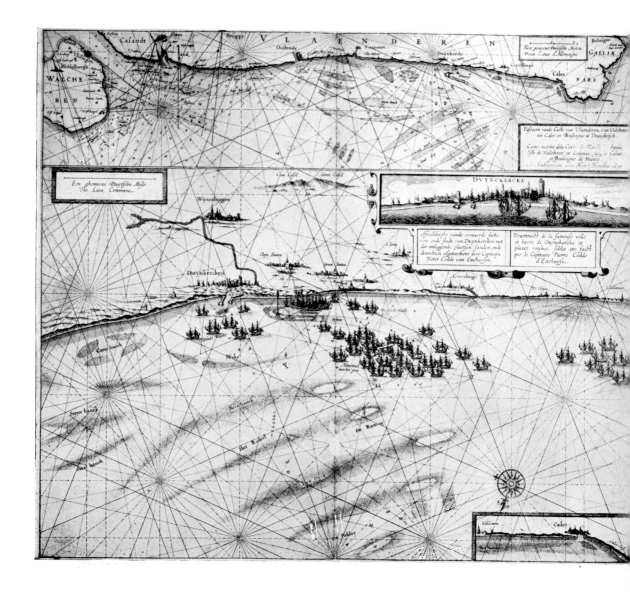

46. H. Hondius: Chart of Dunkirk Roads (1631). *British Library*

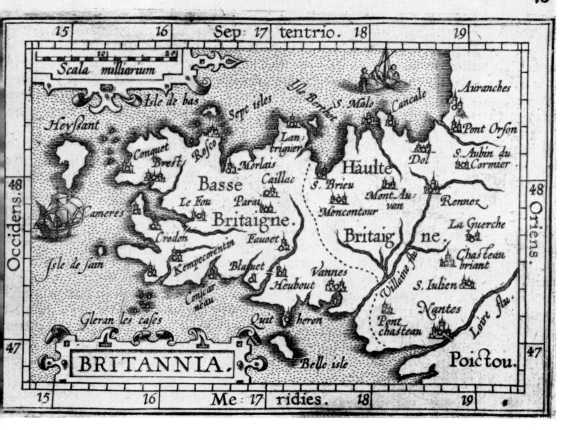

47. Map of Brittany from Abraham Ortelius's *Epitome Theatri Orbis Terrarum* (Antwerp, 1601). 9·9 × 12·2 cm.   *Author's Collection*

Labels visible on the map:

INDIA superior

Cahay

Quinsai

NO VVS

Terra florida

Archipelagus 7448
insularū

Zipangri

Chamaho

Panuco· Inf. Tortucarū

Temiftitan

Iucatána

VBA

Beragua

inf. ydonum

ORENS

Infula Atlantica quam uocant
& Americam

Catigara

Inf. infortu
natæ

Calensuan

Regio Gig

Mare pacificum

48. Western portion of the map of the New World in Münster's edition
of Ptolemy (Basle, 1540). *British Library*

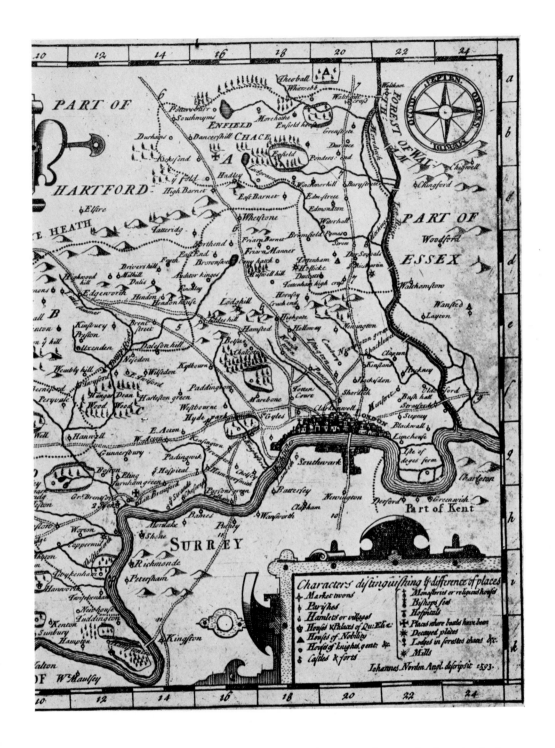

49. Detail from the map of Middlesex from the 1723 edition of John Norden's
Description of Middlesex and Hartfordshire. Engraved by Senex. Size of detail:
17·1 × 13·3 cm.   *Author's Collection*

50. Detail from the map of Switzerland from Ptolemy's *Geography* (1513).
British Library

51. Detail showing lettering from the map of Gt. Britain in Münster's edition
of Ptolemy (Basle, 1540). *British Library*

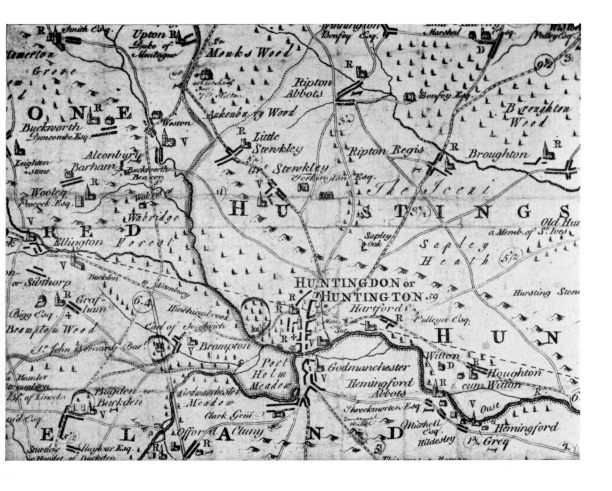

52. Detail from Emanuel Bowen's map of Huntingdonshire (1749).   *British Library*

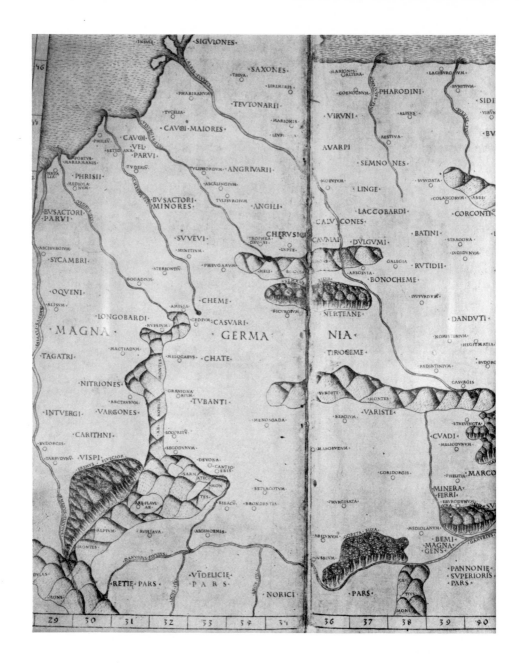

53. Portion of 'Quarta Tabula Europae', showing western Germany, from Ptolemy's *Geography*, published at Rome in 1478. *British Library*

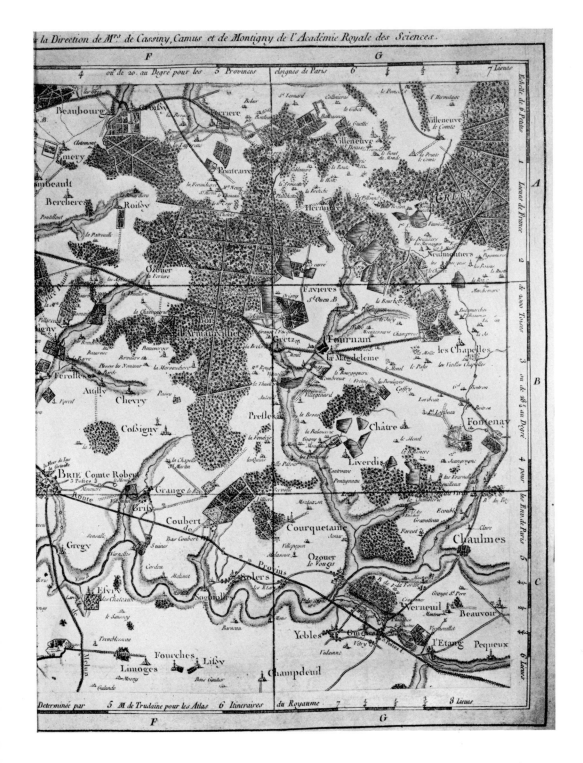

54. Detail from map in Cassini's *Atlas Topographique des Environs de Paris* (Paris, 1768).
British Library

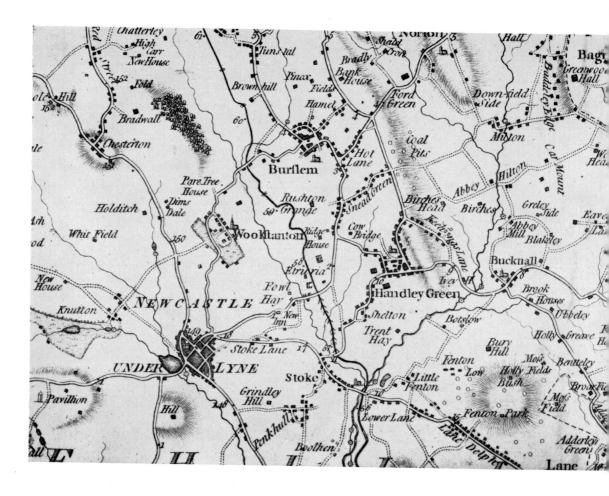

55.   Detail from William Yates's map of Staffordshire (1775).   *British Library*

56.   Detail, showing symbols for towns, etc., from Saxton's map of Norfolk (1579).
British Library

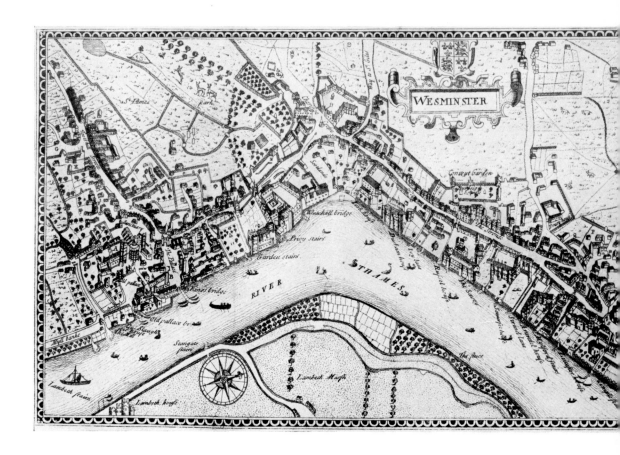

57. Plan of Westminster. From the 1723 edition of John Norden's *Description of Middlesex and
Hartfordshire*. Engraved by Senex. 15·9 × 24·8 cm. *Author's Collection*

58. Group of terrestrial globes. Left to right:
   (i) Nollet (French, 1728: 27·9 cm diameter. Painted and lacquered pedestal
   (ii) Lapié-Langlois (French), 1815: 22·9 cm diameter. Stand supported by black and gold caryatides. The base has brass paw feet, and has a compass fitted
   (iii) Malby (English), 1845: 30·5 cm diameter
   (iv) Klinger (German), 1792: 30·5 cm diameter
   (v) N. Hill (English), c. 1750: 12·7 cm diameter. Fitted with an hour-glass
   (vi) Newton (English), 1836: 45·7 cm diameter. Fitted with a compass

All have meridian circles or segments (vertical); all but no. iii have horizon circles (horizontal).
From The Connoisseur Encyclopædia of Antiques Vol. III, *and by permission of the Trustees of the National Maritime Museum, Greenwich.*

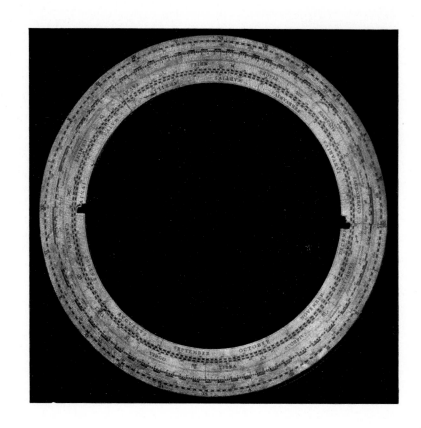

59.   Horizon ring of terrestrial globe by Blaeu, 1602.
By permission of the Trustees of the National Maritime Museum,
Greenwich.

large size, include a large number of manuscript specimens, some of which are of great beauty. But in 1688 he issued engraved terrestrial and celestial gore maps for globes of 106·7 cm diameter, the largest engraved gores issued up to that time. He is known to have made smaller sizes too, but hardly any of them seem to have survived. Coronelli's globes are notable for the way in which they incorporate the maximum of information with the minimum of overcrowding. As on Greuter's globes, names are inscribed in several languages. Ships, oriental people, whales, exotic hunting scenes are all depicted, but there are no loxodromes.

His celestial globes contain thirty-eight constellations in the northern hemisphere, thirty-three in the southern hemisphere, as well as the twelve in the zodiac, and a total of 1,902 individual stars. By far the commonest editions of Coronelli's gores are those issued in 1688, and reissued frequently thereafter with little or no alteration.

In passing, allusion must also be made to a few globe-makers whose output, to judge by what has survived, was limited, but who nevertheless have played their part in the history of the craft. Such were George Christopher Eimmart (1638-1705) of Ravensburg, Giovanni Battista de Cassine (fl. 1700), a Capuchin friar, and Joseph Moxon (1627-1700), the indefatigable English typographer and author of *Mechanick Exercises*, who, in addition to globes priced from 30*s*. upwards and from 15·2 to 66 cm in diameter, advertised 'Astronomical Playing Cards, teaching an ordinary capacity by them to be acquainted with all the Stars of Heaven; to know their places, Colours, Natures and Bignesses. Also the Poetical Reasons for every Constellation. Very useful, pleasant, and delightful for all lovers of Ingeniety. . . . Price plain 1*s*. Coloured 1*s*. 6*d*. best coloured, and the Stars Gilt, 5*s*. Geographical Playing Cards, wherein is described all the Kingdoms of the Earth, curiously engraved. Price Plain 1*s*. Coloured 2*s*. best coloured and Gilt 5*s*. the Pack.'

As with maps, so with globes; Guillaume Delisle's (see p. 30) excellent work had a great influence on their future development. Some of his maps and gores were engraved by Carolus Simonneau, others by the craftsman, Berey. The globes were made of plaster-coated papier mâché, on which the engraved gores were glued. Another designer of globes of high quality was Giovanni Domenico Cassini (see pp. 29–30), discoverer of four of the satellites of the planet Jupiter, and of the rotation periods of Jupiter, Mars and Venus. His celestial globes, which were made by Nicolas Bion (*circa* 1652-1733) of Paris, incorporated an adjustment by means of which the precession of the equinoxes (that is the slow westward retreat of the equinoctial point) could be shown. This was achieved by enclosing the sphere within armillae. Bion was himself a well-known globe-maker in his own right and received the appointment of Engineer to the King for Mathematical Instruments. He made terrestrial and celestial globes and armillary spheres, but they are of considerable rarity.

The Dutchman Gerhard Valck (1651/52-1726) and his son Leonhard (1675-1755) made fine globes. Between them they made beautifully engraved terrestrial and celestial globes of sizes ranging in diameter from 7 to 46 centimetres. Most of the Valck globes were published *circa* 1700; but some, published by the son, are dated 1750. Some of them have among their mountings a band of wood on which is pasted a paper strip engraved with zodiacal and other signs and details.   The cartouches are relatively simple, but contain exquisitely proportioned lettering.   On the terrestrial globes geographical accuracy has in some degree been here and there sacrificed to artistic effect.   Loxodromes radiating from compass roses placed on the meridians of 0, 180 and 270 degrees are featured; colouring is excellent.

John Senex (d. 1740) was a noted English cartographer.   He published atlases, maps and globes, and was sufficiently eminent to be elected a Fellow of the Royal Society in 1728; he dedicated one edition of his globes to its President, Sir Isaac Newton.   His headquarters were at Salisbury Court, London.   The globes of Senex are of various sizes from pocket globes of 7·6 cm diameter upwards.

Georg Matthew Seutter (1678-1757) of Augsburg published maps and globes and manufactured scientific instruments; he became Imperial Geographer to the Emperor Charles VI.   Robert Morden (see p. 42) made globes as well as maps, but apparently only one complete set of gores has remained to tell us of this branch of his activities; it is in the British Museum.   Pietro Maria da Vinchio (fl. 1745), a Franciscan, made some fine manuscript globes; James Ferguson (1710-76), a self-educated Scotsman, made concave spheres of leather, 7·6 cm in diameter, on the inside of which were pasted gore maps of the heavens, and which fitted on to terrestrial globes; they were similar to others made by Senex *circa* 1710; P. D. Pietro Rosini (fl. 1760) an Olivetan monk, constructed a magnificent manuscript terrestrial globe; Jean Antoine Nollet (1700-70), a French scientist, was responsible for the manufacture of terrestrial and celestial globes, which are now rarities; Johann Gabriel Doppelmayr (1671-1750), a mathematician of Nuremberg, made globes consisting of engraved gores pasted on to a wooden ball; one of his terrestrial globes includes portraits of several famous explorers, including Columbus, Magellan, Amerigo Vespucci and others, and their routes are marked.

The de Vaugondys (see p. 31) were Royal Geographers to the King of France. Gilles Robert de Vaugondy, the father of Didier Robert, issued his first maps in 1748. He made preparations for making a 182·8 cm diameter manuscript terrestrial globe for the King, but it does not seem that it was ever completed.

Contemporary with Didier Robert de Vaugondy was L. C. Desnos (fl. 1750), a Danish geographer and Geographical Engineer to the King of Denmark.   He published atlases as well as globes.   The globes of Desnos are well engraved.   The terrestrial ones show many of the routes of the great explorers; one refers to the Russian exploration of the Bering Sea in 1743.   Another Scandinavian cartographer and globe-maker was the Swede, Andreas Åkerman (1723?-78), who received a subsidy from the Stockholm Academy of Sciences to enable him to set up a workshop at Uppsala, from which he issued terrestrial and celestial globes; it is said that he exported them to Denmark, Germany and Russia.   They varied from 30 to 59 centimetres in diameter. Åkerman was succeeded by his engraver, Fredrik Akrel, who reissued the globes in 1779 in a corrected and more detailed form, under the auspices of the Cosmographical Society of Uppsala.

Charles Messier (1730-1817) of Lorraine, an astronomer, published maps and globes.

He worked for Delisle (see pp. 30 and 81) as his secretary and then as his assistant, and in time became a member of the Science Academies of Berlin and St. Petersburg. In his work he collaborated with Jean Fortin (1750-1831), and together they produced terrestrial and celestial globes and armillary spheres, at least one example of the latter, now in the collection of the Hispanic Society of America, being constructed of pasteboard (Fig. 5). It contains a device for the representation of eclipses.

The Adams family were an outstanding group of English globe-makers of the second half of the eighteenth century; they practised their craft in Fleet Street in London. They were George (*circa* 1704-73) the father, who was appointed Optician and Maker of Mathematical Instruments to George III, and his sons, George (1750-95) and Dudley (fl. 1797-*circa* 1825), 'Globe Maker to the King, Instrument Maker to his Majesty & Optician to H.R.H. the Prince of Wales'. The elder George was a writer on cosmography and one of his books *A Treatise describing and explaining the Construction and the Use of New Celestial and Terrestrial Globes,* first published in 1766, had reached its thirtieth edition by 1810, when it was published with a preface and additions by his son Dudley. The Adams globes, the earliest of which now surviving are dated 1772, are made of plaster-covered pasteboard, on which the engraved gores have been pasted. They are all of one size: 45·7 cm diameter. Both celestial and terrestrial globes are quite up-to-date in their information.

Another London map- and globe-maker was Nathaniel Hill (fl. 1746-64); he made small, not very detailed globes of about 7·6 cm diameter. He also made covers for his terrestrial globes, with celestial maps inside, like those of James Ferguson. Little is known about the English globe-maker Michael Lane (fl. 1775). The British Museum owns a small wooden globe of his with line-engraved gores pasted on it, and measuring 7·6 cm in diameter.

Globes of William Cary (1759-1825) dated after 1800 are fairly common; specimens with earlier dates than that are much scarcer. William Cary was a pupil of Jesse Ramsden the optician and mechanician. He had his own business by 1790, and it flourished up to the time of his death. He made many types of instrument, in addition to globes, including microscopes and transit circles. Globes also were made by his brother John Cary (see p. 43), whose sons George and John II (see *List of Cartographers*) carried on William's business after his death in 1825.

Foreign globe-makers of the late eighteenth and early nineteenth centuries included Gian Francesco Costa (fl. 1775) a Venetian engineer, who is of some interest to us in England as he based his celestial globes on the measurements and observations of the celebrated English astronomer, John Flamsteed (see p. 48). Rigobert Bonne (1727-95) and Joseph Jérôme le Français Lalande (1732-1807), were two French cartographers who co-operated in the construction of terrestrial and celestial globes made from engraved gores. French, too, was Charles François Delamarche (1740-1817). Plaster-coated papier mâché globes were made by Vincenzo Rosa (fl. 1790) an Italian; another Italian who made globes, too, at about the same time was Giovanni Maria Cassini.

Nineteenth-century globes, like the maps of the period, became more austere and practical, less decorative, yet some have considerable charm, such as those of the Society for the Promotion of Useful Knowledge, on which were based those of Malby and Co. They have the attractive plainness of early Victorian commercialism.

☆   ☆   ☆

A general word is necessary on the subject of the mountings of globes, their stands or bases, and the circles surrounding them. Of the latter the simplest and most usual are a meridian circle, marking the meridian line of the globe – that is, that line marking midday and dividing the day into two equal parts – and a horizon circle (Plate 58). The meridian circle is the vertical circle which is usually accurately divided into degrees, to give measurements of declination, usually in a quarter circle of 90 degrees in each direction from equator to poles. A globe could thus be adjusted with the polar elevation set for any given latitude. Sometimes the meridian circle was made of wood, but more often of brass or some other metal. It sometimes had attached to it an adjustable quadrant for the measurement of altitude, or for ascertaining azimuths or amplitudes (Fig. 8). Sometimes, too, there was attached to the meridian at the South Pole a

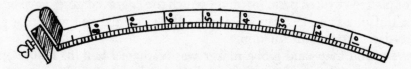

Fig. 8.   Adjustable Quadrant for the measurement of altitude, etc.
From *A Tutor to Astronomy and Geography* by Joseph Moxon (1699)

small hour-circle, divided into twenty-four sections to represent the hours of the day. As the globe was revolved a pointer indicated the hour. This enabled the user to ascertain such things as the rising and setting of the constellations (Fig. 9).

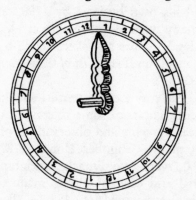

Fig. 9.   Hour Circle.   From *A Tutor to Astronomy and Geography* by Joseph Moxon (1699)

The horizon circle, marking the mathematical horizon, is wider than the meridian circle and usually is of wood, though metal is not unusual, and is notched to allow the meridian circle to pass through it. If it was of metal it had its information engraved on it; if of wood, an engraved paper strip was usually glued to it. The markings consisted of several concentric circles; some had more than ten of these, each affording information of one kind or another, such as zodiacal signs, months and days, church feasts, winds and compass signs (Plate 59). Sometimes a compass needle was let into the horizon circle.

'Those that are *Astrologically* addicted, will want a Circle of Position to their Globes.' So says Joseph Moxon. This is a brass semi-circle divided into degrees and attached to the horizon circle by bearings on which it may be raised or lowered. It was used for finding the astrological 'Houses of Heaven' (i.e. the Zodiac) and positions of other heavenly bodies.

The actual stands in which a globe was set varied enormously as to style and design. Plate 58 gives some typical patterns in vogue during various ages. Some are of such grandeur as to completely submerge the globe as a dominant feature of the piece. Care should be taken in attempting to date a globe by its mountings and stand. The latter in particular is often a later addition. Styles, too, did not necessarily vary in relation to the age in which they were produced. Nevertheless, a knowledge of furniture styles is, in this connexion, an excellent thing to have, and with this in view, I have in the Bibliography mentioned one or two books on the subject worth consulting.[1]

Concerning the dating and identification of globes by conventional signs, geographical features and so forth, the information given in the previous chapters of this book may be used.

Incidentally, the positions of the constellations and the placing of individual stars may sometimes be used for dating celestial globes. The apparent position of constellations varies over the years. Joseph Moxon alludes to this in his *Tutor to Astronomy and Geography*, when he says, in describing celestial globes: '. . . upon which globe I have placed every Star that was observed by Tycho Brahe and other Observers, one degree of Longitude farther in the Ecliptick than they are on any other Globes: so that whereas on other Globes the places of the Stars were correspondent with their places in Heaven 69 Years ago, when Tycho observed them, and therefore according to his Rule want almost a degree of their true places in Heaven at this Time: I have set every Star one degree farther in the Ecliptick, and rectified them on the Globe, according to the true place they had in Heaven in the Year 1671.'

Sometimes a star, or nebula, will appear in the sky, only to disappear after a time. There was one such, for example, discovered in the Cygnus constellation by Tycho Brahe in 1600, which disappeared in 1629, only to reappear in 1659; another one was first discovered in the same constellation in 1670, which has been visible ever since. Obviously, if a celestial globe or map shows a star that was unknown, for example, until 1670, then it cannot have been made before that time. But one should be careful not to use the opposite assumption as a matter of course; that is to say, that if the star is missing, the globe must date from before 1670. It may be that the engraver had never bothered to put it in. On the other hand, it may still be an early globe. I make this observation merely to urge caution.

[1] See entries of Fastnedge and Gloag.

APPENDIX

The Use of Watermarks
in dating Old Maps and Documents[1]

THE idea of authenticating old documents by an examination of the watermarks of the paper is no new one. All who have had to do with early documents, whether as collectors, dealers, archivists or students, have naturally turned to the marks on the paper as a possible clue to the date or place of origin. Even in the eighteenth century Sir J. Fenn, when editing the Paston Letters, took account of the marks of the papers used and gave copies of many of them. In the nineteenth century Sotheby in this country, Midoux and Matton in France, and others elsewhere, attempted a more systematic study; but the vast number of the marks made the collection of data a slow and laborious process, and though much good work was done by Weiner, Heitz, Likachev, Bofarull y Sans and others, it was only on the publication in 1907 of Briquet's monumental work,[2] in which over 16,000 examples of paper-marks are figured, that material at all adequate for the solution of questions of date or authenticity was at last made available. It is needless to say that Briquet's work has been drawn upon to the fullest extent in the preparation of the present paper.

It might be thought that Briquet's labours had left little to be done by others in this field, but nothing could be farther from the truth. The vastness of the subject compelled Briquet to adopt a posterior limit of date for his great collection, and he fixed this at the year 1600, so that for later periods we have only incidental references in his work. Owing to the comparatively late appearance in the field of English paper-makers, he left their work entirely on one side; moreover, he devoted his chief attention to manuscript documents, and allowed that there was still 'infiniment à faire' in the collection and study of marks from printed books. It is hoped that the present paper, while offering little that can be called new, may induce others to co-operate in collecting the data still needed; and that those interested in the subject may be able to supply the answers to some at least of the many questions which to me are still unsolved riddles.

A few general considerations are necessary at the outset in order to gain some idea of

---

[1] Reprinted, by permission of the Royal Geographical Society, from *The Geographical Journal*, vol. LXIII. pp. 391-410 (1924). The author is Edward Heawood.
[2] For details see *Bibliography*.

the degree of precision of the evidence supplied by watermarks. It is not claimed, of course, that any date can be absolutely determined by this method: all that can be said is that it may supply useful collateral evidence in support or refutation of conclusions reached otherwise, or may at least bring the possibilities of date within comparatively narrow limits. Opinions differ on the important question of the length of life, if it may be so called, of individual marks, or of the moulds to which the devices were attached, but there seems little doubt that the latter would wear out pretty rapidly and have to be often renewed. (The date 1742 on many French papers used thirty or more years later proves nothing, as this date was given in obedience to an order of 1741 governing the French paper manufacture.) Briquet, after careful consideration of the statistical evidence, concludes that as a general rule the life of an individual mark may be put down as not over fifteen years. Did we possess only one dated example of a mark occurring in any given document, we could only say that the *probable* date of the latter would lie between limits fifteen years on either side of the dated example. But when, as often happens, a large number of dated examples are available, we may fix the life of the marks much more closely; and if, further, a number of different marks occur in one book or document, the possibilities of the common use of all at one time are narrowed down still more considerably. A test of Briquet's conclusions has been obtained thus. A number of marks from dated books or maps in the Society's Library, of which closely similar examples are given by Briquet, have been taken at random and the divergence of date noted for each. In 37 per cent. of the cases, the divergence proved to be two years or less; in 50 per cent. three years or less; and in 84 per cent. ten years or less. If therefore, the dating of any of these books had depended upon the watermarks, the *probable* error would certainly be less than ten years.

There is of course a bare possibility that paper might be held in stock for some years after making. An exceptional case is recorded of the use of a paper two centuries after it was made, and this might happen with a manuscript document requiring but a small amount of paper: the survival of a stock of old paper sufficient for printing is so improbable that it is hardly necessary to consider it. Briquet's figures are pretty conclusive evidence that no large stocks were held; thus, by comparing the *watermarked* dates of certain papers with the dates of actual use he found that in half the cases the interval was at most four and a half years, and in 92 per cent. of the cases twelve years or less. The great variety of marks often found in a single document points in the same direction. Much of the paper was made in small establishments, whose output was collected by agents for supply to printers at home or abroad, and the accumulation of large stocks would be extremely improbable.

In order to make the best use of our material it is not enough to establish the period in which a given mark was current, or even its place of origin as well. We must try to discover its normal range in space: for example, if a paper can be shown to have, normally, a quite restricted range, we shall feel it necessary to examine more closely than otherwise into the authenticity of a document on such paper, purporting to originate in a widely distant area. It is also desirable to note the usual sources of supply used by any given map-maker, for this may help us to judge of the correctness or otherwise of the ascription to him of an unsigned document. These points have been kept in view in the section of the examples reproduced with this paper[1] which will also illustrate the varying character of the marks used at different periods and in different countries. From time to time special attention will be called to the actual instances in which the marks have supplied evidence of date or authenticity. Some of them are also of interest as taking us behind the scenes, as it were, and giving an insight into the way in which early collections were brought together. Others when occurring on the end-papers of old bindings are useful as helping to establish the date of these. Thus the end-paper of a Ptolemy of 1597 in the Society's Library, in an old stamped leather binding, is marked with the pot, in a form recorded by Briquet once only, and that from the very year 1597; and the somewhat uncommon mark of the arms of Thann in Alsace (17), found on the end-papers of Pontanus's 'History of Amsterdam', of 1611, is recorded elsewhere from 1602 and 1606. Both bindings may therefore with some confidence be pronounced contemporary.

The Earliest Types of Watermarks. Many of the early marks represent common objects of everyday use (sickle (1), hammer, pincers, anvil (48, 49), ladder (57), scissors (4), and even the curry-comb (5), which shows a surprising identity with the modern form). Some even of these may have had an heraldic significance, heraldic charges being commonly chosen as marks. Thus the ladder, found only in Italy, may with some reason be supposed to point to the Scala family, and the anvil may perhaps be taken from the arms of Fabriano – cradle of paper-making in Italy – a smith working at his trade. Weapons, such as the crossbow (2), bow-and-arrow (9), crossed arrows (54, 89), are often met with.

Representations of animals or birds, real or mythical (10-14, 24), are among the commonest marks used by the early makers, especially in Italy, which long supplied paper to the rest of Europe – including even Spain, though this had itself learnt the art from the Arabs. Commonest of all is the bull's head (6, 7), first used in Italy, but soon copied in France and Germany. Its forms are legion, often in conjunction with the cross, crown or other addition, but certain broad types may be localised: thus, in a simplified form, with eyes, but no indication of mouth or nostrils (7), it is always Italian, whilst a type with eyes omitted is always German (used originally, it is thought, by the Holbeins of Ravensburg).

Religious emblems, or marks with a certain symbolic significance, form another group which has been specially dealt with by Mr. Harold Bayley, who holds that even the pot (16, 22, 109), so hackneyed in later times, which gave its name to 'pott' paper, may have originally carried an allusion to the Holy Grail. The Cross is met with in many forms (8. 15, 23, 32, 49, 57, etc.), alone or in combination, in later times mostly in conjunction with the I. H. S. (171). The simple Latin cross in an ovate shield is commonly met with in Spanish documents. The paschal lamb (45), emblem of St. John the Baptist, may have been taken at second hand from the insignia of places or

---

[1] References to these reproductions are given by numbers within brackets. It should be mentioned that a few of the earliest examples, shown for the sake of completeness, are taken from Briquet and other sources, but almost all the rest have been collected by myself from geographical books and maps.

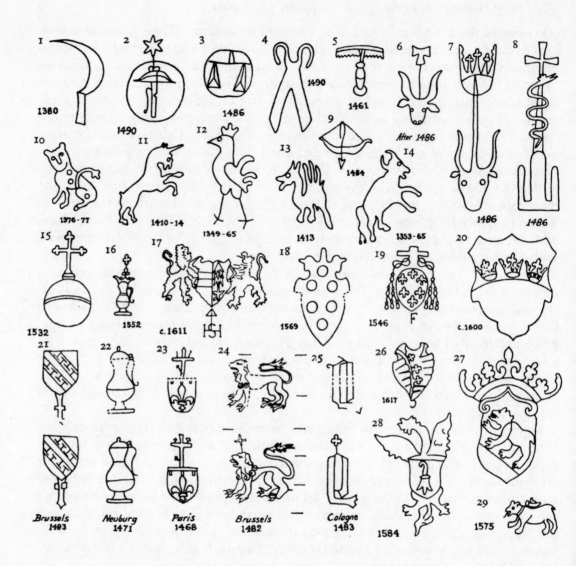

1   1380

2   1490

3   1486

4   1490

5   1461

6   After 1486

7   1486

8   1486

9   1484

10   1376-77

11   1410-14

12   1349-65

13   1413

14   1353-65

15   1532

16   1552

17   c.1611

18   1569

19   1546

20   c.1600

21   Brussels 1483

22   Neuburg 1471

23   Paris 1468

24   Brussels 1482

25   Cologne 1483

26   1617

27

28   1584

29   1575

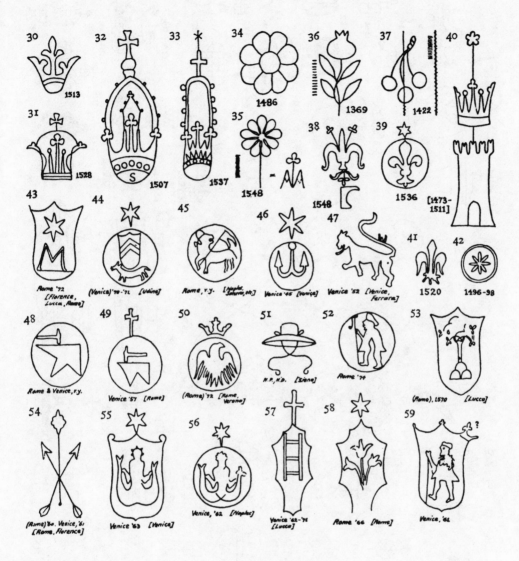

30 1513

31 1528

32 1507

33 1537

34 1486

35 1548

36 1369

37 1422

38 1548

39 1536

40 [1473-1511]

41 1520

42 1496-98

43 Rome '72 [Florence, Lucca, Rome]

44 (Venice) '70-'71 [Udine]

45 Rome, v.y. [Naples, Salerno, etc.]

46 Venice '45 [Venice]

47 Venice '52 [Venice, Ferrara]

48 Rome & Venice, v.y.

49 Venice '57 [Rome]

50 (Rome) '72 [Rome, Verona]

51 N.P., N.D. [Siena]

52 Rome '74

53 (Rome), 1570 [Lucca]

54 (Rome) '60, Venice, '61 [Rome, Florence]

55 Venice '63 [Venice]

56 Venice, '62 [Naples]

57 Venice '62-'71 [Lucca]

58 Rome '66 [Rome]

59 Venice, '61

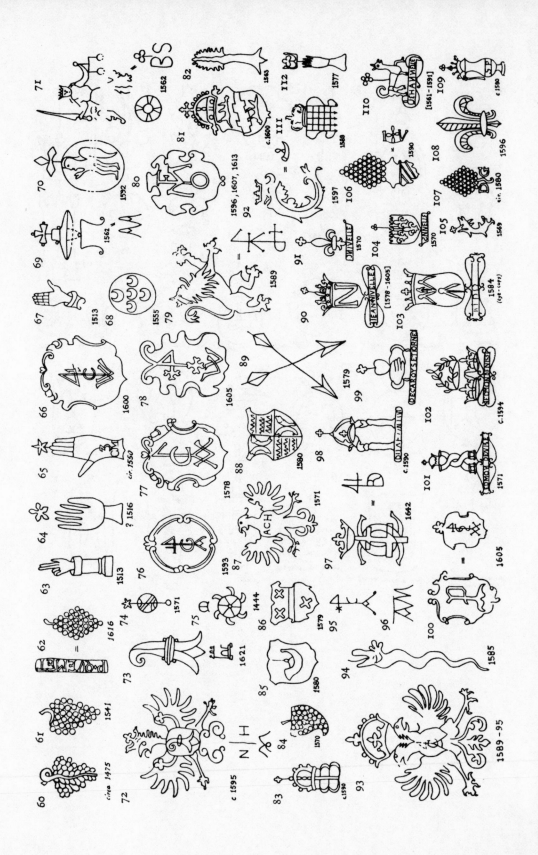

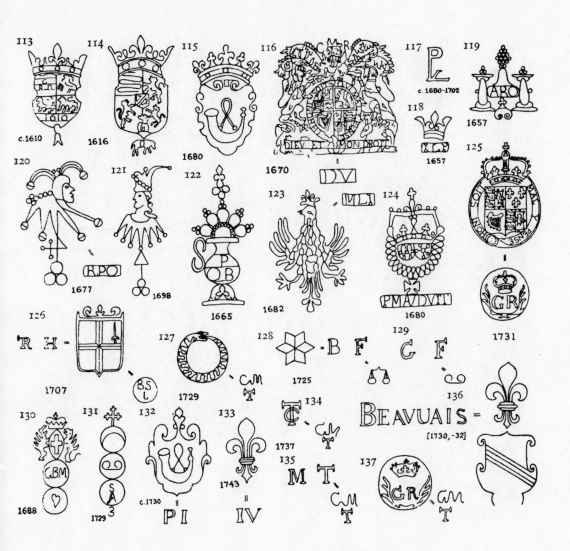

113 c.1610

114 1616

115 1680

116
DIEV ET MON DROIT
1670
DV
VLI

117 c.1680-1702

118 1657

119 1657

120 1677
RPO

121 1698

122 1665
P B

123 1682

124 1680
PMAIDVII

125
HONI SOIT... PENSE
1731
GR

126 1707
R H

127 1729
B.S.L

128 1725

129
B F
G F

130 1688
G.B.M

131 1729

132 c.1730
PI

133 1743
IV

134 1737
TC

135
M T
CM
T

136 BEAVUAIS
[1730, -32]

137
GR
CM
T

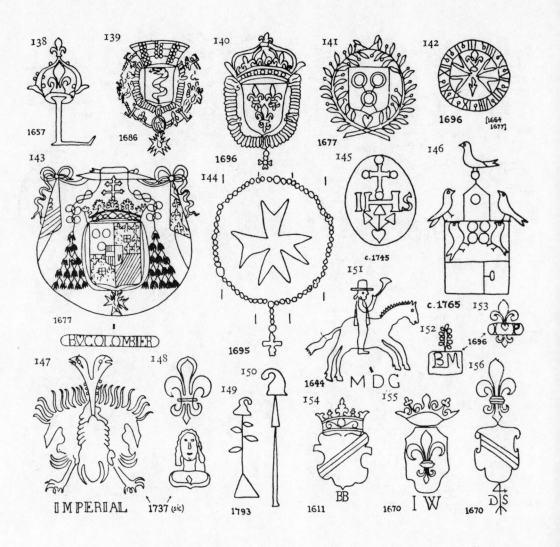

138
1657

139
1686

140
1696

141
1677

142
1696   [1664 1677]

143
1677

BVCOLOMBIER

144
1695

145
c.1745

146
c.1765

147
IMPERIAL

148
1737 (sic)

149
150
1793

151
1644   M D G

152
BM   1696

153

154
1611   BB

155
1670   I W

156
1670   D S

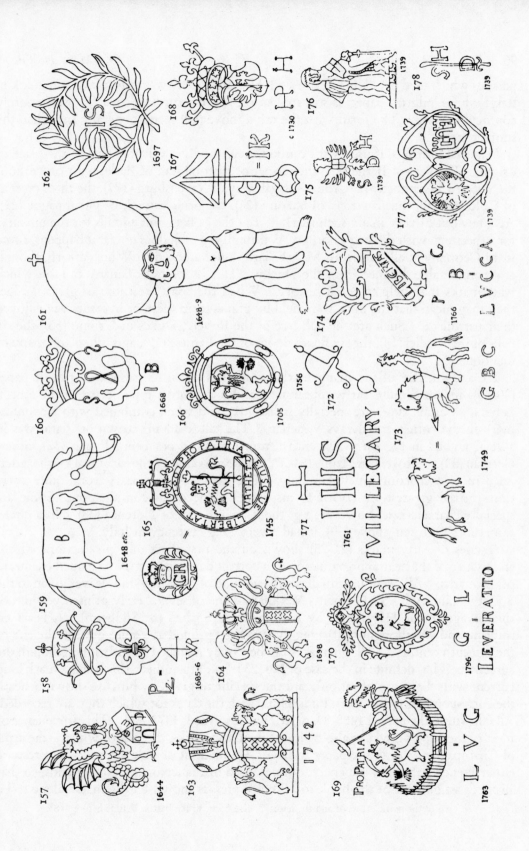

persons which used it.   The pilgrim with his staff (52, 59) was long a favourite mark in Italy, where 'pilgrim paper' is still spoken of, and the cardinal's hat (51) was entirely confined to Italy.   The form with the cross above (69) seems generally later than the simpler form.

Armorial bearings, if used with caution, may supply a valuable clue to the place of origin.   The bear of Berne (27), the crozier or pastoral staff of Basle, with or without the wyvern as supporter (28), the castle gateway of Ravensburg (177), the three crowns of Cologne (20), the *Kranzlein* of Saxony (26), the bend *potencée* of Champagne (21), are a few out of very many such marks.   But they offer some pitfalls for the unwary, for a specimen with the stag's horns of Württemberg (88), does not, as it happens, come from Germany at all, but from Montbéliard, long annexed to Württemberg, though at a considerable distance from the Duchy.   This is one out of many cases in which watermarks may help to teach history.   When the coats are those of great families they may mean that the mills were held by grants from them or in some way enjoyed their patronage.   Such are the oak tree of the Rovere, Dukes of Urbino (53), the six balls of the Medici (18), the six fleurs-de-lys of the Farnese (19), and others too many to mention.

The crown, one of the many marks which has given its name to a size of paper (30–33), makes a quite early appearance, and a primitive type occurs in the Paston Letters.   Certain types are specially Italian (31), and when combined with a counter-mark in the corner are always Venetian.   The tall 'diadem' crown or 'tiara' was in special favour in the early sixteenth century, some types being Italian (32), others German (33).   Flowers and fruits (34–37, 58) are also among the early marks, the latter, once more, occurring chiefly in Italy.   The fleur-de-lys appears from quite early times in the greatest variety of forms, either alone or in combination.   Some are specially characteristic, however, e.g. the *florencée* form of Florence (38).   In a circle, sometimes with star above (39), it had a very long currency in Italy.

A series of early marks (21–25) shown on the first sheet of reproductions affords an instance of the near approximation to a correct date which may be gained by means of the marks.   All occur in a small quarto volume of only fifty-four leaves given to the Society's Library by Sir Clements Markham – one of several early printed versions of the voyage to the Holy Land by Ludolphus de Suchen (or Südheim).   It bears no imprint, and was thought by Brunet to have been printed at Venice quite at the end of the fifteenth century.   All five marks can be closely matched from Briquet (though the agreement is less definite in the case of No. 23 than of the others, this being a mark long current with slight modification), and the specimens given by him are shown beneath the corresponding marks from Ludolphus, with the dates for which they are recorded. All fall within the period 1468–83, or, if 23 be discarded, 1471–83.[1]   The provenance of most of the papers used can also be learnt from the marks, the first of which – the arms of Champagne – shows the paper to have been made at or near Troyes.   The rest are either French papers, or papers commonly used in France or the Netherlands, so that Brunet's assignment of the book to a Venice press is shown by the marks alone to be

---

[1] Two of the marks, the pot and the letter Y, also occur in the Paston Letters before 1483.

most improbable, the use of foreign paper at Venice in early days being almost unheard of. The researches of modern bibliographers have shown that the book was in fact printed at Gouda in Holland in 1484, so that the conclusions drawn from the marks prove to have been quite near the truth.

Less definite but still interesting conclusions may be reached by a study of the marks in the famous Berlinghieri Ptolemy, first printed as indicated by the dedication, and what is known of the printer, about 1480. It was reissued at an unknown date with no change beyond the printing of a title in red on the recto of the first leaf (blank in the first issue), and the addition of a 'registrum' at the end. In both issues most of the text is printed on paper marked with the Cardinal's hat in its earliest form, but as this remained in use through a good part of the sixteenth century it gives little help for dating purposes. The other marks are different in the two issues, the first showing, e.g. the scissors, in a form current about 1480-90, whilst the second has a variety of marks (eight in all) of which those that can be most closely matched elsewhere bear dates lying between 1514 and 1543. Thus the date of 1500 suggested by bibliographers as about that of the second issue seems likely to be somewhat too early. The marks are not here reproduced, but four of them agree closely with Nos. 48, 50, 51 and 57, the others being a ship, bull, crescent in circle, and crossed arrows in circle.

Italian Marks of the Sixteenth Century. Another series of marks on the same sheet (43–59) is given because it not only shows well some of the general characteristics of Italian marks of the sixteenth century, but because all occur in the collection of maps known as Lafreri's Atlas, and (in conjunction with the imprints on the maps themselves) throw some light on the actual way in which the collection was brought together. It has been recognised that this is not to be placed on the same level with the more systematic collections of Ortelius and Mercator, being an assemblage of maps by various authors, many of them drawn and engraved in Venice, whilst Lafreri was established at Rome. But the watermarks bring out more clearly still the haphazard nature of the collection, showing that a large number of the maps must have been not only engraved but printed away from Rome. Nordenskiöld's suggestion that Lafreri may have acquired some of the plates from their original makers seems therefore untenable. The place-names and dates below the copies given are those to be found on the respective maps (the name being indicated in round brackets when only arrived at by deduction), whilst the names added in square brackets are those of places from which the same marks have been otherwise recorded. It will be found that of ten marks associated with maps drawn or engraved at Venice only three are recorded from Rome, while most are known to have been used at or near Venice. Lafreri, therefore, must have merely brought together into one volume (to which he added a title-page) such maps as were contained in his stock-in-trade, or were specially chosen from this by his customers.

In order to make all the maps of a suitable size for binding together it was necessary in many cases to add wide margins to the sheets as originally printed, and it may be noted that the watermarks on these margins (Nos. 48 and 54) also occur, the one on the title-page, the other on certain maps, in the collection itself – a sufficiently clear indication that the Atlas (at least the copy given to the Society by Lord Peckover) has come down to us in the form in which it was originally issued, and that the margins were not added later in binding for the owner of a fortuitous collection of maps.

Among the specially Italian characters shown by these marks we may note (1) the penchant for the circle drawn round the mark proper; (2) the common use of the star either above the circle or shield (as in 44, etc.) or within the shield (43); (3) the typically Italian forms of the shield (43, 57, etc.). Other marks from Italy are shown in Nos. 35-39, 68-71, those with the countermark in the corner (letters with often a trefoil above) being shown by this to be probably Venetian. The mark representing Justice with sword and scales (71, unfortunately rather indistinct) is by no means common, and only a much smaller specimen is given by Briquet. Another Italian mark, the pear-shaped fruit reproduced in the *Journal*, vol. LXII, p. 282, is there shown to be of use in confirming the date of the newly found Contarini map of 1506. Since the account of that map was written I have found still another record of the use of the same mark, in an engraving by Andrea Mantegna of Mantua, who died in 1506; further confirmation of the date of the map, if such were needed, being thus supplied. In the Tower with crown and fleuron (40) we have another rare mark, to be found in Dürer's 'Apocalypse' of 1511, which has been used by Mr. Henry Stevens as evidence for the date of an unique early map, discovered by him some twenty years ago.

Typical French Marks. Marks specially characteristic of France (at least after the earliest period) are the pot, the hand and the grapes. The pot was used chiefly in Auvergne and Champagne. Of the vast numbers of forms of the hand, Briquet has shown that one (64) is characteristic of Genoa, another of north-west France (65), a third of north-east France (67) and a fourth (with crown above) of south-west France (112). Of the grapes (60-62, 84, 107), the earliest forms are decidedly the most artistic, the later ones becoming far too geometrically regular. The mark seems to have originated in north Italy, and to have spread to France, where it became so well established that it gave its name to certain sizes of French paper, still recognised as 'raisin'. Placed near the edge of the sheet, as in the Servetus Ptolemy of 1541, printed at Vienne in Dauphiné (61), it may be localised in southern or south-western France, where it seems to have been a more or less general practice to place the mark in that unusual position. Other marks typically French are the St. Catherine's wheel (used chiefly in Auvergne; 75) and the sphere (in the Angoumois; 74).

Marks used in the Netherlands and Rhine Countries, 1570-1600. With the marks found in the Italian Collection of Lafreri, it is of some interest to compare those met with in the atlases and maps of the great cartographers of the Low Countries, on whom the mantle of their Italian predecessors fell from about 1570 onwards. These marks, with others used in west Germany, are represented by Nos. 66, 72, 73, 76-104, and, as will be seen at once, are of a totally different character from those used in Italy. The pastoral staff or crozier (72, 73) was the special mark of Basle makers, and much paper so marked found its way to various parts of the Rhine basin, and other regions accessible by water-transport. The personal marks of the actual makers help still further to locate the papers with this mark. Thus the N H of No. 72 stands for Nicolas Heusler or Hüssler, whose special mark is added below, and he is also indicated by the house – a play upon his name – in No. 157.   Similarly, the tower below the crozier in No. 73 is a play upon the name Thurneysen, another of the great Basle makers.   A favourite mark in this region is the monogram on an ornamental shield (66, 76–78, 80), and in this and other marks of the present group the frequent occurrence of the symbol resembling a Roman 4 (also much used as a masonic or merchant's trade-mark) is to be noted.   It is held by Briquet to be specially indicative of an origin in the Rhine countries, particularly Lorraine.   In No. 100 it is found in association with the Gothic letter P, a mark of much wider extension, both in time and space, the meaning of which has, however, never been satisfactorily determined, in spite of much discussion.   The displayed eagle (72, 87, 93) either double- or single-headed, is also a mark long in favour in different regions.   The fine specimen (93) occurs in the first edition of Mercator's Atlas, and being found on the map of America sometimes present in this, as well as on the sheet bearing the date of issue (1595) and other maps, lends support to the belief, sometimes challenged, that America was really included in this issue, even if not in all copies.   A careful study of the marks in all copies available would no doubt throw light on the precise manner of issue of the various parts of this first edition.   Other marks on paper used by Mercator are the monograms 76–78 (77 being the earliest to occur), the serpent (94) and the smaller monograms (95 and 96).   The last (96) occurs in all the covers of the set of big Mercator maps once existing in the University Library at Rome (see *Journal*, vol. LXII, pp. 33, 138) and shows that, though those maps were originally published at widely separated dates, the set in question must have been reissued as a whole after all had been published.

The monogram 66 occurs on the excessively rare map by Hondius, showing Drake's route round the world, which bears no date or place of publication.   As the same mark is found in a book and on a map printed respectively at Cologne in 1600 and at Amsterdam in 1604, there seems no reason to adopt the suggestion (lately made to me by an American student of Drake) that the map was printed in England during Hondius's residence there before 1595.   No. 97 is of a type which had a great vogue over a wide area for many years, with or without the interlaced Lorraine cross.   Of its origin in Lorraine there is no doubt, the initials standing for Charles Duke of Lorraine and his wife Claude.   In the simpler form it occurs in Wytfliet's work printed at Douai in 1605, and throughout the Mercator-Hondius Atlas of 1633. Besides the eagle (72, 87, 93) Nos. 80–82 are probably German marks, the F in 80 standing perhaps for Frankfurt.

Turning to marks of paper used for the maps of Ortelius (83–89, 91, 98–99, 101–4) we find a very different series, though including one (the smaller displayed eagle, 87) also used by Mercator.   (The A C H of the Ortelius specimen almost certainly stands for Aachen, or Aix-la-Chapelle.)   Most of the marks indicate a French origin, and the series as a whole seems to show that closer relations were maintained by Antwerp printers with the great paper-making centre of Troyes than with the Rhine countries which supplied Amsterdam.[1]   Of fourteen marks here reproduced from Ortelius,

[1] The Plantijn Press used also a good deal of paper from Frankfurt.

just half are almost certainly from Troyes.   The makers here included (besides Denise, Nivelle, De Caroys and Journé, whose names are to be read in the specimens shown) various members of the family of Le Bé, to which the reversed B's of No. 83 may with some probability be ascribed.   The large crossed arrows (89), much used in some of the earliest editions of the 'Theatrum', are almost certainly from Troyes, as they are found on the paper used for the local records.   Another French paper is that with the arms of Montbéliard (88) already spoken of.   The capital N (90), another mark used by one of the Nivelles, is not from Ortelius, but occurs on the British Museum copy of the rare map, 'veuee et corige par le Seigneur Drach', reproduced in Mrs. Nuttall's volume *New Light on Drake*, published by the Hakluyt Society.[1]   Of the history of this map nothing definite is known, but the mark may perhaps help, in conjunction with other evidence, to throw some light upon it.   Mr. Sprent of the British Museum thinks, from internal evidence, that the map may have been printed at Antwerp before 1590, and it may thus be the very earliest map made to illustrate the first English circumnavigation.   It has been suggested, on the other hand, that it was made much later to accompany a French translation of Drake's voyages; but the known use of Troyes papers at Antwerp about 1580-1600 (as shown above), and the fact that this particular mark was current, according to Briquet, between 1578 and 1602, lends decided support to Mr. Sprent's view.

Still another mark used by Jean Nivelle is the quaint one seen in No. 110.   In connexion with this mark Briquet quotes an old French verse about 'le chien de Jean Nivelle, qui fuit quand on l'appelle', to which there is an evident allusion, although the personage referred to in the verse is not the paper-maker but a French nobleman of earlier days whose son took up arms in a time of disturbance on the side opposed to his father, and resisted all efforts to induce him to return to his duty.

Marks in English Books, Sixteenth Century.   So far we have had no occasion to refer to English papers or their marks, although paper was made in England before the end of the fifteenth century.   No. 42 (borrowed from Briquet) is a mark used by John Tate, who set up a mill at Hertford to which a visit was paid by Henry VII.   It is to be found in the Paston Letters as well as in the 'Golden Legend' printed by Wynkyn de Worde in 1498.   But paper-users in this country long continued to depend largely upon the French mills, as they had from quite early times.   The specimens 105-112, all taken from English geographical works, are mostly French, and other French marks found in England before 1600 include the crossed arrows (in the first issue of Saxton's County maps) and the hand of the type shown in No. 65.   As above mentioned, this is characteristic of north-west France, while the hand with crown (112) came from southwest France.   The specimen given is from Eden and Willes's collection of Voyages, the first general collection to be printed in English.   The marks 108 and 111, however, are probably English – the portcullis as representing the Tudor badge (which now finds a place in the Westminster arms), and the fleur-de-lys because this type, with half of each petal shaded, is commonly found in English books for some time on from this

---

[1] The specimen shown is borrowed from Briquet, as the mark is very indistinct in the map in question.   It is there slightly smaller.

period. The small pot (109) occurs commonly in England from this time on, but the paper probably came from Normandy. The fool's cap (120), so hackneyed a mark in the next century, already appears in England before 1600. Although originating abroad, the mark was adopted by Sir John Spielman (who started a mill at Dartford in the reign of Queen Elizabeth) and is to be found on a rare map by John Blagrave printed in 1596. One of the French marks in this group (106) gives an instance of a near approximation to true date afforded by a mark of somewhat unusual form. It occurs on the star maps published in 1590 by Thomas Hood, a Fellow of Trinity College, Cambridge. On a first rapid inspection of these maps (which are not often met with) the date escaped notice, but was very approximately fixed by a reference to Briquet whose sole specimen of an identical watermark is recorded as from 1588.

Seventeenth and Early Eighteenth Centuries: England. In the seventeenth century we lose Briquet's valuable guidance,[1] apart from occasional historical excursuses, and any conclusions put forward must be regarded as more or less provisional and tentative. English makers now appear more often in the field, but it is not always easy to discriminate between English and foreign makes. A paper much used in England in the first half of the seventeenth century seems to have been of foreign origin, unless indeed the marks were copied by English makers from a foreign original. They represent a coat-of-arms roughly agreeing with one used abroad even before 1600 (114), and like this have apparently a representation of the 'toison d'or' (golden fleece) below the shield (113). Several members of the House of Orange and Nassau (including William the Silent) were members of the Order of the Toison d'or, but the arms, though bearing some resemblance to those of Nassau, are not identical.[2] The mark, or its variations, is found in many English geographical works, including Purchas's 'Pilgrims', Camden's 'Britannia', Fynes Moryson's 'Itinerary', and Foxe's Map of the Arctic Regions (in the last forming a clear link with the foreign examples, as it includes the tower omitted in the others). As Elizabeth, daughter of William the Silent, married Henri de la Tour d'Auvergne, Duc de Bouillon, it might be thought that a mark embodying quarterings both of Nassau and (possibly) La Tour, had a reference to this lady. But it appears, in an example given by Briquet, as early as 1594 or before the date of the marriage. As used in the paper of Captain Smith's map of Virginia (the very rare first issue) it is of special interest as containing the date 1610, and therefore proving that the first issue was *not*, as is sometimes stated, brought out before that year. Other marks specially characteristic of paper used in England in this century (in addition to the ubiquitous fleur-de-lys) are the pot (very common in a small form in the first half of the century, and becoming larger and more elaborate in course of time – 122), the fool's cap in various forms (120, 121), the horn on shield with crown above (in a form identical with that now in use) and the pair of pillars or posts (119) which I have suggested as the real origin of the term 'post' paper, usually attributed to the supposed

[1] But see *Bibliography* for details of Heawood's own book on later watermarks, which was published after this paper.
[2] A foreign specimen from 1591 has *two* lions passant in the fourth quarter, bringing the mark still nearer the arms borne by the Prince of Orange, though the lion and fess of the first and third quarters in these are transposed. The second lion in the fourth quarter was very soon discarded, perhaps from want of space.

post-horn. (That 'horn' and 'post' papers were distinct kinds is shown by a letter from Madras dated 1666 to which my attention has been called by Mr. Foster of the India Office.) Paper with the Royal Arms was used by the House of Commons until tabooed by the 'Rump' Parliament, but of this early type I have not yet secured an example. On the Restoration the mark gained a new lease of life in the elaborate form found in books of about 1670 (116). The displayed eagle, in a somewhat new form (123), is commonly found in English books of the late seventeenth century, and for a limited period a somewhat remarkable mark, taken by Mr. Bayley to represent the five loaves and two fishes (124), was frequently used, but the makers' names attached seem to be all French (Jobert, Durand, Conard, Vaulegard, Mauduit, and perhaps also the singular name Homo which I have found on the fly-leaf of a book bound in 1705 in France for the Polish collector, Count Hoym). Unless, therefore, the names are those of Huguenot refugees the paper would seem to have been imported.[1]

A common mark of the same period is a monogram (117), for which I have as yet found no certain explanation, but which may perhaps be read as the letters I L P to be found in an unmistakable form in association with other marks of this century (118). The monogram is met with in both English and Dutch books, either alone or as countermark, but I am assured by a Dutch student of watermarks that it does not signify a Dutch paper.

In the early eighteenth century the Royal Arms were again used (125), and, being changed on the accession of a new sovereign, may give a useful clue to date, especially as the old form was sometimes retained for a year or two, coupled with the initial letter of the new ruler. The arms of the City of London (126) were another favourite mark, as also was the serpent in a circular coil (127). Both of these are often supplemented by a countermark in the corner of the sheet, and in this respect they show a decided affinity with a group of marks which now makes its appearance, and is generally associated with an excellent crisp paper (128-9, 134-5). The main mark may consist only of a letter or letters (noteworthy in many cases for the emphasis given to the serifs, which are carried above the line), or may be associated with a star in the other half-sheet, in addition to the mark in the corner. This last may be either a minute design (pair of scales, etc., etc.) or another letter or combination of letters, C M with T below being particularly common (127, 134-5, 137). This is one of many instances of two sets of initials appearing on the same sheet, one possibly standing for the name of the proprietor of the mill, the other for the lessee who worked it.[2]

[1] I have a strong suspicion that the mark may be merely a degenerated form of the French Royal Arms (coats of France and Navarre) surrounded by the collar of one of the Royal orders of the St. Esprit and St. Michel. (The chain most resembles that of St. Michel, but the star below belongs to the St. Esprit). The undoubted arms of the French king, with a similar collar and star, were used by the Troyes paper-maker, Journé, before 1600, and continued in use during the first few decades of the seventeenth century. Towards the end of the same century the arms of France only were much used, surrounded by a similar collar (No. 140), which here is undoubtedly that of St. Michel, though still with the star (or cross) as pendant. Examples of similar degeneration at the hands of workmen who did not understand the meaning of the coats-of-arms are not uncommon.

[2] This group of marks appears to have so close an affinity in some ways with certain foreign marks to be dealt with later (130, 131), that we might suppose the papers above spoken of as commonly used in English books to

These marks may supply a clue to the true date of an issue of Petty's Atlas of Ireland, originally printed in 1683, of which a copy was lately acquired by the Society. The only date to be found in it is that below Petty's portrait given as a frontispiece, but as it is there engraved on the plate itself, it would not be changed on a reissue. That such was made is known by the inclusion, in some copies, of a dedication to Lord Shelburne, who only received that title in 1719, but this is not found in the Society's copy. Yet the watermarks (135-7) show pretty conclusively, when compared with the dated examples of this group of marks, that the copy can hardly have been printed before about 1720. The shield with the bend and countermark Beauvais (i.e. the Beauvais mill at Angoulême) points to a date round about 1730, paper with the name of the same mill, with a similar four-line bend (instead of the usual three-line), being found in use in that year, and, in conjunction with other marks, in 1729 and 1732. The G R (note the serif of the G carried above the line) is not conclusive in itself, as it might conceivably stand for Gulielmus as well as Georgius, but as the mark is constantly found otherwise in the reigns of the Georges, but not in that of William, such a supposition is quite improbable.

A paper used a good deal in England, especially for manuscript work, bore the horn on a shield or cartouche of fanciful shape (132), and might be thought to come from Italy, as it sometimes bears the word Vorno, the name of a village near Lucca, but the initials I V T, associated with some examples, seem to indicate that it came from the mill owned by Van Tongheren of Angoulême. This mark, which will be referred to again presently, seems to have held its ground down to the nineteenth century.

Seventeenth and Eighteenth Centuries: France. In France we find many armorial marks in use in the seventeenth century. The Royal Arms as borne by Louis XIV (France and Navarre) occur about 1670 in an elaborate form, with the collars of the two Royal Orders. The arms of prominent men like Cardinal de Bouillon (143) or Colbert (139) would naturally be used mainly when such men were at the height of their influence. As the Cardinal belonged to the family of La Tour d'Auvergne, the use of his arms in that great paper-making district is easily explained. One of the quarterings of his coat – the three *torteaux* of the Counts of Boulogne, was also much used alone, the *torteaux* being sometimes, however, replaced by *annulets* (141). Other common marks were the crowned L (138, probably for Louis[1]), the clock-face (142), and the

---

be also of foreign make. Some of the small corner-marks are to be found also within the circles of the foreign marks, e.g. that resembling a pair of eye-glasses (compare 129 with 131), and another representing a leg or boot. Small corner-marks (e.g. a star) also occur in some of the foreign papers referred to. Another similarity consists in the way the letters (or some of them) are joined together by the use of a continuous wire (in some papers used in Spain the whole of a maker's name is similarly made by one wire). Further, the initials L C used in one case as a corner-mark are also found in an identical form in conjunction with the specially French mark of the grapes.

[1] An earlier series of marks, all of the same style, and all containing a crowned letter (placed between two fleurs-de-lys within a crowned shield), have caused a good deal of perplexity and offer a problem not yet satisfactorily solved. The letters found are B, F, L, and R (or possibly K), but the puzzle is that the name of no French king or regent began with B, and whilst L might stand for Louis, F for Francis, and K (if such it is) for Charles (Karl), the L is found throughout the sixteenth and early seventeenth centuries (from 1509 on), whilst of the six kings who reigned between 1515 and 1610 not one was a Louis. The R or K is found in use from 1493 to 1682.

mace; the grapes and the fleur-de-lys still remaining much in evidence also. Among the great makers were Colombier, Dupuy, and the Sauvades, all of Auvergne.

During the eighteenth century the Auvergne makers produced stout papers of large size suitable for the huge atlases then in vogue, and the marks most used were the rosary and cross (144; already met with on papers made by Colombier and Dupuy before 1700), the two-headed eagle (147, a form used in the Angoumois), the I H S and Cross in an oval (145), and, in the latter half of the century, the dovecot (146). For smaller-sized papers, the grapes remained perhaps the most common mark. The makers' names, now usually given in full on the other half of the sheet, include the Dupuys, Cussons, Richards, Tamiziers, Micolons, and Vimals of Auvergne, and Pigoizard and De Michel of the Angoumois. A special form of the fleur-de-lys, with another mark below (148), was in use in the latter area. An echo of the French Revolution is to be noticed in the cap of liberty as used in Vaugondy's big Atlas of 1793 (149, 150).

Papers used in Holland, Seventeenth to Eighteenth Centuries. The many geographical works printed in Holland in the seventeenth and early eighteenth centuries show us what were some of the papers most used in that country, but many, if not most, were probably imported. Dr. Enschedé of Amsterdam thinks that the manufacture of paper, for which Holland eventually became noted, acquired no great importance there until the next century. The fleur-de-lys on a crowned shield (155, 158), eventually taken as the mark denoting, with slight variations, paper of Imperial, Super-Royal, Medium and Demy sizes, had a great vogue, but the initials W. R. attached to many examples indicate that some was made by Wendelin Riehel of Strasbourg. This mark was much copied, and Briquet quotes a document showing that paper so marked was made at Nersac, near Angoulême, for the Netherlands market. The bend as an armorial charge, originally representing the arms of Strasbourg, occurs in various forms (154, 156), but soon settles down into that still used for 'Royal' paper (156). For a time the fool's cap found almost as much favour as in England, but generally in a varying form. The huntsman with horn is often found (151), also the horn on a shield (160) and the crozier of Basle (157), whence much paper was sent down the Rhine to Holland. The imposing mark (161) showing Atlas upholding the World, is found in various editions of Blaeu's great atlas, at least from 1648, indicating that the paper for the same was probably made specially to Blaeu's order. Other marks found in the atlas show the letters I B (e.g. the elephant, 159), which might be thought to stand for Blaeu's name, were not the initials fairly common among paper-makers, one of the most notable being Jerome Blum of Basle. Another fine mark found in the productions of the firm of Blaeu is the laurel wreath with the initials I S within (162), which may perhaps stand for Jacques Salmon, proprietor of a paper mill at Angoulême about this time. The Dutch lion with sword, and darts representing the seven independent provinces (166), might be thought to indicate a Dutch paper, but it too may have been used by French makers who supplied the Dutch market. Towards the 'seventies of this century the arms of Amsterdam, with lions as supporters, began a career which lasted over a century at least, but this paper is known to have been made very largely in the Angoulême district for the Dutch market at mills kept going largely by Dutch capital. Paper with this mark was used in England, France and Germany, as well as Holland, and a certain chronological sequence may be traced in its varying forms, so that it is not quite valueless for dating. The forms with mantling (164) seem to have been generally the earliest (one occurs in Dapper's 'Asia' of 1672), whereas the eighteenth-century forms had mostly the plain crown. The date 1742 on one example (163) proves unmistakably its French origin, being inserted in obedience to the French ordinance of 1741, and the makers' names or initials, often given as countermarks, are a further guide. The Amsterdam arms are to be found on the paper of a Dutch MS. in the British Museum – a copy of Tasman's original journal kept during the voyage of 1642 – and, in conjunction with other marks in the same MS., give a fairly close indication of the date of writing, about which there has been considerable uncertainty. The arms are of the style of No. 163 (though without the date) and the countermark gives the name of I. Laroche, a member of which family is known to have worked a mill at Angoulême about 1750. Another of the marks is the horn within a shield of the pattern shown in No. 132, and the associated initials I V T seem to point to one of the Van Tongherens, owners of one of the Angoulême mills in the early eighteenth century; they are to be found also in one of the volumes of Valentyn's great collection of documents relating to the Dutch colonies, published in 1724-6. A third mark is of the type of No. 115, so long used with hardly any variation, but the initials L V G below and the I V of the countermark, are those of Dutch makers of the eighteenth century, which begin to occur commonly from about 1730 onwards. These indications point consistently to a date for the writing of the manuscript between 1725 and 1750, and confirm the view of Professor Heeres that it was perhaps written after 1700, as against that of others who thought it nearly contemporary with Tasman's voyage.

Some German Marks. It is impossible within the limits of this paper to deal adequately with the marks of German paper-makers (some of which have been already referred to incidentally), though some of the German mills, as those at Ravensburg, Augsburg, Bautzen and various places in Baden and Saxony, were noted from quite early times. Coats-of-arms were much used, as in other countries, some being extremely elaborate. One of the simpler forms, which seems to imply a connexion with Ravensburg, is given in No. 177, and another in No. 168. The style of 167 seems to imply a German origin, and Nos. 175-6 and 178 are all from the same German book as 177.

Later Dutch and English Marks. In the early eighteenth century the Dutch lion appears in a new form within a ring fence (169), in company with an allegorical female figure bearing the hat of liberty on a spear, with the words 'Pro Patria' above. I am told that 'pro patria' is a term still used in Holland for a class of paper. It was especially used for MS. work (e.g. legal documents), or for the fly-leaves of bound books, and had a great vogue in this country as well as in Holland.

It is often exceedingly difficult to say definitely whether a given paper is of Dutch or English make. The Dutch papers were now coming to the fore, and the reputation in which they were held in the eighteenth century is shown by the mark 'All' Olandese' which I have found on an *Italian* paper of 1764, showing that, even in the early home of paper-making, to be made after the Dutch style was thought a recommendation.   In England the Dutch lion holding the hat of Liberty on a pole or spear (165) was for a time, it seems, the recognised mark of 'foolscap' paper, until replaced by the figure of Britannia now current.   With this and the plain fleur-de-lys (133), the present-day marks for 'Imperial', etc. (roughly similar to the old mark 158), 'Royal' (cf. 136, 156), and 'post' (cf. 115) held the field at the expense of many of their former rivals, so that something at least of the pleasing variety of the old marks became lost.   Even in England these marks are commonly associated with the initials of the great Dutch makers, Villedary and Van Gerrevink (I V and L V G), both sometimes occurring on the same sheet, or one of them in conjunction with the name of an English maker (e.g. J. Bates) so that one might almost think that they had become common property among paper-makers.   Another Dutch firm – that of the Honigs – does not seem to have catered largely for the English market.   On the English side the famous firm of Whatman was already a serious rival of the Dutch makers.

Later Spanish and Italian Marks.   In dealing with the later Spanish and Italian marks, it is not easy to disentangle them completely, so much Italian paper being used in Spain.   In the late seventeenth and early eighteenth centuries two closely allied forms are commonly met with in Spanish books and documents.   One is an obvious development from the three moons or three circles of early Italian paper (131), and the other shows the arms of Genoa combined with two of the three circles (130).   But here again caution is needed in assigning a place of origin, for there is documentary evidence (quoted by Briquet) that in the eighteenth century the paper with the 'three O's' or 'three rounds' was commonly made in south-west France after the 'façon de Gênes' for export to Spain and the Indies.   In another form – three crescents placed horizontally, with unusually coarse wire-marks – the mark does apparently denote an Italian paper, often exceedingly stout and firm, such as was used for collections of engravings, and almost to be identified by the touch.   The fleur-de-lys in a circle (cf. 39) continues to be a common Italian mark, and we find a late variety of the crossbow (172), showing an entire change of style from the early form.   It supplies the single instance I have met with where a tentative suggestion of Briquet's seems mistaken, as it appears in his collection as a solitary intruder among the older forms, with the suggested date 1602. It was found by him on the fly-leaf of a book of that date in what he thought a contemporary binding, but where the fly-leaf at least would appear to be much later. Elaborate designs with scrollwork (170) found considerable favour in Spain, but the most popular mark there in the eighteenth century appears to have been the bullfight (173).   The arms of Lucca (174) are an unmistakable instance of an Italian paper used in Spain.

Besides the marks themselves there are other helpful indications, such as the position of the mark or its countermark on the sheet.   We have seen that a position near the edge points generally to the south of France; similarly, one in the very centre of the sheet was, according to Briquet, a characteristic of Genevan paper, whilst a countermark in the corner was, in old days at least, a mark of Venetian papers.   The spacing of the two sets of wires to which the 'chain' and 'laid' marks are due may also be helpful, since in early days two periods of fine spacing were separated by one of wide spacing (see Nos. 36, 37 for examples of different types). But this part of the subject cannot now be enlarged upon.

Another branch of the subject which, by reason of its definitely geographical bearing, it would be interesting to pursue further did space permit, is that of the course of early trade as evidenced by the extent of country over which individual paper-marks were current. When it is possible to determine the centres at which papers bearing given marks were made, the great number of records brought together by Briquet permit us to lay down what we may call the 'spheres of influence' of such papers, and this has been done for two marks of fairly wide diffusion in the accompanying two sketch-maps (p. 108). The first shows by means of small circles the principal places from which the crozier of Basle has been recorded, the second doing the same for the crowned hand of south-west France. The close clusters of circles around the respective districts of origin would suffice in themselves to indicate roughly whence the papers issued, and many of Briquet's conclusions are based upon a consideration of such regional distribution of occurrences. The importance of water-transport is particularly well brought out in the case of the Basle mark, while it will be seen that the two maps are more or less complementary, the spheres of the two papers hardly overlapping at all, except in the case of such a noted emporium of trade as Hamburg. It must suffice here to point out the interest of such a line of study.

It has been impossible in this paper to refer to more than a small proportion of the vast number of marks current in old days, even though attention has been confined chiefly to the three centuries, 1500–1800. Enough has, however, been said to show that even a general knowledge of the subject may be of some help in authenticating documents or disposing of unfounded claims, and there are cases where such a general knowledge may be of more help than even a thorough knowledge of a limited period only. Thus the mere fragment of a mark shown to me as occurring on an engraving, the date of which was in question, could at once be identified as part of the shield and crown of the type seen in No. 115, which was certainly not in use in the fifteenth century, when the original engraving was made. Again, a minute patch used in repairing a leaf of an Ortelius when last re-bound could at once be seen to contain a small part of the crown and 'mantling' in the arms of Amsterdam, as found elsewhere from about 1680. It is thus important that students should acquire such a general knowledge of the whole period in which watermarking has been practised, before specialising on more limited periods, which may also be advisable in view of the tens of thousands of marks that have to be dealt with. In proportion as the data accumulates, the precision of the results will increase. If it be thought that the positive results here mentioned are of no very great importance, they at least show something of what may be done, and if we learn to use the help offered with insight and discretion, we may no doubt find that really important points can occasionally be settled in this way.

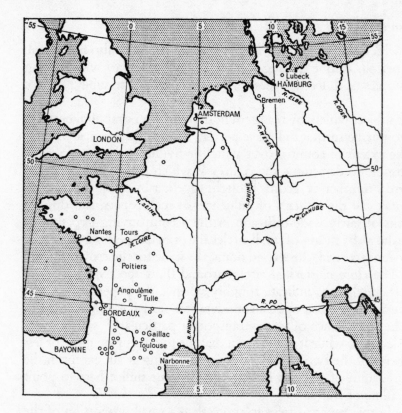

° Occurrences of Paper
marked with Hand and
Crown, as recorded by
Briquet

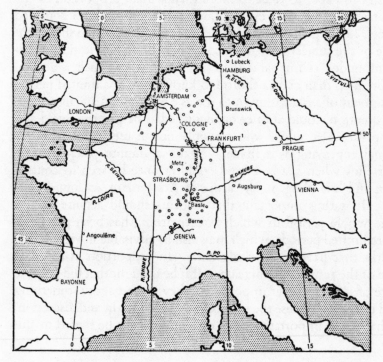

° Occurrences of Paper
marked with Crozier, as
recorded by Briquet

Select Bibliography

GENERAL AND HISTORICAL

ACTA CARTOGRAPHICA: 1967 onwards. Series reprint of monographs and cartographical-historical studies.

ALMAGIÀ, R.: *Monumenta Italiae Cartographica.* 1929.

ATKINSON, G.: *La littérature géographique française de la renaissance.* 1927.

BAGROW, L.: *Imago Mundi* (1953 onwards). Annual periodical devoted to the history of maps, edited by Bagrow.

—— *Die Geschichte der Kartographie.* 1951. New edn. 1964 (with R. A. Skelton), entitled *Meister der Kartographie.* English translation *The History of Cartography, 1964.*

—— *Abrahami Ortelii catalogus cartographorum.* 1928-30.

BEAZLEY, C. R.: *The dawn of modern geography.* 3 vols. 1897-1906.

BERTHAUT, H. M.: *La carte de France 1750-1898.* 1898-9.

BEVAN, W. L., and PHILLOTT, H. W.: *Mediaeval geography. An essay in illustration of the Hereford Mappa Mundi.* 1873.

BINYON, LAURENCE, and SEXTON, J. J. O.'B.: *Japanese colour prints.* 1960.

BONACKER, W.: *Kartenmacher aller Länder und Zeiten.* 1966.

BRITISH MUSEUM: *Catalogue of printed maps, charts and plans,* 15 vols., 1967, and *Ten-Year Supplement 1965-74.* 1978.

BRITISH MUSEUM: *Prince Henry the Navigator and Portuguese marine enterprise.* 1960.

BRITISH MUSEUM QUARTERLY: Articles.

BROWN, BASIL J. W.: *Astronomical atlases, maps and charts.* 1932.

BROWN, LLOYD A.: *The story of maps.* 1949.

CHAPIN, H. M.: *Check lists of maps of Rhode Island, Providence.* 1918.

CHUBB, THOMAS: *A descriptive list of the printed maps of Somersetshire, 1575-1914.* 1914.

—— *The printed maps in the atlases of Great Britain and Ireland. A bibliography.* 1927. An important book. Reprinted 1974.

—— and STEPHEN, G. H.: *Descriptive list of the printed maps of Norfolk, 1574-1916: Descriptive list of Norwich plans, 1541-1914.* 1928.

CLARK, J. W., and GRAY, A.: *Old plans of Cambridge 1574-1798.* 2 parts. 1921.

CLOSE, SIR CHARLES: *The early years of the Ordnance Survey.* 1926, reprinted 1969 with new Introduction by J. B. Harley.

COLBY, T.: *Ordnance survey of the county of Londonderry: memoir of city and N. W. liberties.* 1837.

COLLINDER, PER: *History of marine navigation.* 1954.

COLONIAL OFFICE: *Catalogue of maps, plans and charts.* 1910.

CORTESÃO, A. *Cartografia e cartografos Portuguese dos séculos XV-XVI.* 1935.

CRONE, G. R.: *Maps and their makers. An introduction to the history of cartography.* 1953. Second, revised, edition, 1962. Excellent short introduction.

CURNOW, I. J.: *The world mapped.* 1930.

DENUCÉ, J.: *Oudnederlandsche kaartmakers in betrekking met Plantijn.* 1912.

DOPPELMAYER, J. G.: *Historische nachricht von den Nürnbergischen mathematicis und künstlern.* 1730.

DRAYTON, MICHAEL: *Poly-Olbion.* Edited by J. William Hebel. 1933.

ECKERT, M.: *Die Kartenwissenschaft.* 1921-5.

EMMISON, F. G. (editor): *Catalogue of maps in the Essex record office 1566-1860.* 1947.

FASTNEDGE, RALPH: *English furniture styles from 1500 to 1830.* 1961. An excellent survey. Recommended as an aid for identifying globe stands, etc.

FIORINI, M.: *Sfere terrestri e celesti.* 1898.

FITE, E. D., and FREEMAN, A.: *A book of old maps delineating American history.* 1926, reprinted 1969.

FOCKEMA, ANDREAE S. J., and VAN'T HOFF, B.: *Geschiedenis der Kartografie van Nederland,* 1947.

FORDHAM, SIR HERBERT GEORGE: *Hand-list of catalogues and works of reference relating to carto-bibliography and kindred subjects for Great Britain and Ireland 1720 to 1927.* 1928. Very useful.

—— *John Cary.* 1910 and 1925, reprinted.

—— *Maps, their history, characteristics and uses.* 1927.

—— *Notes on a series of early French atlases 1594-1637.* 1921.

—— *Notes on British and Irish itineraries and road books.* 1912, reprinted.

—— *Notes on the cartography of the counties of England and Wales.* 1908.

—— *Road-books and itineraries of Great Britain.* 1924.

—— *Road-books and itineraries of Ireland 1647-1850.* 1923.

—— *Some notable surveyors and map-makers of the 16th, 17th and 18th centuries and their work.* 1929.

—— *Studies in carto-bibliography, British and French.* 1914, reprinted.

FRERE-COOK, G.: *The decorative arts of the mariner.* 1966.

GALLOIS, L.: *Les géographes allemands de la renaissance.* 1890.

GEOGRAPHICAL JOURNAL: Articles.

GEOGRAPHICAL MAGAZINE: Articles.

GEOGRAPHICAL REVIEW: Articles.

GERISH, W. B.: *The Hertfordshire historian, John Norden, 1548-1626(?): a biography.* 1903.

GLOAG, JOHN: *A short dictionary of furniture.* 1952. For an aid in the identification of globe mountings, etc.

GLOBUSFREUND, DER: A multilingual annual published since 1952 at Vienna by Societas Coronelliana Amicorum Globorum. It is of the greatest value to collectors of globes. Several issues contain detailed lists of globes and their makers, as well as articles on such subjects as 'The restoration of old globes', 'Carto-bibliography of old globes', etc., etc.

GOUGH, R.: *British topography.* 1780.

GRACE, F.: *Catalogue of maps, plans and views of London.* 1878.

GRIMALDI, A. B.: *A catalogue of zodiacs and planispheres, originals and copies, ancient and modern . . . from B.C. 1320 to A.D. 1900.* 1905.

HALLIDAY, F. E. (editor): *Richard Carew of Antony: The survey of Cornwall.* With the maps of John Norden. 1953.

HARLEY J. B.: *Christopher Greenwood county map maker and his Worcestershire map of 1822.* 1962.

HARMS, HANS: *Künstler des Kartenbildes.* 1962. Contains portraits of many famous cartographers.

HARRISSE, H.: *The discovery of North America . . . with an essay on the early cartography of the New World.* 1892, reprinted 1969.

—— *Notes pour servir à l'histoire, à la bibliographie et à la cartographie de la Nouvelle France 1545-1700.* 1872, reprinted 1975.

HARVEY, P. D. A., and THORPE, H.: *The printed maps of Warwickshire, 1576-1900.* 1959. A model of its kind.

HASTINGS MUSEUM: *Catalogue of maps and plans in the exhibition of local maps.* 1936.

HEAWOOD, EDWARD: *English county maps in the collection of the Royal Geographical Society.* 1932.

HERRMANN, A.: *Die ältesten Karten Deutschlands.* 1940.

HILLIER, J.: *The Japanese print. A new approach.* 1960.

HINKS, A. R.: *Maps and survey* (5th ed., 1944).

HODSON, D.: *Printed maps of Hertfordshire.* 1974.

HOLMDEN, H. R.: *Catalogue of maps, plans and charts in the map room of the Dominion archives.* 1912.

HOWSE, DEREK (with MICHAEL SANDERSON): *The sea chart.* 1973.

HUMPHREYS, ARTHUR L.: *Old decorative maps and charts.* 1926. Later reissued with new text by R. A. SKELTON (q.v.), 1952.

—— *A hand-book to county bibliography.* 1927.

IMAGO MUNDI: *See under* BAGROW, L.

INDIA: *Catalogue of maps, plans, etc., of India and Burma and other parts of Asia.* Published by Order of H.M. Secretary of State for India in Council, 1891.

INGLIS, H. R. G. (and others): *The early maps of Scotland.* 2nd ed., 1936.

ISCHL, T.: *Die ältesten Karten der Eidgenossenschaft.* 1945.

JOHNSTON, W. and A. K.: *One hundred years of map making.* 1924.

JOURNAL OF THE AMERICAN GEOGRAPHICAL SOCIETY: Articles.

KIMBLE, GEORGE H.: *The Catalan world map of the R. Biblioteca Estense at Modena.* 1934.

—— *Geography in the Middle Ages.* 1938.

KOEMAN, C.: *Atlantes Neerlandici,* Vols. I to V. 1967-71.

—— *Collections of maps and atlases in the Netherlands.* 1961.

LANE-POOLE, E. H.: *The discovery of Africa . . . as reflected in the maps in the collection of the Rhodes-Livingstone Museum.* 1950.

LE GEAR, CLARA EGLI: *United States atlases . . . in the Library of Congress.* 1950. *See also under* PHILLIPS, P. L.

LEHMANN, E.: *Alte deutsche landkarten.* 1935.

LOWERY, WOODBURY: *The Lowery Collection. Descriptive list of maps of the Spanish possessions within the present limits of U.S.* 1912.

LYNAM, EDWARD: *The first engraved atlas of the world, the Cosmographia of Claudius Ptolomaeus Bologna, 1477.* 1941.

—— *The map of the British Isles of 1546.* 1934.

—— *The mapmaker's art.* 1953. Of great value.

—— *Middle level of the fens and its reclamation . . . with maps of fenland.* 1936.

—— *British maps and map-makers.* 1944.

LYNAM, EDWARD: *Early maps of Scandinavia and Iceland.* 1934.

MacFADDEN, C. H.: *A bibliography of Pacific area maps.* 1941.

MAGGS BROS. LTD.: *Cartography. Maps and atlases. Books of cartographical importance and the use of spheres* (Catalogue No. 693). 1940. Valuable reference material.

MAGGS, F. B. *Voyages and travels in all parts of the world.* 5 vols. 1942-63. Bound catalogues. Valuable reference material.

MAP COLLECTOR: Quarterly magazine edited by R. V. Tooley. 1977 onwards.

MAP COLLECTORS' CIRCLE (ed. R. V. Tooley). 1963-75. Ten volumes a year.

MARCEL, G.: *Cartographie de la Nouvelle France.* 1885.

MARINELLI, G.: *Saggio di cartografia della regione veneta.* 1881.

—— *Saggio di cartografia italiana.* 1893.

MILLS, COL. D.: *University of Manchester. Catalogue of historical maps.* 1937.

MITTON, E. E. *Maps of Old London.* 1908.

MOXON, JOSEPH: *A tutor to astronomy and geography. Or, an easie and speedy way to know the use of both the globes, celestial and terrestrial.* 1699.

NORDEN, JOHN: *Speculi Britanniae Pars. A description of Hartfordshire.* Reprinted 1903 with a bibliography by W. B. Gerish.

NORTH, F. J.: *The map of Wales before 1600 A.D.* 1935.

—— *Geological maps, their history and development, with special reference to Wales.* 1928.

—— *Maps, their history and uses, with special reference to Wales.* 1933.

OEHME, H.: *Old European cities.* 1965.

PALMER, CAPT. H. S.: *The Ordnance Survey of the kingdom.* 1873.

PHILLIPS, PHILIP LEE: *A list of geographical atlases in the Library of Congress.* Vol. I and Vol. II (1909); Vol. III (1914); Vol. IV (1920); Vol. V (compiled by Clara Egli Le Gear, 1958), Vol. VI (ditto, 1963). Of the greatest importance. An indispensable work of reference for all those who want really detailed information about maps and atlases. Vols. I-IV reprinted 1971.

PRAESENT, H.: *Beiträge zur deutschen Kartographie.* 1921.

PRITCHARD, JOHN E.: *Exhibition of old Bristol plans, etc.* 1906.

RADFORD, P. J.: *Antique maps.* 1971.

RAISZ, E.: *General cartography.* 2nd edn., 1948, 1971.

RAVENSTEIN, E. G.: *Martin Behaim: his life and his globe.* 1908.

REEVES, E. A.: *Maps and map-making.* 1910.

ROBINSON, A. H. W.: *Marine cartography in Britain. A history of the sea chart to 1855.* 1962.

RODGER, ELIZABETH M.: *The large-scale county maps of the British Isles.* 1960. 2nd edn., 1972. An excellent listing.

ROYAL SCOTTISH GEOGRAPHICAL SOCIETY: *The early maps of Scotland.* 3rd edn., revised, 1936.

SANDLER, C.: *Die Reformation der Kartographie um 1700.* 1905.

SANFORD, W. G.: *The Sussex scene in books and maps.* 1951.

SHEERER, J. E.: *Old maps and map-makers of Scotland.* 1905.

SHEPPARD, T.: *William Smith, his maps and memoirs.* Hull, 1920.

—— *The evolution of topographical and geological maps.* Cardiff, 1920.

SKELTON, R. A.: *Decorative printed maps of the 15th to 18th centuries.* London, 1966. Reissue of A. L. Humphreys's book (q.v.) with new text.

—— *Explorers' maps.* 1958. Reprinted 1970.

STEVENS, HENRY N.: *Ptolemy's Geography. A brief account of all printed editions down to 1730.* London, 1908, reprinted 1972.

—— *Notes biographical and bibliographical on the Atlantic Neptune.* 1937.

STEVENSON, EDWARD LUTHER: *Portolan charts.* New York, n.d.

—— *Terrestrial and celestial globes.* 2 vols. Hispanic Society of America. 1921, reprinted 1971.

—— *William Janszoon Blaeu 1571-1638.* New York, 1914.

—— *Maps illustrating early discovery and exploration in America.* 1903-6.

—— *A description of early maps, originals and facsimiles, 1452-1611.* New York, 1921.

TAYLOR, E. G. R.: *Tudor geography 1485-1583.* 1930.

—— *Late Tudor and Early Stuart geography 1583-1650.* 1943.

—— *The mathematical practitioners of Tudor and Stuart England.* 1954.

TIELE, P. A.: *Nederlandsche bibliographie van land – en volkenkunde.* 1884.

TOOLEY, R. V.: *Maps and map-makers.* London, sixth edition, 1978. Of great value. Contains extensive bibliographies.

—— *Collectors' guide to maps of the African continent and Southern Africa.* 1969.

—— *Dictionary of map-makers.* 1964-75 (to Powell).

—— (with BRICKER and CRONE): *Highlights of cartography.* 1968.

—— (with BRICKER): *A history of cartography,* 1969.

—— —— *Landmarks of mapmaking.* 1976.

TYACKE, SARAH: *London map-sellers 1660-1720.* 1978.

UZIELLI, G. and P., AMAT DI SAN FILIPPO: *Studi biograficie bibliografici sulla storia della geografia in Italia.* 1882.

VICTORIA & ALBERT MUSEUM: *Tapestry maps 16th and 17th century.* 1915.

WAGNER, H. R.: *The cartography of the northwest coast of America to the year 1800.* 1937, reprinted 1968.

WALLIS, HELEN (with SARAH TYACKE, joint eds.): *My head is a map.* 1973.

WALTERS ART GALLERY (Baltimore, Maryland): *The world encompassed. An exhibition of the history of maps.* 1952. Valuable.

WAUWERMANS, H. E.: *Histoire de l'école cartographique belge et anversoise du XVIᵉ siècle.* 1895.

WEISZ, L.: *Die Schweiz auf alten Karten.* 1945.

WHITAKER, H.: *The Harold Whitaker collection of county atlases, road-books and maps presented to the University of Leeds. A catalogue.* 1947.

WIEDER, F. C.: *Nederlandsche historisch-geographische Documenten in Spanje.* 1915, reprinted 1976.

WINTERBOTHAM, H. S. L.: *A key to maps.* 1947. A useful little guide for those who wish to know more of the background of the 'mechanics' of cartography.

HISTORY OF ENGRAVING
ORNAMENT, TECHNIQUE AND WATERMARKS

BEANS, G. H.: *Some sixteenth-century watermarks found in maps prevalent in IATO atlases.* 1938.

BRIQUET, C. M.: *Les filigranes* (i.e. Watermarks [to 1600]). 1907, reprinted.

BRITISH MUSEUM: *A guide to the processes and schools of engraving.* 1923.

CHURCHILL, W. A.: *Watermarks in paper in Holland, England, France, etc., in the XVII and XVIII centuries.* 1935.

COLVIN, SIR S.: *Early engraving and engravers in England 1545-1695.* 1905.

DOSSIE, ROBERT: *The handmaid to the arts.* 2 vols. 1764 and other edns. Contains a chapter 'Of colouring or washing maps, prints, etc.'

FAIRBANK, A.: *A book of scripts.* 1949.

FLETCHER, F. MORLEY: *Wood-block printing.* 1916. A clear exposition of Japanese wood-block printing.

HEAL, SIR A.: *The English writing-masters and their copy-books 1570-1800.* 1931.

HEAWOOD, EDWARD: *Watermarks mainly of the 17th & 18th centuries.* 1950.

HIND, A. M.: *A history of engraving and etching.* 3rd edn. 1923.

—— *An introduction to a history of woodcut.* 2 vols. 1935, reprinted 1968.

—— *Engraving in England in the sixteenth and seventeenth centuries.* Vol. I, *The Tudor Period,* 1952. Vol. II, *The Reign of James I,* 1935 (other volumes in preparation).

—— *Early Italian engraving.* 1938-48.

JESSEN, P.: *Der Ornamentstich.* 1920.

—— *Meister der Schreibkunst.* 1923.

JOURNAL OF THE BIBLIOGRAPHICAL SOCIETY. Many useful articles on printing, paper, etc.

LE CLERT: *Le Papier.* 1926.

LYNAM, E.: 'Period ornament, writing, and symbols on maps, 1250-1800.' *Geographical Magazine,* xviii, 1945-6. *See also* LYNAM'S *The mapmaker's art* under the General and Historical Section, which contains much useful information on ornament, much of it incorporated from the above article.

NAGLER, G. K.: *Allgemeines Künstler-lexikon.* 1835-52.

PLANT, M.: *The English book trade.* 1939. Useful for the background of map and print retailing.

SALMON, WILLIAM: *Polygraphice.* 1685 and other edns. Contains chapters 'Of Colours simple for washing of maps' and 'Of Colours compound for washing of maps'.

SMITH, JOHN: *The art of painting in oyl.* 1701 and other edns. Contains a valuable chapter 'The whole art and mystery of colouring maps, and other prints, in water colour.'

THIEME, V., and BECKER, F.: *Allgemeines Lexikon der bildenden Künstler.* 1907-47. Valuable.

VON WURZBACH, A.: *Niederländisches kunstlerlexikon.* 1886.

WALLER, J. G.: *Biographisch Woordenboek van Noord-Nederlandsche Graveurs.* 1938.

WALPOLE, HORACE: *A catalogue of engravers.* 1763.

WILME, B. P.: *A hand-book for mapping, engineering and architectural drawing in which maps of all descriptions are analysed and their several uses fully explained.* 1846.

A SELECTION OF FACSIMILES

A few facsimiles are included here. Many more are available. They are valuable for making comparisons.

ALBA, DUQUE DE (and others): *Mapas españoles de América, siglos XV-XVI.* 1951.

ALMAGIÀ, R.: *Monumenta cartographica Vaticana.* (1944 on).

—— *Monumenta Italiae cartographica.* 1929.

BRITISH MUSEUM: *Six early printed maps.* 1928.

CARACI, G.: *Tabulae geographicae vetustiores im Italia adservatae.* 1926-32.

COOTE, C. H.: *Facsimiles of three mappemondes 1536-50.*   1898.

CORTESÃO, ARMANDO AND TEIXIERA DE MOTA, AVELING: *Portugaliae monumenta cartographica.*
   From 1960.   7 vols. (5 folios, 2 quartos).

DESTOMBES, M.: *La mappemonde de Petrus Plancius.*   1944.

FISCHER, J.: *Claudii Ptolemaei.   Geographiae Codex Urbinas Graecas.*   1932.

FITE, E. D., and FREEMAN, A.: *A book of old maps delineating American history . . . to the close of
   the Revolutionary War.*   1926, reprinted 1969.

GUILLÉN Y TATO, J. F.: *Monumenta cartographica Indiana.*   1942, etc.

JOMARD, E. F.: *Les monuments de la géographie.*   1842-62.

KRETSCHMER, K.: *Die Entdeckung Amerikas.*   1892.

MARCEL, G.: *Reproductions de cartes et globes relatives à la découverte de l'Amérique.*   1893.

MERCATOR, G.: *Drei Karten von G. Mercator.*   1891.

MÜLLER, F.: *Remarkable maps of the XVth, XVIth and XVIIth centuries reproduced in their
   original size.*   1894-7.

NORDENSKIÖLD, A. E.: *Facsimile atlas to the early history of cartography.*   1889, reprinted 1973.

—— *Periplus.   An essay in the early history of charts and sailing directions.*   1897.

ROYAL GEOGRAPHICAL SOCIETY: *Map of the world by J. Hondius.*   1927.

—— *English county maps.* n.d.

—— *Catalan world map, R. Bibla. Estense at Modena.*   1934.

—— *Hereford World Map.*

—— *Gough Map.*

—— *Early Maps of the British Isles A.D. 1000-1579.*   1961.

SAXTON, CHRISTOPHER: *An Atlas of England and Wales.*   British Museum, 1936.   One of the
   finest facsimile atlases ever made.   The maps were also issued singly.   A new series of
   facsimiles of Saxton maps is in the process of being issued at present.   About 18 are so
   far available.

SPEED, JOHN: *John Speed's England.*   5 vols.   1953.   Edited by John Arlott.

STEVENSON, E. L.: *Maps illustrating early discovery and exploration in America.*   1903-6.

—— *Maps selected to represent the development of map-making from the first to the seventeenth
   centuries.*   1913.

THEATRUM ORBIS TERRARUM.: A series of fine reproductions of maps by Ptolemy, Ortelius,
   Waghenaer and others.   Edited by R. A. Skelton and Alexander O. Victor.   From 1963.

TELEKI, P.: *Atlas zur Geschichte der Kartographie der Japanischen inseln.*   1909.

WIEDER, F. C.: *Monumenta cartographica.*   5 vols.   1925-33.

YOUSSOUF, KAMAL, PRINCE: *Monumenta cartographica Africae et Aegypti.*   1926 on.

A List of Cartographers, Engravers, Publishers and Printers Concerned with Printed Maps and Globes from circa *1500* to circa *1850*

THIS list is not exhaustive, but it contains most of the names likely to be encountered by the average collector and dealer.

The names and dates of the various people listed are in each case followed by details of the category for which they are usually known (e.g. cartographer, engraver) and of the types of work they produced or were associated with (e.g. maps, globes); abbreviations, a list of which is given below, denote these categories and products. In some cases, particularly in those of important figures, further details are given, including the titles of works and collaborators. In one case (that of John Speed) various editions of a work have been dealt with more elaborately; this has been done to illustrate the difficulties of cartographic identification, and to demonstrate typical variations. Similar problems exist in many others of the works described, but it is obviously impossible in one volume to mention them all.

In many cases variations in name are given, especially where the Latinised version shows considerable variation from the original (e.g. Visscher=Piscator). It should be remembered, too, that Latin names are subject to case; thus the nominative Hondius or Hondium (=Dutch, de Hondt) becomes, in the genitive, Hondii (i.e. *of* Hondius); in some entries the names are given in the cases in which they usually appear. Also letters are sometimes exchanged; one of the most frequent being I for J (Iohn=John).

The following Latin words from map inscriptions are worth noting:

AUCTORE DELINEAVIT DESCRIPSIT INVENIT	Indicate the cartographer or draughtsman
CAELAVIT FECIT INCIDIT INCIDENTE SCULPSIT SCULP. SC.	Indicate the engraver

APUD ⎫
EXCUDIT ⎪
EXCUD. ⎪
EXC.   ⎬ Indicate the printer or
EX OFFICINA ⎪ publisher
FORMIS ⎪
SUMPTIBUS ⎭

Finally, two observations on the subject of atlases.   The student should distinguish between atlases, which have been intended as such, and printed and bound up in that form, and collections of maps, merely assembled and bound up for convenience; examples of both are given under John Speed's entry.

In many cases two dates are given for the publication of an atlas.   In the case, for example, of an atlas with the dates 1715-30, it means that it consists of maps bearing individual dates within that period, the earliest being dated 1715, the latest 1730.

ABBREVIATIONS

Amer.	America, American		Ger.	German, Germany
Amster.	Amsterdam		incl.	includes, including
Astron.	Astronomer		Ital.	Italian, Italy
b.	born		Lond.	London
Brit.	British		M	Maps, charts, atlases, plans
c.	*circa* = about		n.d.	no date
Cart.	Cartographer		n.p.	no place
cent.	century		O	Armillary spheres
d.	died		Port.	Portugal, Portuguese
Dan.	Danish		Pub.	Publisher, published
Dut.	Dutch or Netherlandish		q.v.	*quod vide* = which (or whom) see
edn.	edition		Russ.	Russia, Russian
Eng.	Engraver, engraving, engraved		Span.	Spanish
fl.	*floruit* = flourished		subs.	subsequent
Fr.	France, French		Surv.	Survey, surveyed, surveyor
G	Globes		Swed.	Sweden, Swedish
G (C)	celestial globes		Topg.	Topographer
G (T)	terrestrial globes		trs.	translated, translation, translator
Geog.	Geographer		X	Astronomical charts, maps

IMPORTANT. In order to make identification easier, most of the names in the atlases and collections of maps in this list have not been standardised, but have been copied literally as they appear on the maps.

Aa *Same as* van der Aa (q.v.)

Abel, Gottlieb Friedrich 18th cent. Eng.
  Ger. G

Acerlebout, J. 17th cent. Cart. Brit. M
  Hatfield Chase W. Dugdale, 1662, 1772.

Adair, John d. 1722 Surv., Cart. Brit. M
  Maps published posthumously.
  East-Lothian R. Cooper, 1736.
  West-Lothian 1737 Eng. by R. Cooper.
  Perthshire c. 1690 Eng. by J. Moxon.
  Description of the sea coast of Scotland 1703.
    Only Part I (with 6 charts) was published.
  Lothians 1745 Eng. by R. Cooper; 1745,
    A. Millar, eng. by T. Kitchin.
  Midlothian 1735 Eng. by R. Cooper.

Adam et Giraldon *See under* Lapié, P.

Adami, C. 1838-52 Military instructor
  Ger. G (C)

Adams, Clément c. 1519-87 Eng. Brit.
  M
  Eng. the English edn. of Cabot's map, 1549.

Adams, Daniel 1773-1864 Geog. Amer.
  M
  *Atlas to Adams' Geography for the use of schools
    and academies* ... Boston, West and Blake,
    1814.
  School atlas to Adams' Geography . . .
    Boston, Lincoln and Edmands, 1823.
    Maps eng. by Annin and Smith.

Adams, Dudley fl. 1797- c. 1825. Cart.
  Optical-instrument maker Brit. G (C
  and T)

Adams, George, Senr. 1704-73 Cart.
  Brit. G (C and T)

Adams, George, Junr. 1750-95 Cart.
  Brit. G (C and T)

Adams, John fl. 1670-96. Cart. Brit. M
  Constructed on 12 sheets a map based on
  Saxton:
  *Angliae totius tabula cum distantiis notioribus
    in itinerantium usum accomodata.* Pub. by
  Philip Lea (q.v.). Distances between towns
  and villages are shown by straight lines.

Adams, Robert fl. 1588 Eng. Brit. M
  Armada charts.
  See also under Ryther, A.

À Daventria, Jacob} d. 1575 *See under*

À Daventer, Jacob }   Ortel, A.

Addison, John, and Co. fl. 1825. Cart.
  Brit. G (C and T)

À Doetecum *See* Doetecum

Adrichom, Christian}
Adrichem   } fl. 1593 Cart.
Adrichomius   } Dut. M
  Theatrum terrae sanctae. An early atlas of
  the Holy Land, copied by many later carts.
  See also under Hondius, H.; Horn, G.;
    Jansson, J.; Mercator, G.

Aedgerus, Cornelius 16th cent. Cart.
  Dut. M
  Coloniensis dioecesis typus 1583.

Aernhofer, Hans d. 1621. Sculptor Ger.
  G

Aeuer, H. fl. 1827-8 Cart. Ger.? G
  (C and T)

Afferden *Same as* de Afferden (q.v.)

Agas, Ralph 1540?-1621 Surv. Brit. M

Agnese, Battista fl. 1527-64 Cart. Ital.
  M

À Horn, Joannes} fl. 1526 Cart. Dut.
van Hoirne, Jan } M

Aiken, John}
Aikin   } fl. 1790 Cart. Brit. M
  England Delineated 1790 and subs. edns.

Ainslie, John 1745-1828 Surv., Eng.
  Brit. M
  Various Scottish maps.
  A new atlas ... 1814, A. Constable and Co.,
    Edinburgh. *See also under* Blackadder, J.

Ainslie, Mrs. *See under* Blackadder, J.

Aitkin, W. 18th cent. Pub. Brit. M

Aitzing } *Same as* von Aitzing (q.v.)
Aitzinger}

Akademiia Nauk Russian Academy
  Atlas russicus ... Petropoli, 1745.
  Plan de la ville de St. Petersbourg . . . St.
    Petersburg, 1753.

Akamidzu, Demito 18th cent. Cart.
  Japanese M

Åkerman, Andreas 1723?-78 Eng.
  Swed. M G (C and T)
  Atlas juvenilis ... Uppsala, 1789?
  Atlas hydrografica ... 1768

Akhmatov, Ivanov 19th cent. Cart. Russ. M
Geographical atlas of Russia 1845 (in Russ.)
Historical atlas of the Russian Empire 1831 (in Russ.)

Akrel, Fredrik 1748-1804 Eng. Swed. G

Alagna, J. Giacomo fl. 1767 Cart. Ital. M
Complete set of new charts of Portugal and the Mediterranean Lond., Mount and Page, 1760?

À Langren *Same as* VAN LANGREN (q.v.)

Alberti, Gian Battista fl. 1675 Instrument-maker Ital. G O

Alberts, R. C. *See under* BLAEU, J.

Albertus, Leander 1479-1553 Cart. Dominican Span. M
Isole appartenanti alla Italia Venice, 1567 (with 7 maps).

Albin, John 18th cent. Cart. Brit. M
Map of the Isle of Wight 1795, 1805, 1807, 1823 (pub. by L. Albin) Eng. by S. J. Neele.

Albin, L. 18th cent. Pub. Brit. M
See also under ALBIN, J.

Albrecht, Ignatius *See under* VON REILLY, F. J. J.

Albrizzi, G. *See under* DE VALLEMONT, PIERRE LE LORRAIN

Alcmar, W. J. }
Alcmarianus, G. J. } *Same as* G. BLAEU (q.v.)

Aleni, Giulio 1582-1642 Cart., Jesuit Ital. M
World map printed in China, *c.* 1630

Algoet, Lieven}
Panagathus } fl. 1562 Cart. Dan. M
Extremely rare copper-engraved map on 9 leaves:
Terrarum septentrionalium. It was used also by DE JODE (q.v.) in his atlas (1578, etc.).

Allard, Abraham *See under* DANCKERTS, J.; VALCK, G.

Allard, Karel or Carolus} c. 1648-1709

Allardt } Pub. Dut. M
Atlas Minor Amster. 1682?, 1696?
Atlas Major 3 vols.
Nova tabula India Orientalis Amster. *c.* 1680.
Magnum theatrum belli . . . Amster. 1702.
Orbis habitabilis oppida et vestitus Amster. 1698? Carts. and engs. incl. A. Meijer, T. Doesburgh. He also pub. editions of maps by JANSSON (q.v.). *See also under* BEEK, A.; BRAAKMAN, A.; BROSTENHUYZEN, J.; DANCKERTS, J.; DE LA FEUILLE, J.; DE WIT, F.; OTTENS, R.; VALCK, G.; VISSCHER, N., III.

Allardt, Hugo }
      Hugh } 1628-66 Cart. Dut. M
      Huych}
Dioecesis Leondiensis accurata tabula n.p. n.d. *See also under* SELLER, JOHN

Allen, George *See under* DE RAPIN-THOYRAS, P.; EDWARD, B.; LAURIE, R.

Allen, J. *See under* BAKER, R. G.

Allen, W. fl. 1798 Surv., Cart., Pub. Irish M
Maps of Co. Carlow, 1798, 1824, and Co. Wicklow, 1834 (eng. C. E. Maguire). *Ireland*, Dublin, 1815 [1825?].

Allen and Co., Wm. H. 19th cent. Pub. Brit. M
See also under HORSFIELD, T.

Almada *See under* TASSIN, N.

Almeda *Same as* DE ALMEDA (q.v.)

Almon, J. *See under* POWNALL, T.

Alphen *Same as* VAN ALPHEN (q.v.)

Altheer, J. *See under* VAN WIJK ROELANDSZOON, J.

Alting, Menso (1637-1713) **and Schotanus à Sterringa, B.** Carts. Dut. M
Uitbeelding der heerlijikheit Friesland . . . Leeuwarden, F. Halma, 1718.

Altmütter, Georg 1787-1858 Scientist Austrian G

Alvarez, Fernando 16th cent. Cart. Port. M
Map of Portugal, 1560.

Amato, Bell fl. 1554. Cart. Ital. M

À Merica, G. and P. *Same as* À MYRICA (q.v.)

Ameti, Giacomo Filippo *See under* DE ROSSI, G. G.

Amman, Johannes 1695-1751 Eng. Ger. G

Amman, Jost } 1539-91 Eng. Ger.
Jodokus} M
Eng. Ph. Apian's map of Bavaria, 1568.

À Myrica, Gaspard } fl. 1537 Cart.,
and Peter } Eng. Dut. M G
Van der Heyden }
Collaborated with G. MERCATOR and GEMMA FRISIUS (qq.v.).

Ancelin and Le Grand, P. 18th cent.
Cart. Fr. M
Atlas général et élémentaire de toutes les Russies . . . Moscow and St. Petersburg, 1795.

Anders Lebiano, Adolario Ericho *See under* BLAEU, G.

Anderson, A. *See under* D'ANVILLE, J. B. B.; WINTERBOTHAM, W.

Anderson, John, Junr. fl. 1837 Cart. Brit. M
See also under PAYNE, J.

André, P. *See under* CHAPMAN, J.; MENTELLE, E.; RAYNAL, G. T. F.

Andreae, Johann Ludwig fl. 1717-26 Cart. Ger. G (C and T)

Andreae, Johann Philipp 1720-57 Mathematician Ger. (C)

Andrée *See under* BOTERO, G.

Andrews, John fl. 1766-1809 Surv., Geog., Eng., Mapseller Brit. M
Worked with P. ANDREWS, A. DURY and W. HERBERT (qq.v.).
A set of plans and forts in America (with P. Andrews), 1765.
County maps of Herts., 1766; Wilts., 1773 and subs. edns. (both with A. Dury); Kent, 1769 and subs. edns. (with A. Dury and W. Herbert).
A map of the country sixty-five miles round London (with A. Dury) Lond., 1776.
A collection of plans of the most capital cities of every empire . . . Lond., J. Hand, 1772, 1792?
Plans of the principal cities of the world . . . Lond., J. Stockdale, 1792?

A collection of plans of the capital cities of Europe, and some remarkable cities in Asia, Africa and America . . . Lond., 1771.

Andrews, P. fl. 1765 Eng. Brit. M
Worked with J. ANDREWS (q.v.) on *Plans and forts in America*.
See also under ROCQUE, J.

Andriveau, J. *See under* ANDRIVEAU-GOUJON

Andriveau-Goujon (a firm) 19th cent.
Carts., Pubs. Fr. M G
Names on maps variously appear as J. Goujon, J. Andriveau, J. Andriveau-Goujon or E. Andriveau-Goujon. Eugène Andriveau-Goujon (1832-97) was a geog. and cart., and Gilbert Gabriel Andriveau-Goujon (1805-94) was a globe-maker and seller.
Atlas classique et universal de géographie . . . Paris, 1843-4.
Atlas élémentaire simplifié de géographie . . . (with E. Soulier) Paris, 1838.
Atlas de choix . . . Paris, 1841-62.
Engs.: Erhardt, Flahaut, Gérin, Erhard Schieble, Ch. Smith.
Lettering eng. by Arnoul, P. Rousset, Ch. Simon.
Carts. incl.: H. Nicollet, E. Soulier, Hase, A. H. Dufour, A. Vuillemin, Dufrénoy et Elie de Beaumont, E. Stanford, Ch. Picquet, A. Brué.

Angelo, S. *See under* JULIEN, R. J.

Anich, Peter 1723-66 Cart. Eng.
Austrian M G (C and T)
Worked with Huber or Hueber, Blasius (fl. 1774-1804) on *Atlas Tyrolensis* Vienna, 1774.

Annin, W. B. *See under* CUMMINGS, J. A.

Annin and Smith *See under* ADAMS, DANIEL; EDMANDS, B. F.

Anse *See under* OTTENS, R.

Anselin *See under* DE JOMINI, A. H.; DE GOUVION SAINT-CYR, L.; LE BOUCHER, O. J.; LEGRAND, P.

Anselin et Carilian-Goeury *See under* POUSSIN, G. T.

Anson. *See under* WEILAND, C. F.

Anthonisz, Cornelis } 1499?-1556?
Anthoniszoon, Cornelis} Cart. M

Caerte van Oostlandt 1543. No copy survives but a later edition (1560) survives. His work was much used by carts. who succeeded him, among them Ortelius and Mercator.

Charts of Zuyder Zee, etc.

Antillon y Marzo *Same as* DE ANTILLON Y MARZO (q.v.)

Apianus, Petrus}
Apian, Peter   } 1495–1552 Cart. Ger.
Bienewitz   } M G (C and T) X

Sometimes signed his work 'Petrus Apianus Mathematicus'.

World map (woodcut) in—

Joannis Camertis Minoritani . . . by C. S. Solinus (Vienna, Joannes Singrenius, 1520); it is the second map to bear the name 'America' (the earliest is by WALDSEE-MÜLLER) (q.v.).

Apianus, Philip} 1531–89 Cart. Ger. M
Bienewitz   } G (C and T)

Son of Peter.

Maps by the Apians are of extreme rarity, only single examples being known in most cases.

See also under AMMAN, J.; DE Wale, P.

À Porta, Hugo *See under* PTOLEMAEUS, C.

Applegate, H. S. and J. *See under* CONCLIN, G.

Aragon, Madame Anne Alexandrine *See under* PERROT, A. M.

Araos *Same as* DE ARAOS (q.v.)

Archer, J. fl. 1841–61 Eng. Brit. M
Made 17 ecclesiastical maps for *The British Magazine*, 1841.

See also under SOCIETY FOR THE PROPAGATION OF THE GOSPEL.

À Regibus, S. *Same as* DI RE, S. (q.v.)

Arenhold *See under* HOMANN, J. B.

Aretinus, Paulus 17th cent. Cart. Ger. M
Map of Bohemia, 1619.

Argaria, Gasparo 16th cent. Cart. Ital. M
Executed maps of Naples (1538) and Messina (1567) for Lafreri. Sometimes signed with the initials 'G.A.'

Argelander, Friedrich Wilhelm August
fl. 1843–63 Astron. Ger. X

Argoli, Andrea fl. 1644 Astron. Ital. X

Arias Montanus, Benedictus *c*. 1527–98
Theologian Span. M
Compiled a Bible atlas, 1571.

Arigoni, Fra Bono} fl. 1509–11 Cart.
Harigoni   } Ital. M

Armstrong, Andrew fl. 1768–81 Cart.,
Surv. Brit. M.
Father of MOSTYN JOHN A. (q.v.), with whom he traded as Capt. Armstrong and Son, A. Armstrong and Son, or Capt. A. and M. Armstrong. Their joint imprint appears on many of the following maps.

Map of Durham. Eng. and pub. by T. Jefferys, 1768 and subs. edns.

Map of Lincolnshire. Eng. S. Pyle, 1779 and subs. edns.

Reductions were pub. in 1781 (R. Sayer and J. Bennett) and 1794 (R. Laurie and J. Whittle).

Map of Ayrshire, 1775.

Map of Northumberland, 1769. Eng. by T. Kitchin. Reductions were pub. in 1770, 1781 (R. Sayer and J. Bennett), 1787 (R. Sayer), 1796 (R. Laurie and J. Whittle).

Map of Rutland, 1781. Eng. J. Luffman, 1781.

Map of Argyllshire. Eng. S. Pyle, 1775. Reduction pub., 1774 (eng. A. Baillie).

Map of Berwickshire. Eng. A. Bell, 1771. Reduction pub., 1772 (eng. H. Gavin).

Map of Lothians. Eng. T. Kitchin, 1773.

Armstrong, Mostyn John fl. 1769–91
Geog., Pub. Brit. M
Son of ANDREW A. (q.v.).
Map of Peebleshire, 1775. Eng. S. Pyle.
Other county maps of Scotland.
A Scotch Atlas 1777, 1787, 1794. Pub. by Laurie and Whittle.
Map of Cambridgeshire, 1778.
Map of Norfolk, 1778.
See also under LAURIE, R.

Arnoldi *Same as* DE ARNOLDI (q.v.)

Arnoul *See under* ANDRIVEAU-GOUJON

Arnz, J. 19th cent. Pub., Cart. Ger. M
Atlas der Alten Welt . . . Dusseldorf, 1829. 16 lithographed maps.

À Rotenhan, Sebastian b. 1478 Knight, Humanist Ger. M

Arrowsmith, Aaron 1750-1823 Cart., Pub. Brit. M
Pub. very many large-scale maps, e.g. U.S.A. (1796 and subs. edns.); Persia, 1813; Pacific Ocean, 1798; Panama Harbour, 1806, etc.
Various maps of American counties and localities.
A map exhibiting all the discoveries in the interior parts of North America Lond., Faden, 1802.
Maps of Dalmatia, Lond., 1812; Upper Egypt and Lower Egypt, Lond., 1807; Holland, Lond., 1815; India, Lond., 1804; Hindostan (with R. Wilkinson), Lond., 1804, 1806.
Map (of) . . . *the great post roads* . . . *of Europe* Lond., 1810.
Map of the Physical Divisions of Germany Lond., 1812.
A Sketch of the countries between Jerusalem and Aleppo Lond., 1814.
Chart of the Cape of Good Hope . . . Lond., 1805.
A new map of Mexico . . . Lond., 1810.
Chart of the West Indies . . . Lond., 1803.
Map of the world . . . Lond., 1794, 1799.
A new general atlas . . . Edinburgh, Lond., A. Constable and Co., 1817.
A new and elegant general atlas . . . (with S. Lewis) Philadelphia, J. Conrad and Co., 1804; Boston, Thomas and Andrews, 1805, 1812, 1819.
Nouvel atlas universel-portatif de géographie ancienne et moderne Paris, H. Langlois, 1811. Maps eng. by Ambroise Tardieu, Bondeau, Semen.
Maps by Arrowsmith, J. B. Poirson, Bonne, Lorin, d'Anville.
A pilot from England to Canton . . . Lond., Arrowsmith and Blacks and Parry, 1806, Each map bears Arrowsmith's name, but in addition James Johnstone's appears on some.

See also under BLACKADDER, J.; BOSSI, L.; BURNS, LIEUT. ALEX.; DELAMARCHE, F.; GARDEN, W.; PIMENTEL, M.

Arrowsmith, Aaron, Junr. 19th cent. Cart. Brit. M
Son of AARON, Senr.
'Hydrographer to his Majesty.'
Orbis terrarum veteribus noti descriptio . . . Lond., 1828, 1830 (pub. for use at Eton College).
Arrowsmith's comparative atlas . . . (with S. Arrowsmith) Lond., 1829, 1830.
Outlines of the world . . . Lond., S. Arrowsmith, 1828. All maps except no. 1 (which has the imprint of A. Arrowsmith) bear the imprint of Samuel Arrowsmith.

Arrowsmith, John 1790-1873 Cart., Pub. Brit. M
Nephew of AARON (1750-1823).
The London atlas Lond., 1834 and various other edns.
The London atlas of universal geography Lond., 1840 and various other edns.
Flinder's and Light's *Maritime portion of S. Australia* Lond., 1839.
Light's *District of Adelaide* Lond., 1839.
See also under BURR, D. H.

Arrowsmith, Samuel fl. 1839 Cart. Brit. M
Son of AARON, Senr.
The Bible atlas . . . Lond., 1835.
See also under ARROWSMITH, A., JUNR.

A.S. Signature sometimes used by A. SALAMANCA (q.v.).

Ascensio, José (fl. 1759) **and Martinez de la Torre, F.** Carts. Span. M
Plano de la villa y corte de Madrid . . . Madrid, J. Dobaldo, 1800.

À Schagen, Gerardo *See under* DE LA FEUILLE, J.

Ashby, H. *See under* FANNIN, P.

Ashby, Robert 16th/17th cent. Cart. Brit. M

Ashley, Sir Anthony 1551-1627 Pub. Brit. M Pub. the English edn. of Waghenaer's *Mariner's Mirrour* (1588), which contained 45 charts.

Aspin, J. fl. 1814–40 Astron. Cart. Brit.
M X
See also under LAVOISNE, C. V.

Aston, John fl. 1816 Cart. Brit. M
Map of Warwickshire, Coventry, 1816;
J. Turner, 1830; H. Merridew.

Atkinson, J. fl. 1677 Pub. Brit. M
Pub. a section of *The English Pilot* by JOHN
SELLER (q.v.).

Auber, L. *See under* MENTELLE, E.

Aurigarius *Same as* WAGENHAER (q.v.)

Ausfeld, J. C. *See under* VON SCHLIEBEN,
W. E. A.

Austen, Stephen T. fl. 1732 Pub. Brit.
M H. Popple's *Map of the British Empire in
America* Lond., 1732.
*A map of the king's roads . . . in the Highlands of
Scotland* 1746, Willdey and S. Austen.
See also under TANNER, H. S.; WILLDEY, G.

À Varea, L. *Same as* SAVANAROLA, R. (q.v.)

Aventinus } 1477–1534 Historian,
Turmair, Johannes} Cart. Ger. M

Avinea, Antonio fl. 1557. Cart. Ital.?
M

Ayrouard, Jacques} 18th cent. Cart. Fr.
Ayrouart } M
Recüeil de plusieurs plans des ports . . . Paris,
1732–46.
Eng. by Louis Corne and H. Coussin.
See also under HEATHER, W.

Azara *Same as* DE AZARA (q.v.)

Aznar, Pantaleon *See under* VAZQUEZ, F.

Baalde, S. J. 18th cent. Pub. Dut. M
See also under BRENDER À BRANDIS, G.;
ROUSSET DE MISSY, JEAN.

Bachiene, Willem Albert 1712–83 Cart.
Dut. M
Atlas Amster., M. Schalekamp, 1785.

Bachmeier, Wolfgang 17th cent. Cart.
Ger. M

Bäck, E. *See under* SEUTTER, G. M.

Backer, Remmet Tennisse *See under*
DANCKERTZ, J.

Backhouse, Thomas 18th cent. Cart.
Brit. M
*New pilot for the south-east coast of Nova
Scotia . . .* Lond., Laurie and Whittle,
1798.

Badeslade, Thomas (fl. 1719–45) **and Henry**
(fl. 1742) Engineers, Survs. Brit. M
Chorographia Britanniae 1742, 1743, 1745,
1747. Eng. William Henry Toms.

Baeck, Elias 1679–1747 Cart. Ger. M
Atlas geographicus . . . Sachs-Weimar, 1710.

Bagge, T. 18th cent. Cart. Brit. M
Map of the fens of e. England, 1796.

Bagnoni, Jacopo *See under* DUDLEY, SIR R.

Bahre, Hans Georg 17th cent. Eng. Ger.
M
See also under HOLZWURM, I.

Bailey, J. *See under* GILL, V.

Bailleul *See under* JULIEN, R. J.

Bailley, F. and R. *See under* SCOTT,
JOSEPH.

Baillie, Alexander *See under* ARMSTRONG,
A.; LAURIE, J.; ROSS, LIEUT. C.

Baillieu, Gaspard 18th cent. Cart. Fr.
M *See under* BEEK, A.; DESNOS, L. C.

Bailly, Robertus *Same as* DE BAILLY (q.v.)

Baily, Francis 1774–1844 Astron. Brit.
X

Baker, Benjamin fl. 1780–1824 Eng.,
Pub. Brit. M
Universal magazine 1791–7 and subs. edns.
(44 maps).
See also under FADEN, W.; DONN, B.;
LINDLEY, J. and C.; SHERRIFF, J.; WIL-
KINSON, R.

Baker, Richard Grey fl. 1821 Cart.
Brit. M
County map of Cambridgeshire, 1821 and
subs. edns. Eng. J. Allen.

Baker, T. 18th cent. Cart. Brit. M
Map of Isle of Wight, 1795 (eng. T.
Bowles), 1806 (with Fletcher).

Bakewell, Thomas fl. 1730–64 Bookseller, Printseller, Mapseller Brit. M
Map of Africa, 1745.

Bald, William fl. 1830 Cart. Irish M
County map of Mayo, 1812–14 (eng. J.
Basire), 1830 (eng. P. Tardieu).

Baldwin and Cradock } *See under* KOCH,
Baldwin, Cradock and Joy} C. G.; RUSSELL,
J.; SOCIETY FOR THE DIFFUSION OF USEFUL
KNOWLEDGE; THOMSON, JOHN, AND CO.

Baldwin, E. *See under* Hoxton, W.

Baldwin, Richard d. 1770 Bookseller Brit. M

Ballard, Christophe *See under* Buache, P.; du Caille, L. A.

Ballière, H. *See under* Prichard, J. C.

Ballino, Giulio fl. 1560-9 Geog. Ital. M
De' desegni dell piu illustri città et fortezze del mondo . . . Bologna, Zalterii, 1569.

Balugoli, Alberto 16th cent. Cart. Ital. M
Map of Modena, 1571.

Banister, T. C. fl. 1820 Cart. Brit. M
County maps.

Bankes, Thomas, Blake, E. W., and Cook, A. 18th/19th cent. Carts. Brit. M
A new royal, authentic and complete system of universal geography . . . Lond., J. Cook, 1787-1810?.

Bansemer, J. M. (19th cent.) **and Zaleski Falkenhagen, Piotr** (1809-83) Carts. Polish M
Atlas, containing ten maps of Poland . . . Lond. J. Wyld; J. Ridgway and Sons, 1837.

Baranca, Joseph 17th cent. Cart. ? G

Barber, J. fl. 1776 Eng. Brit. M

Barbié du Bocage, Jean Denis 1760-1825 Cart. Fr. M G
Recüeil de cartes géographiques . . . Paris, Sanson and Co., 1791.

Barbou, J. *See under* Brion de la Tour, Louis.

Barclay, Rev. James 19th cent. Lexicographer Brit. M
A complete and universal dictionary of the English language . . . Lond., George Virtue, various edns., *c.* 1840/52. Some edns. contain maps from Moule's *English Counties Delineated*, with eng.'s name and other imprints deleted. Other edns. incl. maps 'Engraved by the Omnigraph E. P. Becker & Co. Patentees' and with the imprint 'LONDON. GEO. VIRTUE, IVY LANE'. These maps are outlined in colour by hand. The omnigraph was obviously a form of mechanical lithography.

Bardin, William, J. M. and R. fl. 1800-14 Explorer, Surv. (William); Carts. Brit. G (C and T)
See also under Edkins, S. S.

Barents, William }
Barentszoon } 1550-97 Cart.
Barendsz } Dut. M

Bernard, Guillaume }
Nieuwe beschryvinge ende caertboek vande Midlandtsche Zee . . . Amster., C. Clae, 1595.
Description de la mer Méditerranée . . . Amster., C. Nicolas, 1607.
Sea-charts, maps of Scandinavia in Waghenaer's *Spieghel der Zeevaert* (1596, Dut. edn.), viz. 'Tabula hydrographica tam maris Baltici' and 'Hydrographica septent. Norvegiae'.

Barfield, J. *See under* Lavoisne, C. V.

Barkeley, Sir Francis fl. 1601 Surv. Brit. M
Maps of Ulster, Tyrone and Sligo, 1601.

Barker, Robert d. 1645 Pub., Printer Brit. M
Sea-chart of the English coast. Lond., 1604.

Barker, William *See under* Carey, Mathew; Guthrie, W.; Payne, J.

Barläeus, Caspar } 1584-1648 Cart. Dut.
von Baarle } M
Atlas of Italy, 1627 (with Jodocus Hondius).
Descriptio Indiae occidentalis (with 17 maps) Amster., 1622.

Barnalet, I. A. *See under* Faden, W.

Barocci, Giovanni Maria 1560-70 Cart. Instrument-maker Ital. G

Baronovskii, Stepan Ivanovich b. 1817 Cart. Russ. M
Geographical atlas of the ancient world . . . (title in Russ.) St. Petersburg I.A. Iungmeister, 1845.

Barreo, A. *See under* Wolfgang, A.

Barres *Same as* des Barres (q.v.)

Barrière Frères *See under* Delamarche, F.

Barrow, J. and T. *See under* Staunton, Sir G. L.

Barthelemi *See under* Gendron, P.

Bartlett and Welford   *See under* FADEN, W.

Bartsch, Jacob   fl. 1661 Astron. Ger.   X

Bartsch, Johann Gottfried   fl. 1665   Eng.
Ger.   G

Bascarini, Nicolo   fl. 1548   Pub.   Ital.   M
Pub. an edn. of Ptolemy at Venice in 1548
with 60 copper-plate maps.

Basire, I.   *See under* BALD, W.; KIRBY, J.;
LEMPRIÈRE, C.

Bassaei, N.   *See under* ROMANUS, A.

Basset, Thomas}   *c.* 1659-93   Bookseller
Bassett,   **}** Brit.   M
Publishers of some later editions of Speed's
maps.   *See under* CHISWELL, RICHARD, with
whom he collaborated, and SPEED, J.

Basso, Francesco} fl. 1560-70   Cart.,
Pilizzoni   **}** Instrument-maker   Ital.
G (T)

Bate, R. B.   fl. 1810   Cart.   Brit.?   G
(C and T)

Batelli and Fanfani   *See under* ROSSI, L.

Bates, T. M.   *See under* WALKER, J.

Bath, N.   fl. 1811   Cart.   Irish   M
Map of County Cork (1811).

Battista, Giovanni de Cassine   fl. 1560
Cart.   Ital.   G (C and T)

Baudoin, Gaspar   *See under* TASSIN, N.;
TAVERNIER, M.

Baudouyn, P.   *See under* LÜBIN, A.

Bauer, Johann Bernhard   d. 1839
Mechanic, Turner   Ger.   G (T)

Bauernkeller   *c.* 1840   Maker of relief maps
Fr.   M   G

Baugh, Robert   fl. 1808   Cart., Eng.
Brit.   M
County map of Shropshire (1808).
See also under EVANS, REV. J.

Baum, Heinrich   fl. 1573-5   Astron.,
Jesuit priest   Ger.   G

Baum, Johann Christoph   18th cent.   Cart.
Ger.   M

Baumeister, I.   *See under* HOMANN'S HEIRS

Bayer, Johann   **}**
Bayr   **}** 1572-1625   Astron.   Ger.
Bayeri, Iohannis}   X

Bazeley, Charles William   18th/19th cent.
Cart.   Amer.   M
The new juvenile atlas   Philadelphia, 1815.

Beach and Beckwith   *See under*
WOODBRIDGE, W. C.

Beaufort, Daniel Augustus   18th cent.
Cart.   Brit.   M
Diocese of Meath   1797, 1816.   Eng. S. J.
Neele.

Beaulieu   *Same as* DE BEAULIEU (q.v.)

Beaumont   *Same as* DE BEAUMONT (q.v.)

Beaurain   *Same as* DE BEAURAIN (q.v.)

Beautemps-Beaupré, C. F.   1766-1854
Cart.   Fr.   M
Pilote français. Environs de Brest   Paris,
1822.
Atlas de voyage de Bruny-Dentrecasteaux . . .
Paris, 1807.

Beauvais   18th cent.   Cart., Pub.   Fr.   M
Atlas ecclésiastique . . .   Paris, 1783.

Becker, E. P., and Co.   19th cent.   Litho-
graphers   M
Patentees of the 'Omnigraph', a machine
for lithographing.
See also under BARCLAY, J.

Beckher, Joh.   fl. 1726   Globemaker
Ger.?   G (C)

Beckmann, Johann Christoph   1641-1717
Theologian   Ger.   G

Beek, Anna   17th cent.   Cart.   Ger.   M
Plans of fortifications and battles. Some-
times bound up with similar plans by
Allard, Baillieu, Blaeu, Browne, Cartier,
Gournay, Fricx, Gutschoúen, Hooge,
Husson, Lafeuille, Lea, Morden, Mortier,
Mosburger, Mee, Plouich, Skynner,
Sprecht, Visscher, N. J., Wit.

Beeldesnijder   *Same as* DE BEELDESNIJDER
(q.v.)

Beer, Wilhelm   fl. 1834   Astron.   Ger.   X

Begbie, P.   *See under* DAY AND MASTERS;
GARDEN, W.

Begule, L. G.   *See under* DE FER, N.

Behaim, Martin   *c.* 1459-1506   Geog.
Ger.   G (T)
Made the oldest surviving modern terrestrial
globe (MS.).

Beighton, Elizabeth 18th cent. Pub.
Brit. M

Beighton, Henry, F.R.S. 1686-1743
Surv., Cart. Brit. M
Map of Warwickshire, 1725 and subs. edns.,
from Dugdale's *Warwick*.

Beins *Same as* DE BEINS (q.v.)

Beke *Same as* VAN DER BEKE (q.v.)

Belch *See under* LANGLEY

Belga, Guilielmus Nicolo fl. 1600
Cart. Dut. G (T)

Belga, Jacobus Bossius fl. 1558 Eng.
Ital. M

Belidar *See under* JEFFERYS, T.

Belknap and Hamersly *See under* SMILEY,
T. T.; WOODBRIDGE, W. C.

Bell, A. *See under* ARMSTRONG, A.; HORSLEY,
J.

Bell, Allan, and Co. 19th cent. Pub.
Brit. M
A new general atlas of the world . . . Lond.;
New York, J. K. Herrick, 1837-8.
Although pub. jointly with an American
firm, the maps have Bell's imprint and
the date, 1837.

Bell, Andrew fl. 1694-1715 Pub. Brit.
M

Bell, James fl. 1833-6 Topg. Brit. M
*New and comprehensive gazetteer of England
and Wales* 4 vols., 44 maps, which were
eng. by Gray and Son, J. Neele and
Co., and B. Scott and were used again in
*The parliamentary gazetteer of England and
Wales* (1843 and other edns.).

Bell, John Thomas William 19th cent.
Surv. Brit. M
Map of Northumberland and Durham
coalfields, 1843-61. Eng. by M. and
M. W. Lambert. A reduction was pub.
in 1850.

Bell, Peter 18th cent. Cart. Brit. M
Map of London district, 1769 (with A.
Dury; 2 edns., 1 eng. by I. Caldwal),
1771 (pub. C. Bowles and A. Dury),
1775 (pub. C. Bowles).

Bellamie, Iohn *See under* WOOD, W.

Bell'Armato, Girolamo 1493-c. 1560
Cart. Ital. M
Map of Tuscany, 1536.

Bellero, Ioan 16th cent. Cart. Ital.? M
Woodcut map in Richard Eden's *Decades of
the newe worlde* . . . Lond., William
Powell, 1555.

Bellin, Jacques Nicholas} 1703-72 Cart.,
Belin } Pub. Fr. M
Le petit atlas maritime Paris, 1764.
Atlas maritime 1751.
Neptune français 1753.
Corsica 1769.
*Carte de cours de fleuve de Saint Laurent depuis
Quebec* Paris, 1761.
*Partie occidentale de la nouvelle France, ou du
Canada* Paris, 1755.
*Partie orientale de la nouvelle France, ou du
Canada* Paris, 1755.
Carte réduit de l'océan oriental . . . Paris, 1740.
Recüeil des villes, ports d'Angleterre (with J.
Rocque) Paris, Desnos, 1766 etc.
See also under DELAMARCHE, C. F.; HOMANN,
J. B.; HOMANN'S HEIRS; PONCE, N.;
RAYNAL, G. T. F.; ROCQUE, J.

Bellius *Same as* DOMENICO MACHANEUS
(q.v.)

Bellizard, Dufour et Cie} *See under* HELLERT,
Bellizard, F. et Cie } J. J.

Bellue, P. 19th cent. Cart. Fr. M
*Atlas ou Neptune, des cartes de la mer Méditer-
ranée* . . . Toulon, 1830-4.

Bembo, Pietro 1470-1547 Cart. Ital.
G (T)

Benard, J. F. *See under* DE FER, N.

Benci, Carlo 1616-76 Cart. Ital. G
(C and T)

Benedicht, Laurentz fl. 1568 Cart. Dan.
M

Beneventanus, Marco 16th cent. Geog.
Ital. M

Benincasa, Grazioso and Andrea 15th/
16th cent. Cart. Ital. M

Bennet, R. G. *See under* VAN WIJK
ROELANDSZOON, J.

Bennett, John *See under* SAYER, R.

Berchem, N. P. *See under* VISSCHER, N., III

Berckenrode *See under* Jansson, J.;
    Mercator, G.

Berey, C. A. *See under* Blaeu, G.; Placide
    de Saint Hélène, Père

Berey, Nicolas 17th cent. Cart., Map-
    colourist Fr. M
Father-in-law of A. H. Jaillot (q.v.).
Plan of Paris, 1645.
See also under Delisle, G.; Tassin, N.

Berg *Same as* van der Berg (q.v.)

Bergano, Georgius Jodocus 16th cent.
    Cart. Ital. M

Berghaus, Heinrich Karl Wilhelm 1797–
    1884 Cart. Ger. M
*Dr. Heinrich Berghaus' Physikalischer schul-
    atlas . . .* Gotha, J. Perthes, 1850.
Physikalischer atlas Gotha, J. Perthes, 1850,
    1852, 1892.
Atlas von Asia Gotha, J. Perthes, 1832–43.
There was also a Hermann Berghaus (1828–
    90) who pub. a school atlas of the Austrian
    monarchy in 1855.

Bergman, Johann. *See under* Verardus,
    Carolus

Berlinghieri, Francesco di Niccolo *See
    under* Ptolemy, C.

Bernard *See under* Mentelle, E.

Bernhardt, L. *See under* Weiland, C. F.

Berry fl. 1836 Pub., Bookseller Brit. G

Berry, C. 18th cent. Pub. Brit. M

Berry, William fl. 1669–1708 Geog.
    Eng., Bookseller Brit. M G
Pub. maps based on those of Sanson (q.v.).
A mapp of all the world 1680.
Europe 1680. *Africa* 1680. *North America*
    1680. *South America* 1680.

**Bertelli, Fernando,
Francesco, Donato,
    Petrus, Luca** } 16th/17th cent.
Berteli } Carts. Ital. M
Bertelius }
The Bertellis sometimes signed with initials,
    e.g. 'F.B.', 'D.B.'
Theatrum urbium Italicorum (Petrus Bertelius)
    Venice, 1599. Copies of Gastaldi's oval
    world map, 1562, 1565 (F.) and 1568
    (D. and L.)
See also under Lafréry, A.

Bertius, Petrus}
Berts, Pierre } 1565–1629 Cart., Pub.
Bertii, Petri } Belgian M
Cosmographer to Louis XIII.
Tabularum geographicam 1600, 1602, 1606;
    Amster., J. Hondius, 1616. Some maps
    eng. by S. Rogiers.
*Geographischer eyn oder zusammengezogener
    tabeln* Frankfurt, 1612.
Petri Bertii Beschreibung der gantzen welt . . .
    Amster., J. Janssonio, 1650.
La géographie racourcie . . . Amster.,
    Iudoci Hondij, 1618. Some maps eng. by
    S. Rogiers.
Variae orbis universi . . . Paris, 1628? Maps
    eng. by Melchior Tavernier.
See also under Blaeu, G.; Tassin, N.;
    Tavernier, M.

Bertrand, A. *See under* Collot, V.; de
    Bourgainville, H. Y. P. P.; Duflot de
    Mofras, E.; Duperrey, L. I.; Joly, J. R.;
    Mallat de Bassilan, J.; Voogt, C. J.

Berts *Same as* Bertius (q.v.)

Besson, J. *See under* Blaeu, G.

Best, Thomas fl. 1578 Cart. Brit. M
Maps of Frobisher's Strait and of the world.

Beste, George d. c. 1584 Geog. Brit.
    M
*A true discourse of the late voyages . . . of Martin
    Frobisher* Lond., Henry Bynnyman, 1578.

Bestehorn, C. B. *See under* Homann's
    Heirs

Betts, W. *See under* Hoxton, W.

Betulius, Sigismund 17th cent. Cart.
    Ger. M

Bevis, John 1695–1771 Astron. Brit. X

Bew, John d. 1793 Cart., Bookseller
    Brit. M
Australia 1784.

Beyer, Johann 1673–1751 Cart. Ger.
    G (C and T)

Beyer, L. *See under* Gaspari, A. C.;
    Weiland, C. F.

B.F. Signature sometimes used by F.
    Bertelli (q.v.).

Bianchini, Francesco 1662–1729 Astron.
    Ital. G

Bickham, George, Junr. d. 1771 Eng.
Brit. M
Made bird's-eye views (semi-maps) of
English counties for *The British Monarchy*.
Complete work first issued 1754, but
there is a title page dated 1743, and the
views are dated from 1750 to 1754.
Another edn. entitled *A curious antique
collection of birds-eye views . . .* was pub.
by Laurie and Whittle in 1796.

Bienewitz *Same as* APIAN (q.v.)

Bigges, Walter *See under* BOAZIO, BAPTISTA

Bill, John d. 1630 Pub. Brit. M

Biller, Bernhard, the Elder 1790–1838
Eng. Austrian G

Biller, Bernhard, the Younger 1802–40
Eng. Austrian G (T)

Billings, T. *See under* YATES, W.

Bindoni *Same as* DI BINDONI (q.v.)

Binet *See under* MONIN, V.

Bion, Nicolas *c.* 1652–1733 Instrument-
maker Fr. G (C and T)
Globe celeste . . . Paris, 1708.
Globe terrestre . . . Paris, 1728.

Bironius, Gallus *Same as* DE BRION, M.
(q.v.)

Birt, S. *See under* MELA, POMPONIUS;
OSBORNE, T.; WELLS, E.

Bishop, Captain fl. 1794–6 Eng., Cart.
Brit.
*Charts of the gulf and windward passage, old
straits of Bahamas* (6 charts) 1794–6.

Bishop, George d. 1611 Pub. Brit. M

Bishop, George 1785–1861 Astron.
Brit. X

Black, Adam (1784–1874) **and Charles**
Pubs. Brit. M
General atlas Edinburgh and Lond., 1840
and many subs. edns.
County atlas of Scotland 1848.
Atlas of Australia 1853.
See also under HALL, S.; HUGHES, W.

Blackadder, John 18th cent. Surv. Brit.
M
Berwickshire Mrs. Ainslie, W. Faden and A.
Arrowsmith. 1797. Eng. J. Ainslie.

Blackie, Wallis Graham d. 1906 Cart.
Brit. M
The imperial atlas of modern geography Lond.,
1860

Blacks and Parry *See under* ARROWSMITH, A.

Blackwood, William, and Sons fl. 1839–53
Pub. Brit. M
Atlas 1838 and subs. edns. Eng. by LIZ-
ARS (q.v.)

Blaeu, Cornelis d. 1642 Cart. Dut. M
Younger son of GUILJELMI, B. (q.v.).
See also under BLAEU, J.; DE LA FEUILLE, J.

Blaeu, Guiljelmus } 1571–1638 Surv.

Blaeu, Willem Janszoon} Cart. Dut. M

Blaeuw, Guiljelmum } X   G (C and T)

Janssonius, Guilielmus } His Latin work

Jans Zoon, Willems } usually signed thus,
causing confusion
with the work of
JAN JANSSON, his
rival (q.v.).
In English his name was shown as 'William
Johnson'. Sometimes he used his Christian
names only, combined with that of his birth
place: e.g. Willem Jansz. Alcmar or
Guilielmus Janssonius Alcmarianus.
Father of CORNELIS and JOHANNES B. (qq.v.).
One of the most important cartographers
and map publishers of the 16th and 17th
centuries.
Maps of Holland (1604), Spain (1605) and a
series of atlases, culminating in the *Atlas
maior* of 9–12 vols. in various languages.
Was appointed in 1633–4 cart. to Dutch
East India Co.
Acquired reproduction rights of over 35
maps from JODOCUS HONDIUS (q.v.).
Le flambeau de la navigation . . . Amster.,
Jean Jeansson, 1620.
The light of navigation . . . Amster.,
William Johnson, 1622.
Le théâtre du monde . . . (with J. Blaeu)
Amster., 1635 and various edns.
Tooneel des aerdriicx (with J. Blaeu) Amster.,
1635 and various edns.
The light of navigation . . . Amster., 1622.

Novus atlas . . . 2 vols. Amster. (with Bleau, J.) 1635 and other edns. Carts. incl. (according to edn.): G. Mercator, Eilhardo Lubino, Ioanne Laurenbergio, Ioanne Gigante, Ioanne Mellingero, Christiano Mollero, E. Sÿmonsz, Ioanne Westenberg, Tilemanno Stella, I. A. Comenio, Wolfgang Lazio, Philip Cluvero, Fortunato Sprechero à Berneck, Michaele Florentio a Langren, Iohannes Surhonio, Damien de Templeux, Ab. Fabert, I. B. Nolin, P. Coronelli, N. de Fer, P. du Val, G. Delisle, N. Sanson, I. Besson, Isaaco Massa, Adolario Ericho, Anders Lebiano, Bathol, Scutelto, Jona Scutelto, Abraham Ortelius, Aegidio Martini, Captain Peter Codde, Martino Doué, Adriano Metio, Gerardo Freitag, Bartholdo Wicheringe, Corneliu Pynacker, P. Petit Bourbon, Petro Bertio.

Eng. incl.: Josua van der Ende, S. Rogiers, F. D. Lapointe, C. A. Berey, C. Inselin, R. Cordier, Ioannes Somer Pruthenus, A. Peyrouin, P. Starckman, Desrosiers, Hesselum Gerardrem, Hamersveldt.

See also under GOULART J.; MARTINI, M.; VISSCHER, N., III; WOLFGANG, A.

Blaeu, Johannes }
Johann }
Iohannem } 1596-1673 Pub. Dut.
Joan } M X G (C and T)

Occasionally took the Latin name CAESIUS = blue = Blaeuw (Dut.) = bleu (Fr.).

Son of GUILJELMI, B. (q.v.).

Extremely prolific.

Brasilia Amster., *c.* 1664.

Nova et accurata Brasiliae totius tabula Amster., *c.* 1660.

Praefecturae Paranambucae pars borealis Amster., 1664.

Extrema Americae . . . Amster., 1660.

Archevesché de Cambray Amster., *c.* 1660.

Magnia mogolis imperium (with C. Blaeu) Amster., *c.* 1660.

Paraquaria . . . Amster., 1664.

Nova Belgica et Anglica Nova Amster., 1655. Some copies have Latin text, some Dutch text on reverse of maps.

Virginiae partis australis . . . Amster., 1655; 1658 (Dut. text on reverse, ending 'van te sien en is'); 1664 (Dutch text on reverse, ending 'van te sein is').

Atlas maior sive cosmographiae Blaviana . . . Amster., 1654-72. Various edns. Number of vols. varies from about 9-11, according to contents.

Grooten atlas . . . Amster., 1664-5.

Le grand atlas . . . Amster., 1667.

Het nieuw stede boek ven Italie 4 vols. Amster., P. Mortier, 1704-5.

Nouveau théâtre d'Italie . . . Amster., P. Mortier, 1704.

Novum Italiae theatrum 2 vols. Hagae comitum, R. C. Alberts, 1724.

Nouveau théâtre du Piémont et de la Savoye . . . 2 vols. Le Haye, R. C. Alberts, 1725.

Theatrum civitatum et admirandorum Italiae . . . Amster., 1663.

Toonel der Statien van de vereenighde Nederlanden . . . 2 vols. Amster., 1649.

America nova tabula Amster., 1655.

Asia noviter delineata Amster., *c.* 1660.

Mappa aestiuarum insularum, alias Barmudas Amster., 1665.

Theatrum orbis terrarum 6 vols. (G and J. Blaeu) Amster., 1646-55 and other edns.

Le théâtre du monde ou nouvelle atlas (G. and J. Blaeu) Amster., 1645.

Chili Amster., 1664.

Terra firma et novum regnum Granatense et Popayan (Colombia) Amster.; 1664.

Le royaume de France Amster., 1660.

Comitatuum Boloniae et Guines descriptio Amster., 1660.

Le pais de Brie Amster., 1660.

Champagne (G. and J. Blaeu) Amster., 1660.

Bourbonnais (G. and J. Blaeu) Amster., 1660.

Insulae divi Martini et Uliarus . . . Amster., *c.* 1660.

Duché de Berri Amster., *c.* 1660. *Utriusque Burgundiae* Amster., *c.* 1660 *Le souveranité de Dombes* Amster., *c.* 1660. *Gastinois et Senonis* Amster., *c.* 1660. *Le Beausse* Amster., *c.* 1660. *Languedoc* Amster., *c.* 1660.

Linnois, Forest, Beauiolais et Masconnois Amster., *c.* 1660.

Lotharingia Ducatus (i.e. Lorraine) Amster., *c.* 1660.

Le pais de Poictou Amster., *c.* 1660. *Savoye* Amster., 1660. *Valois* Amster., 1660. *Le souverainetez de Sedan* Amster., 1660. *Yucatan . . . et Guatimala* Amster., *c.* 1665. *Guiana sive Amazonum regio* Amster., 1664. *Nova Hispania et Nova Galicia* Amster., 1665. *Terra Sancta . . .* Amster., 1629. *Paraguay . . .* Amster., 1640. *Peru* Amster., 1664. *West Africa* Amster., *c.* 1660.

Insula S. Laurentii, vulgo Madagascar Amster., 1665.

Tabula Magellicana . . . Amster., 1655. *Aethiopia inferior . . .* Amster. *c.* 1660. Many maps of Spain, e.g.: *Arçobispado de Caragossa* Amster., *c.* 1665. *Arragonia regnum* Amster., 1665. *Regnorum Hispaniae nova descriptio* Amster., *c.* 1660. *Nova Virginiae tabula* Eng. D. Grijp. Amster., 1655, 1662.

Venezuela . . . Amster., *c.* 1660. *Canibales insulae* Amster., 1665. *Insulae Americanae . . .* 1665.

See also under BEEK, A.; BLAEU, G.; DE BEINS, JEAN; DE LA FEUILLE, J.; DE WIT, F.; FABERT, AB.; JUBRIEN, JEAN; MARTINI, M.; OTTENS, R.; TASSIN, N.; VALCK, G.; VAN KEULEN, J. I.; VISSCHER, N., III.

Blaeu, W. J. *Same as* BLAEU, G. (q.v.)

Blagrave, John 1558?-1612 Mathematician Brit. M
Map of the forest and manor of Feckenham, 1591.
World map, 1596. Eng. Benjamin Wright.

Blake, E. W. *See* BANKES, T.

Blake, John L. 19th cent. Cart. Amer. M
A geographical, chronological and historical atlas . . . New York, Cooke and Co., 1826.

Blancard, Nicolaus}
Blancardus } *See under* HORN, G.

Blanchard and Lea *See under* BUTLER, S.

Blanco, Manuel 19th cent. Cart. Span. M
Mapa general de las almas . . . en estas islas Filipinas . . . Manila, M. Sanchez, 1845.

Bland, J. *See under* WARBURTON.

Blaskowitz, Charles fl. 1780 Cart. Amer. M
A topographical chart of the bay of Narraganset . . . Lond., Faden, 1777.
See also under FADEN, WM.; LE ROUGE, G. L.

Blau, Georg fl. 1579 Astron. Ger. X

Bleauw *same as* BLAEU (q.v.)

Bleuler, Johann Ludwig 1792-1850 Surv. Ger. G

Blome (or **Bloome**), **Richard** d. 1705 Cart., Topog. Brit. M
Britannia 1673 (printed by Tho. Roycroft) and subs. edns. (50 maps of poor quality).
Speed's maps epitomised 1681, 1685. The maps were later used in Thomas Taýlor's *England exactly described,* 1715.
Cosmography and geography Lond. Printed by T. Roycroft, 1682, 1693. Engs.: Thomas Burnford, W. Hollar, Francis Lamb.
See also under HOLLAR, W.; TAYLOR, THOMAS.

Blondeau *See under* DELAMARCHE, F; MENTELLE, E.

Blondel *See* BULLET ET BLONDEL

Blunt, E. and G. W. 1802-78 Cart., Pub. Amer. M
Blunt's charts of the north and south Atlantic oceans . . . New York, 1830.
In this pub. certain maps are based on maps by des Barres and others.
See also under FURLONG, L.

Boazio, Baptista} fl. 1588-1606 Cart. Ital.
Boazius } M
Irelande 1599.
Maps in *A summarie and true discourse of Sir Francis Drake's VVest-Indian voyage* by Walter Bigges Lond., 1589.

Bobrik, Hermann 1814–45 Cart. Ger. M
Atlas zur Geographie des Herodot . . . Königsberg, August Wilhelm Unzer, 1838.

Bode, Johann Ehlert 1747-1826 Astron. Ger. G (C and T)

Bodenehr, Gabriel 1664?-1758? Cart. Ger. M

Atlas curieux . . . Augsburg, 1704?.

Curioses staats und kriegs theatrum . . . Augsburg, 1710-30?.

Dritter theil des tractats genañdt Europeas pracht . . . Augsburg, 1737?

See also under KILIAN, G. C.

Bodenehr, Hans Georg 1631-1704 Cart. Ger. M

Sac. imperij Romano Germanici geographica descriptio Augsburg, 1677, 1682 (contains 3 maps).

Bodley, Sir Josias 17th cent. Surv. Brit. M

Map of Northern Ireland, 1609.

Bogardi, Johannis *See under* WYTFLEETE, C.

Bohemio, A. G. *See under* HOMANN'S HEIRS

Böhm, Josef Georg 1807-68 Astron. Ger. G (C)

Bohn, Henry George *See under* CAMDEN, W.

Boisseau, Jean fl. 1642 Pub. Fr. M

'*Enlumineur du roi pour les cartes géographiques.*'

Reissued the maps of BOUGUEREAU (q.v.) in 1642 as *Théâtre des Gaules* (75 maps).

Trésor des cartes géographiques . . . Paris, 1643.

Le clef de la géographie générale . . . (with Isaac Dumas de Fores) 1645, Paris, I. Boisseau; L. Vendosme, 1645.

Reissued by L. BOISSEVIN (q.v.) in 1654, whose name replaced that of Boisseau in the imprint. *See also under* TASSIN, N.

Boissevin, Louis 17th cent. Cart. Fr. M

Trésor des cartes géographiques . . . Paris, 1653.

See also under BOISSEAU, J.

Boiste, Pierre Claude Victoire 1765-1824 Cart. Fr. M

Dictionaire de géographie universelle, ancienne, du moyen âge et moderne . . . Atlas Paris, Desray, 1806.

Bokel, Martin fl. 1568/9 Cart. Ger. M

Bompare, Petrus Johann} 16th cent.

Bomparius } Cart. Fr. M

Drew a map of Provence for Ortelius.

See also under MERCATOR, G.

Bonaldo, Dolfin fl. 1519 Cart. Ital. M

Bonardel *See under* GENDRON, P.

Boncompagni, Girolamo 1622-84 Cardinal Ital. G

Boncompagni, Hieronymo fl. 1570 Cart. Ital. G (C)

Bondeau *See under* ARROWSMITH, A.

Bonifacio, Natoli di ⎫ 1550-92 Cart.,

Girolamo ⎬ Eng. Ital. M

Bonifacius, Natalis ⎭ G (T)

Map of Scotland, 1578, pub. in Rome. No impression now known. Signed 'N.B.'

Bonne, Rigobert 1727-95 Cart., Engineer Fr. M G (C and T)

Recüeil de cartes sur la géographie ancienne . . . Paris, Lattré, 1783.

Atlas de toutes les parties connues du globe terrestre . . . Geneva, J. L. Pellet, 1780.

Atlas portatif Paris, c. 1785.

Atlas encyclopédique (with N. Desmarest) Paris, 1787-8.

Atlas maritime Paris, Lattré, 1762.

Petit tableau de la France. . . Paris, Lattré, 1764.

See also under ARROWSMITH, A.; BOSSI, L.; DESNOS, L. C.; RAYNAL, G. T. F.; SCHRAEMBL.

Bonner, John 1643-1726 Cart., Sailor Brit. M

Bonvalet, H. *See under* MALLAT DE BASSILAN, J.

Bonwicke, J. and R. *See under* WELLS, E.

Boom, Henry and T. 17th cent. Carts. Dut. M

Atlas françois . . . 2 vols. Amster., 1667.

Boone, T. and W. *See under* SIBORNE, W.

Booth, ? fl. 1810 Cart., Eng. M

Plan of the settlements of Australia 1810.

Borch, G. *Same as* TERBORCH, G. (q.v.)

Borcht *Same as* VAN DER BORCHT (q.v.)

Bordee *See under* LAURIE, R.

Bordone, Benedetto 1460-1531 Cart., Ital. M

Work incl. a map of the British Isles in his *Isolario* (1528 and subs. edns. They contain numerous woodcut maps.)

See also under D'ARISTOTILE, N.

Bordone, Girolamo 16th cent. Cart., Courtier Ital. M

Borel, Petrus fl. 1656 Astron. Dut. X

Borghi, Bartolommeo 1750-1821 Cart. Ital. M
Atlante generale . . . Florence, 1819.

Borgonio, Giovanni Tommaso 1620-83 Cart. Ital. M

Borough fl. 1576 Cart. Brit. M

Borsari, Bonifacius fl. 1760 Instrument-maker Ital. O

Borven, M. *See under* SAYER, R.

Bory de St. Vincent, Jean Baptiste Genevière Marcellin, Baron 1778-1846 *See under* DESMAREST, N.

Bos, Bossius *See under* VAN BOS, J. B.

Bosch. *Same as* VAN DEN BOSCH (q.v.)

Boschini, Marco 1613-78 Cart. Ital. M

Bossi, Luigi 1758-1835 Cart. Ital. M
Nuovo atlante universale . . . dei signori Arrowsmith, Poirson, Sotzmann, Lapié, d'Albe, Malte-Brun . . . d'Anville e Bonne Milan, P. and G. Vallardi, 1824.

Bostwick, Henry 1787-1836? Cart. Amer. M
A historical and classical atlas . . . New York, 1828.

Boswell, Henry fl. 1786 Topg. Brit. M
See also under CONDER, T.; KITCHIN, T.

Botero, Giovanni 1540-1617 Cart. Ital. M
Map of Africa.
Mundus imperiorum . . . Ursel, Cornelium Sutorium, 1602.
Theatrum oder schawspiegel . . . Cologne, Andrée, 1596.

Botley, Thomas 18th cent. Cart. Brit. M
Map of Surrey, 1765-9

Boudet *See under* COURTALON, L'ABBÉ

Bougainville *Same as* DE BOUGAINVILLE (q.v.)

Bougard, R. 17th cent. Cart. Fr. M
Le petit flambeau de la mer 1684 and subs. edns.
The little sea torch . . . Lond., J. Debrett, 1801; 'Corrections and additions' by J. T. Serres.

Bouguereau, Maurice 16th cent. Cart. Fr. M
Le théâtre françois. Tours, 1594. The first French atlas of local geography.

Bouillon *Same as* DE BOUILLON (q.v.)

Boulenger, Louis d'Albi ⎱ fl. 1515 Cart.,
or d'Alby ⎰ Mathematician
Boulengier ⎱ Fr. G (T)
Boullanger *See under* DE FREYCINET, L. C. D.; JULIEN, R. J.

Boulton, S. late 18th cent. Cart. Brit. M
Africa with its states, kingdoms, republics, regions, islands, etc. Lond., Robert Sayer, 1787.

Bourgoin *See under* NOLIN, J. B.

Bourgoing *Same as* DE BOURGOING (q.v.)

Bourne, E. *See under* WILKINSON, R.

Bourne, Nicholas *See under* DRAKE, SIR F.

Bourtruche, A. 19th cent. Cart. Fr. M
Atlas chronologique et synchronique d'histoire universelle . . . Paris, Daubrée, 1837.

Bouttats, G. fl. 1690 Cart. Dut. M

Bowen, Emanuel d. 1767 Eng., Pub., Printseller Brit. M
Eng. of maps to George III and Louis XV. Collaborated with T. Kitchin, R. Sayer and J. Bennett.
Prolific.
Britannia depicta or Ogilby improved (with JOHN OWEN, q.v.) 1720 and subs. edns.
Complete system of geography 1744-7 (70 maps).
Complete atlas or distinct view of the known world 1752 (68 maps).
Maps for Harris's *Voyages* 1744.
The Royal English atlas 1762 and subs. edns.
An accurate map of the island of Barbadoes Lond., 1747.
The large English atlas (with T. Kitchin) Lond., Bowles, c. 1760, 1767.
The natural history of England . . . Lond., 1759-63.
Atlas Anglicanum . . . Lond., 1767, 1785 (contains reduced versions of maps in the *Large English atlas* of 1760).
Atlas minimus illustratus (with J. Gibson) 1779.
See also under CORBRIDGE, J.; HINTON, J.; JEFFERYS, T.; TINNEY, J.

Bowen, Thomas d. 1790 Eng. Brit. M
See also under De Rapin-Thoyras, P.;
Russell, P.

Bowles, Carrington 1724–93 Pub. Brit. M
Bowles's new medium English atlas 1785
(44 maps).
Bowles's pocket atlas 1785 (57 maps).
Also the 1766 edn. of *Ellis's English atlas.*
Bowles's post-chaise companion . . . Lond.,
1782.
See also under Bell, P.; Dury, A.; Ogilby,
J.; Palairet, J.; Schraembl, F. A.

Bowles, John 1701–79 Pub., Mapseller
Brit. M
Traded also as John Bowles and Son, and
Smith and J. Bowles.
Brother of Thomas B. (q.v.)
Prolific.
Reissued Saxton's *Britannia* 'Insularum in
oceano maxima', *c.* 1763.
Map of Surrey (in collaboration with T.
Bowles) 1733.
North part of Britain called Scotland pub. by
H. Moll (q.v.) in 1714 and compiled,
according to issue, in collaboration with
Thomas Bowles, D. Midwinter and
John King (qq.v.), etc.
A Plan of London as in Queen Elizabeth's days
Lond., *c.* 1715.
A new and exact plan of ye city of London
Lond., 1731.
Battle of Culloden London., 1746.
Sea chart of all the sea ports of Europe Lond.,
c. 1715.
See also under Bowen, E.; Brooking, C.;
Moll, H.; Overton, P.

Bowles, Thomas, I fl. 1683–*c.* 1714 Mapseller. Brit. M

Bowles, Thomas, II 1712–67 Eng., Map-
seller Brit. M
Brother of John B. (q.v.).
See also under Baker, T.; Moll, H.; Ogilby,
J.; Overton, P.; Senex, J.

Bowles and Carver *See under* Palairet, J.

Boyer, Abel 1667–1729 Cart. Brit. M
*The draughts of the most remarkable fortified
towns of Europe* . . . Lond., I. Cleave and
J. Hartley, 1701.

Boyton, G. W. *See under* Edmands, B. F.;
Goodrich, S. G.; Tanner, H. S.

Braakman, Adriaan 17th/18th cent. Pub.,
Cart. Dut. M
Atlas minor . . . Amster., 1706. Contains
maps by Cantelli, Allard, de Wit, Jansson,
Ortelius, Schenk, Valck, van Loon,
Visscher, N. J.
See also under Schenk, P.

Bradford, S. F. *See under* Rees, A.

Bradford, Thomas Gamaliel 19th cent.
Cart. Amer. M
*Atlas designed to illustrate the abridgement of
universal geography* . . . Boston, W. D.
Ticknor; New York, F. Hunt and Co.,
1835.
A comprehensive atlas . . . Boston, W. D.
Ticknor; New York, Wiley and Long,
1835.
An illustrated atlas . . . *of the United States*
Philadelphia, E. S. Grant and Co., *c.* 1838.
A universal, illustrated atlas . . . (with S. G.
Goodrich) Boston, C. D. Strong, 1842.

Brahe, Tycho 1546–1601 Astron. Dan.
G (C and T) O

Brahm *Same as* de Brahm (q.v.)

Bransby, John 1762–1837 Astron. Brit.
X

Brassier, William 18th cent. Cart. Brit.
M
A survey of lake Champlain . . . Lond.,
Sayer and Bennett, 1776.
See also under Sayer, R.

Braun, A. *See under* Taitboret de Marigny,
E.

Braun, Georg and Hogenberg, R.} 16th cent.

Bruin, Joris   } Carts.
                              Ger. M
Noted for bird's-eye view maps of towns.
Civitates orbis terrarum (6 vols.) 1573 on-
wards.
Théâtre des cités du monde . . . Brussels,
1564–1620.
See also under Hoefnagel, J.; Hogenberg, R.

Breese, Samuel 19th cent. Cart. Amer.
M
The cerographic atlas of the United States New
York, S. E. Morse and Co., 1842.
Morse's North American atlas New York,
Harper and Bros., 1842-5.

Bremond *See under* MICHELOT; HEATHER, W.

Brender à Brandis, G. 18th cent. Cart. Dut. M
Nieuwe natuur-geschied-en handelkundige zaken reis-atlas Amster., S. J. Baalde, 1788.

Bret, William 18th cent. Surv. Brit. M
Worked on maps of India.

Bretez fl. 1739 Cart. Fr. M

Breydenbach *Same as* VON BREYDENBACH (q.v.)

Briet, Père fl. 1650 Cart. Fr. M
Map of Japan.
See also under HORN, G.

Brietius, Philipp fl. 1641-9 Cart. Ger.? M

Brigham, John C. 19th cent. Cart. Amer. M
Nuevo sistema de geografía (with 5 maps) New York, White, Gallagher and White, 1827-8.

Brinhauser, A. *See under* LOBECK, T.

Brion, Benjamin *See under* GOURNÉ, P. M.

Brion, M. *Same as* DE BRION, M. (q.v.)

Brion de la Tour, Louis 18th cent. Cart. Fr. M
Atlas ecclésiastique . . . Paris, Desnos; De Lalin, 1766.
Atlas général . . . Paris, 1766, 1767.
Coup d'œil général sur la France . . . (based on Desnos) Paris, Grangé, Guillyon and others, 1765.
Atlas et tables élémentaires de géographie, ancienne et moderne . . . Paris, J. Barbou, 1777. Maps unsigned by author, eng. or pub.
See also under DESNOS, L. C.; GENDRON, P.; MACLOT, J. C.

Briot, B. J. fl. 1660 Pub. Fr. M
See also under MARIETTE, P.; VAN LOCHOM, P. and M.

Brochard, Bonaventura 16th cent. Cart. Fr. M

Brockenrode *See under* HONDIUS, H.

Broeck *Same as* VAN DEN BROECK (q.v.)

Broedelet, G. 18th cent. Pub. Dut. M
See also under RUYTER, B.

Bróen *Same as* DE BRÓEN (q.v.)

Brognolus, Bernardus fl. 1573 Cart. Ital. M

Brooke, Henry fl. 1820 Cart. Brit. X

Brooking, Charles fl. 1728 Cart. Irish M
City and suburbs of Dublin Lond., J. Bowles, 1728.

Brose, W. *See under* TANNER, H. S.

Brostenhuyzen, Johan fl. 1646. Eng. Dut. M
Worked for Allard.

Brouckner, Isaac 1686-1762 Cart. Swiss M
Der erste preussische seeatlas . . . Berlin, 1749.
Nouvel atlas de marine . . . Berlin, 1749.
Nieuwe atlas; of zee en wereld neschryving . . . The Hague, Pieter van Os, 1759.

Brouwer, Hendrick fl. 1640 Cart. Dut. M

Brown, George fl. 1790-1813 Surv., Engineer Brit. M

Brown, Thomas fl. 1800 Pub. Brit. M
Atlas of Scotland (26 county maps).
A general atlas . . . Edinburgh, 1801. Maps eng. by Gavin and Son, McIntyre, I. Menzies, S. I. Neele, A. G. Whitehead.

Brown(e), Thomas fl. 1627-53 Instrument-maker, Cart. Brit. M

Browne, Christopher fl. 1684-1712 Cart., Pub. Brit. M
Map of Ireland, 1691.
A new map of England 1693 and subs. edns.
Nova totius Angliw tabula 1700 and subs. edns.
See also under BEEK, A.; DE WIT, F.; SPEED, J.; VISSCHER N., III.

Browne, D. *See under* NORDEN, J.; OSBORNE, T.

Browne, John 17th cent. Eng. Brit. M
Eng. map of Staffordshire in Plot's *Staffordshire.*
See also under SPEED, J.

Browne, John fl. 1583-92 Surv. Brit. M
Provincial maps of Ireland.

Brownson, W. M. *See under* PALMER, R.

Bruce, Graf Jacob 1670-1735 Soldier, Cart. Russ. M

Bruckner, Isaac 1689-1762 Mechanic, Cart. Swiss M

Brué, Adrien Hubert 1786-1832 Cart. Fr. M

Atlas . . . 1821, 1826 (revised edn.) Paris, J. Goujon; Manheim, Artaria.

Atlas classique de géographie . . . ancienne et moderne . . . 1826 and subs. edns. Paris.

Atlas universel . . . Paris, 1825 and subs. edns.

Grand atlas universel . . . Paris, Desray, 1816.

See also under ANDRIVEAU-GOUJON.

Bruin, J. *Same as* BRAUN, G. (q.v.)

Bruneau, Robert 17th cent. Pub. Dut. M

Brunet et Fruger *See under* BULLOCK, W.

Bruns, C. fl. 1832 Astron. Ger. X

Brus, Caspar} 1518-*c.* 1550 Cart. Ger.?
Bruschius } M

Bruslins *Same as* DES BRUSLINS (q.v.)

Bruun, Malthe Conrad *Same as* MALTE-BRUN, C. (q.v.)

Bruyst, Jan *See under* RENARD, L.

Bry *Same as* DE BRY (q.v.)

Bryant, A. 19th cent. Surv., Cart. Brit. M

Maps of Bedfordshire, 1826; Bucks., 1824-5; Gloucestershire, 1824; Herefordshire, 1835; Hertfordshire, 1822 (eng. Davies); Lincolnshire, 1828; Norfolk, 1826; Northants., 1827; Oxfordshire, 1824; Suffolk, 1826; Surrey, 1823; Yorks., E. Riding, 1829.

Buache, Philippe 1700-73 Cart. Fr. M G (C and T)

Atlas géographique de quatres parties du monde 1769-99.

Atlas géographique et universelle 1762.

Cartes et tables de la géographie physique 1754.

An edn. of De Lisle's *Carte d'Amerique* 1745.

Considérations géographiques . . . Paris, Ballard, 1753-4.

See also under HOMANN'S HEIRS; JEFFERYS, T.; JULIEN, R. J.; NOLIN, J. B.

Buchanan, G. *See under* THOMSON, JOHN, AND CO.

Buchon, Jean Alexandre C. 1789-1846 Cart. Fr. 'M

Atlas géographique . . . Paris, J. Carez, 1825.

Bucius *Same as* PUTSCH, J. (q.v.)

Budgen, Richard fl. 1824-79 Cart. Brit. M

Map of Sussex, 1724, eng. by JOHN SENEX (q.v.). Republished with corrections from a survey by J. SPRANGE (q.v.), 1779.

Buehler, James A.} fl. 1790-1800 Cart.
Bühler   } Ger. G (T)

Buisson, F. *See under* MALTE-BRUN, C.

Bulifon, Antoine d. 1649 Cart. Fr. M

Accuratissima e nuova delineazione del regno di Napoli . . . Naples, 1692.

Maps eng. by D. F. Cassianus de Silua.

Bull, Gascoign, Bryan and de Brahm 18th cent. Cart. Brit. M

A map of south Carolina . . . Lond., Jefferys, 1757.

Caroline méridionale . . . Paris, 1777.

See also under LE ROUGE.

Bulla *See under* LEGRAND, A.

Bullet et Blondel fl. 1676 Cart. Fr. M Plan of Paris, 1676.

Bullock, William 19th cent. Cart. Brit. M

Atlas historique pour servir au Mexique en 1823 Paris, A. Eymery; Brussels, Brunet et Fruger, 1824.

Bully, Le Sieur *See under* JEFFERYS, T.

Bünau, Henry fl. 1507 Cart. Ger. G (T)

Bünting, Heinrich 16th cent. Cart. Ger. M

Buonem, I. *See under* CLUVER, P.

Buonsignori, Stefano}
Oliveti, Stephanus } 16th cent. Cart.
Monachus Montis} Ital. M

Bürck, A. *See under* GASPARI, A. C.

Burdett, P. P. fl. 1777-94 Cart. Brit. M

Maps of Cheshire, 1777, 1794; Derbyshire, 1767, 1791. The Derbyshire map eng. by T. KITCHIN (q.v.).

Bure, Anders } fl. 1626 Cart. Swed.
Buraeus, Andreas} M
  Map of Sweden, *Orbis arctoi nova et accurata delineatio* 1626

Burgdorfer, J. J. *See under* Wyss, J. R.

Burgess, D., and Co. *See under* Smith, R. C.

Burghers, Michael fl. 1672-1700 Eng. Dut. M
  Map of Oxfordshire in Plot's *Oxfordshire* 1677, 1705.

Burght *Same as* van der Burght (q.v.)

Bürgi, Jost 1552-1663 Astron. Ger. G (C and T) O

Burgklehrner, Matthias 1573-1642 Cart. Austrian M

Burgkmair, Hans 16th cent. Eng. Ger. M

Burn, George *See under* Stephenson, J.

Burnett, Gregory 19th cent. Surv. Brit. M
  Map of Sutherland (with W. Scott), 1833, 1835; a reduction was pub. in 1833.

Burnett, William *See* Dugdale, Thomas.

Burnford, Thomas 17th cent. Eng. Brit. M
  Eng. a map in *Cosmography* by Richard Blome (q.v.).

Burns, Lieut. Alex 19th cent. Cart. Brit. M
  Central Asia Lond., Arrowsmith, 1834.

Burr, David H. 1803-75 Cart. Amer. M
  The American atlas . . . Lond., J. Arrowsmith, 1839. (Maps bear Arrowsmith's imprint.)
  An atlas of the State of New York . . . New York, 1829, 1838.
  A new universal atlas . . . New York, D. S. Stone, 1835?

Busch, Andreas fl. 1656-74 Instrument-maker Ger. G (C and T) O

Busch, Georg fl. 1573 Astron. Ger. X

Busius, Albertus 16th cent. Pub. Ger. M

Busius, J. B. *Same as* van Bos, J. B.

Bussemacher, Johann 16th cent. Printer, Pub. Ger. M

Bussemacher, Quad fl. 1600 Cart. Ger.? M
  Map of Africa.

Butler, E. H., and Co. *See under* Mitchell, S. A.

Butler, M. *See under* Cummings, J. A.

Butler, Philip and Richard 18th cent. Cart. Brit. M
  Map of Co. Carlow, 1789.

Butler, Samuel, Bishop of Lichfield and Coventry 1774-1839 Cart., Antiquary Brit. M
  An atlas of antient geography . . . Philadelphia 1831 (Carey and Son); 1834 (Carey, Lea and Blanchard); 1855 (Blanchard and Lea) and many other edns. Maps eng. by P. E. Hamm and J. Yeager, and in some edns. stereotyped by J. Howe.
  An atlas of modern geography . . . Lond., Longman, Brown, Green and Longmans, 1844 and other edns.
  See also under Laurie, Robert.

Butterfield, Michael *c.* 1635-1724 Instrument-maker Brit. G
  Worked in Paris.

Butters, R. fl. 1803 Pub. Brit. M
  An atlas of England 1803.

Buxemacher *Same as* Bussemacher. (q.v.)

Buy de Mornas, Claude d. 1783 Cart. Fr. M
  Atlas méthodique et élémentaire . . . Paris, 1761-2, 1783 (Desnos).

Bye, Joseph *See under* Playfair, J.; Rees, A.; Smith, Charles

Bylandt Pasterkamp *Same as* de Bylandt Pasterkamp (q.v.)

Bynnyman, Henry *See under* Beste, G.

Byrne, G. *See under* Nevill, A. A.; Nevill, J.

Byron, S. fl. 1785 Cart. Irish M
  Plan of Dublin, 1785.

Cabot, Giovanni, or } *c.* 1475-1557 Navi-
John, and Sebastian } gators, Carts. Ital.
Caboto }M G
  Father and son.

Cabrera, Fran. *See under* Maestre, M. R.

Cacciatore, Leonardo 1775-1830 Cart.
Ital. M
Nuovo atlante istorico . . . 3 vols. (with 33
maps and plans) Florence, 1832-3.

Cadell, T. fl. 1820 Pub. Brit. X

Cady and Burgess *See under* SMITH, R. C.

Cagno, P. *Same as* CANIUS (q.v.)

Cahill, D. 19th cent. Cart. Irish M
Map of Queen's County, 1806. Eng. by
J. Ford.

Caille *See under* JULIEN, R. J.

Calamaeus, I. *Same as* CHAMEAU, J. (q.v.)

Calapoda, Georgio } fl. 1537-65
Callapoda Sideri, Giorgio} Cart. Ital. M
Portolans.

Caldwal, I. *See under* BELL, P.

Callender, B. *See under* MALHAM, REV. J.

Camararius, Joachim fl. 1559 Astron.
Ger. X

Camden, William 1551-1623 Antiquary,
Historian M
Britannia
1607 57 maps. Latin text on reverse of
maps.
1610 57 maps. Reverse of maps blank,
separate eng. text.
1637 57 maps. Reverse of maps blank,
separate eng. text.
Subs. edns. incl. 1695 (50 maps, eng.
by SUTTON NICHOLLS and JOHN STURT,
qq.v.), 1715, 1722 (2 vols.), 1737, 1755,
1772, 1789 (3 vols. 60 maps eng. by J.
CARY, q.v.). There was also an abridgement issued in 1626 (52 maps). The
earlier edns. eng. by JOHN NORDEN,
WM. KIP, WM. HOLE and GEORGE
OWEN (qq.v.). Other edns. worthy of
note are those of 1617 (abridged edn.
with Latin text. Illustrated by reduced
copies of Saxton eng. by Peter Keer);
1626 (printed by John Bill. Abridged
edn. 51 small maps).
In 1805 the maps used for the 1789
edn. were used for Stockdale's *New British
atlas*, and in 1845 was pub. *Camden's
Britannia epitomised and continued* by Samuel
Tymms, Lond., Bohn, 6 vols.
See also under SPEED, J.; STOCKDALE, J.

Camerarius, Elias} 16th cent. Mathematic-
Kammermeister } ian, Cart. Ger. M

Cameron, G. *See under* ROSS, LIEUT. C.

Camotto, Giovanni Francesco}
Camocio, Gian Francesco   } fl. 1560
Camocius   } Cart. Ital.
Camotii   } M
Isole famose Venice, 1563.
Map of Ireland, 1572.
Edns. of Gastaldi's oval world map, 1569,
1581.
See also under LAFRÉRY, A.

Campbell, I. C. 19th cent. Cart. Brit.
M
Map of Pembrokeshire, 1827.

Campbell, Lieut. *See under* SAYER, R.

Campi, Antonius 16th cent. Artist, Cart.
Ital. M

Campiglia *See under* DE ROSSI, G. G.

Campii, F. and D. *See under* REICHARD,
C. G.

Canerio, Nicolay fl. 1502 Cart. Port.
M

Canina, L. 19th cent. Historian Ital. M
Pianta tópografica di Roma antica Rome,
1850.

Canius   } fl. 1582 Cart. Ital. M
Cagno, Paolo}

Cantel *See under* JULIEN, R. J.

Cantelli de Vignola, Giacomo 1643-95
Geog. Ital. M G
See also under BRAAKMAN, A.; OTTENS, R.

Cantellius, Jacobius *See under* DE ROSSI,
G. G.

Cantemir, Prince Dimitrie 1673-1723
Cart. Russ. M

Cantino fl. 1502 Cart. Port. M
Sea-charts.

Cantzler fl. 1795 Cart. Brit. M
Map of Australia.

Capitaine, Louis fl. 1789 Cart. Fr. M
Made an abridged edn. of C. F. Cassini's
Carte géométrique de la France 1789.

Capper, Benjamin Pitts fl. 1808 Topg.
Brit. M
*Topographical dictionary of the United King-
dom* (with 44 maps) various edns. from
1808 to 1829.

Capriolo, Elia 16th cent. Cart. Ital. M

Caraffa, Giovanni} fl. 1561 Cart. Ital.
Carafa } G (T)

Carcía de Cespedes, Andres 16th/17th cent. Navigator Span. M
Regimicato de Navigacion . . . Madrid, 1606 (numerous charts).

Cardono *See under* DE ANTILLON Y MARZO, I.

Carey, Henry Charles (1793-1879) **and Lea, Isaac** (1792-1886) Pubs. Amer. M
A complete historical, chronological and geographical American atlas . . . 1822: Philadelphia, 1823, 1827. Maps drawn by J. Finlayson, F. Lucas, F. Lucas, Junr., J. Yeager (eng.), Kneass, S. H. Long, E. Paguenard; the map of South Carolina is 'Reduced by J. Drayton from the state map by J. Wilson'.
Family cabinet atlas Philadelphia, 1832, 1834.
See also under BUTLER, S.; WEILAND, C. F.

Carey, Mathew 1760-1839 Cart. Amer. M
Carey's American atlas . . . Philadelphia, 1795, 1809. Maps are drawn by Samuel Lewis, W. Barker (eng.), J. Smither (eng.), Amos Doolittle, Harding Harris, Vallance (eng.), Genl. D. Smith, J. T. Scott (eng.), Thos. Hutchins.
Carey's American pocket atlas . . . Philadelphia, 1796, 1801, 1805, 1813, 1814. Maps eng. by W. Barker, J. H. Seymour, A. Doolittle. In the 1813 and 1814 edns. the map of New Hampshire is by Saml. Lewis.
Carey's general atlas . . . Philadelphia, 1796, 1802, 1814, 1817, 1818 and other edns. Some of the maps are by Samuel Lewis, Harding Harris, A. Doolittle, Genl. D. Smith.
A scripture atlas . . . Philadelphia, 1817.
A general atlas for the present war . . . Philadelphia, 28 January 1794. Engs.: C. Tiebout, Joseph T. Scott, W. Barker.

Carey and Hart *See under* TANNER, H. S.

Carez, J. 19th cent. Pub. Fr. M
Atlas géographique by J. A. Buchon, Paris, 1825.

Carleton, Osgood *See under* NORMAN, J.

Carnan, Thomas d. 1788 Pub. Brit. M
See also under GIBSON, J.

Carniero, Antonio de Mariz 17th cent. Cart. Moorish M

Carolus, Joris fl. 1634 Cart. Dut. M
Het nieu vermeerde de licht Amster., Janssen, 1634.

Caroly *Same as* DE CAROLY (q.v.)

Carranza, D. G. 18th cent. Cart. Span. M
Chart of coasts of Span. W. Indies and plans of towns there. Lond., Caleb Smith, 1740.

Cartaro, Mario}
Kartaro } fl. 1575 Cart. Ital. M
Karterus } G (C and T)

Cartier *See under* BEEK, A.

Cartilia, Carmelo fl. 1720 Instrument-maker Ital. O

Cartwright, Samuel fl. 1635 Printer, Cart. Brit. M
Historia mundi, or Mercator's atlas, with new maps and tables by the studious industry of Judocus Hondy, Englished by W. S. [Saltonstall], 1635 (in collaboration with Michael Sparke).

Carver *See* BOWLES AND CARVER

Carver, Captain W. 18th cent. Surv., Cart. Brit. M
Executed maps for Thomas Jefferys.
American atlas 1776.
A new map of the province of Quebec . . . Sayer and Bennett, Lond., 1776.

Carwitham, J. *See under* GORDON, W.

Cary, George (d. 1859) **and John II** (c. 1791-1852) Pubs., Carts. Brit. M G (C and T)
Very prolific. A list of their publications is to be found in *Catalogue of maps, atlases, globes and other works published by G. and J. Cary,* 1800. *See also* CARY, JOHN; CARY, WILLIAM.

Cary, John c. 1754-1835 Eng., Pub. Brit. M X G (C and T)
Brother of WILLIAM C. (q.v.).
Very prolific.
Eng. maps for the 1789 edn. of Camden's *Britannia.*
General atlas of the world.

New universal atlas Lond., 1808 and subs. edns.

World map.

Cary's pocket globe . . . Lond., 1791.

Cary's new map of France Lond., 1814.

Inland navigation . . . *throughout Great Britain* Lond., 1795.

New and correct English atlas various edns., 1787-1862—the last pub. by Cruchley.

Cary's English atlas various edns., 1809-34.

Cary's travellers' companion various edns., 1790-1828.

New British atlas (in collaboration with John Stockdale) 1805.

A new map of North America . . . Lond., 1824.

Cary's actual survey of the country fifteen miles round London . . . *preceded by a general map of the whole* . . . Lond., 1786.

Cary's survey of the high roads from London . . . Lond., 1790.

See also CARY, G. AND J.; CARY, WILLIAM; SAYER, R.; STOCKDALE, J.; WALLIS, JOHN.

Cary, William 1759-1825 Cart. Brit. G (C and T) M
Brother of JOHN CARY (q.v.).
Maps of Africa, 1805, 1811, 1828.
See also CARY, G. AND J.; CARY, JOHN; SMITH, W. (1769-1839).

Case, N. *See under* LOTHIAN, J.

Casimirus *See under* VALCK, G.

Cassell, Peter and Galpin fl. 1864-7 Pub. Brit. M
Cassell's British atlas 1864-7.

Cassianus de Silua, D. F. *See under* BULIFON, A.

Cassine *Same as* DE CASSINE (q.v.)

Cassini, Giovanni Domenico (Jean Dominique) 1625-1712 Astron., Geog. Ital. M X
Worked in France. Ancestor of César François C.
Planisphere terrestre . . . Paris, J. B. Nolin, 1696. This eng. map is accepted as the first scientific map of the world.
Nieuwaerdsch pleyn Amster., *c.* 1703, eng. by Cornelis Danckerts.

Cassini, Giovanni Maria 18th cent. Cart. Ital. M O G (C and T).

Nuovo atlante geografico universale . . . 3 vols. Rome, 1790-2, 1792-1807.
Maps signed 'Gio. Ma. Cassini somco. incise'.

Cassini, Jean Dominique, Comte 1748-1845 Astron., Geog. Fr. M X

Cassini, M. fl. 1710 Cart. Fr. M
Hydrographia Gallia: the sea coasts of France 1710, Lond., Morden and Lea.
See also under GUEDEVILLE, N.

Cassini de Thury, César François 1714-84 Geog., Astron., Cart. Fr. M
Descendant of Giovanni Domenico C.; son of Jacques C. de T.
Déscription géométrique de la France 1783. Contains a copper-plate map, dated 1744, eng. by Dheulland; it was the first complete topographic map in outline, based on triangulation and astronomical observation.
Atlas topographique, minéralogique et statistique de la France . . . Paris, H. Langlois, 1818.
See also under COVENS, J.; JEFFERYS, T.; SCHRAEMBL, F. A.; VON REILLY, F. J. J.

Cassini de Thury, Jacques 1677-1756 Cart. Fr. M
Father of César François C. de T.

Castaldo, Jacobo *Same as* GASTALDI (q.v.).

Castlemain, Earl of (Roger Palmer) 1634-1705 Cart. Brit. G (C and T).

Castro, João *Same as* DE CASTRO (q.v.)

Caucigh, R. P. Michael fl. 1725 Cart. Ger. G (T)

Cay, R. *See under* HORSLEY, J.

Caylus *Same as* DE CAYLUS (q.v.)

Celi, Fran. Math. *See under* JEFFERYS, T.

Cella, Philipp fl. 1831 Cart. Ger. G (T)

Cellarius, Andreas b. *c.* 1630 Astron. Dut. M X
Maps were copied for Philip Marent's *Geographia antiqua et nova* Lond., 1742.
Celestial Atlas.

Cellarius, Christophorus *Same as* KELLER, CHRISTOPH (q.v.)

Cellarius, Daniel fl. 1578-93 Pub. Ger. M

Celle, Benedetto *See* MARINO, GEROLAMO

Celtes, Konrad}
Celtis } 1459–1508 Cart. Austrian
Pickel } G (C and T)

Cenus *Same as* ZENO, NICOLO (q.v.)

Cerruti *See under* HOMANN'S HEIRS

Certa, J. *See under* GIUSTINIANI, F.

Chafrion, Joseph fl. 1685 Cart. Fr.? M

Chainlaire, Pierre Grégoire 1758–1817
Cart. Fr. M
Atlas national de la France . . . Paris, 1810–12.
Atlas national portatif de la France (with
Dumez) Paris, 1792.
See also under MENTELLE, E.

Chambon *See under* DESNOS, L. C.

Chameau, Jean} 16th cent. Advocate,
Calamaeus, Io } Cart. Fr. M
Carte du Berry Lyons, 1566.
Reproduced by Bouguereau and Ortelius.

Chamouin, Jean Baptiste Marie b. 1768
Eng. Fr. G

Champion, J. N. *See under* GALLETTI,
J. G. A.

Changuion, D. J. *See under* RAYNAL,
G. T. F.

Chaplin, T. *See under* D'ANVILLE, J. B.

Chapman, John fl. 1777 Cart. Brit. M
Maps of Nottinghamshire, 1776 and subs.
edns., and Essex, 1777. The latter in
collaboration with P. André.

Chapman and Hall *See under* GORTON, J.;
HALL, S.; SHARPE, J.; SOCIETY FOR THE
DIFFUSION OF USEFUL KNOWLEDGE

Charcornac, M. fl. c. 1850 Astron. Fr.
X

Chardon 19th cent. Printer, Pub. Brit.
M

Charle, J. B. L. 19th cent. Cart. Fr. M
*Nouvel atlas national de la France avec augmen-
tations par Darmet* Paris, Dauty et Roret,
1833.

Chase, W. *See under* GODDARD, J.

Chassereau, Peter 18th cent. Cart.
Brit.? M
Plan of Carthagena harbour (America),
Lond., 1741. (Based on William Law's
reports.)

Chassignet, Daniel 17th cent. Cart. Fr.
G (C and T)
See also under DE MONGENET, F.

Châtelain, Henri Abraham 1684–1743
Cart. Fr. M
Atlas historique Amster., F. l'Honoré et
Châtelain, 1705–29; Z. Châtelain, 1732–9.

Châtelain, Z. *See under* CHÂTELAIN, H. A.

Chattilion *Same as* DE CHATTILION (q.v.)

Chauchard, Capt. 18th cent. Cart. Brit.
M
A general map of the empire of Germany . . .
Lond., J. Stockdale, 1800.

Chaudiere, Guillaume *See under* THEVET,
ANDRÉ

Chaves *See* DE CHAVES.

Chetwind fl. 1666 Cart. Brit.? M

Chevalier, Michel 1806–79 Cart. Fr.
M
*Histoire et description des voies de communica-
tion aux États-Unis* (with 4 maps) Paris,
1841.

Chevenau *See under* CHIQUET, J.

Chiaves, Hieronymo 16th cent. Cart.
Dut. M
Hispanensis conventus delineatio Antwerp,
Plantin, c. 1584.

Chiesa, Andrea 18th cent. Cart. Ital.
M

Chieze *Same as* DE CHIEZE (q.v.)

Childe, T. *See under* MOLL, H.

Chiquet, Jacques 18th cent. Cart. Fr.
M
Nouveau atlas françois . . . Paris, Cheveanau,
1719.
Le nouveau et curieux atlas . . . Paris,
Chevenau, 1719.

Chiswell, Richard} 1639–1711 Bookseller,
Chiswel } Pub. Brit. M
Editions of Speed's *Theatre*, pub. by Bassett
and Chiswell in 1676. Two issues were
made: (*a*) maps with plain backs, (*b*) maps
with text on backs.
Speed's *Africae described* 1676.
F. Lamb's *A new map of East India* 1676.

Chkatoff *See under* PIADYSHEV, V. P.

Christoph, Meister *Same as* STIMMER, C. (q.v.)

Christophori, Joann *See under* MATAL, J.

CHS Signature sometimes used by C. STIMMER (q.v.)

Chukei (Ino Tadutaka) fl. 1800-16 Surv. Japanese M

Churchill, Awnsham d. 1728 Pub., Bookseller Brit. M
See also under MOLL, H.

Churchill, John d. *c.* 1714 Pub., Bookseller Brit. M
Pub. an edn. of Camden's *Britannia*.

Cimerlino, Giovanni } 16th cent. Eng.
Cimerlinus, Johannes } Ital. M *See also*
          Paulus } *under* FINÉ, O.

Cingolani, Giovanni Battista 17th/18th cent. Cart. Ital. M

Claesz, Cornelis }
Nicolai, Cornelius } fl. 1599 Pub. Dut.
Nicolas, C. } M
Issued an edn. of B. Langenes's *Caert thresoor* Amster., 1599.
See also under BARENTSZOON, W.; LANGENES, B.; PTOLEMAEUS, C.; WAGHENAER, L. J.

Clarici, Paolo Bartholomeo 18th cent. Cart. Ital. M

Clark, James 17th cent. Eng. Brit. M
Eng. charts for Capt. Greenville Collins's *British coasting pilot*, 1693.
See also under SELLER, J.; THORNTON, J.; VAN KEULEN, J.

Clark, Matthew 18th cent. Cart. Amer. M
Charts of the coast of America . . . Boston 1790.

Clarke, Benjamin fl. 1852 Topg. Brit. M
Compiled, with H. G. Collins, *British Gazetteer*, 1852 (with 43 maps).

Clarke, J. 19th cent. Cart. Brit. M
Map of Isle of Wight, 1826.

Classun *Same as* DE CLASSUN (q.v.)

Claudianus, Nicolaus *Same as* KLAUDIAN, M. (q.v.)

Cleave, I., and Hartley, J. *See under* BOYER, A.

Cleef *Same as* VAN CLEEF (q.v.)

Clericus, Johannes *Same as* LE CLERC, J. (q.v.)

Clerk, James *Same as* CLARK, J. (q.v.)

Clerk, Thomas fl. 1811-35 Eng. Brit. M
See also under THOMSON, J., AND CO.

Cleynhens, Bernardus 18th cent. Cart. Dut. M
Accuraat geographisch kaart-boekje of zakatlas . . . Haarlem, n.d.

Cloppenburg, Everhardus 17th cent. Cart. Dut. M

Cloppenburgh, Jean
          Evertsz } 17th cent. Pub.
Cloppenburgij, Johannis } Dut. M
See also under LINSCHÓTEN, J.

Clouet, Jean, Baptiste Louis, L'Abbé b. 1730 Cart. Fr. M
Géographie moderne avec introduction Paris, 1780, 1787 (Mondhare St. Jean), 1791.

Clusius, Carolus } 1525-1609 Cart.
de L'Escluse, Charles } Fr. M
Gallia Narbonensis.
Map of Spain. Only one impression known.

Cluverius, Philip }
Cluver } 1580-1623 Geog.
Clüver } Danzig M
Introductionis in universam geographicam Amster., Jansson-Waesberg, 1661, 1676, 1682; Joannem Pauli, 1729; Brunswick, G. Muller, 1641; Amster., Elzevir Press, 1659; Amster., J. Wolters, 1697; Wolfenbüttel, I. Buonem, 1667?.
Germaniae antiquae ubri tres Leyden, Ex officina Elzeviriana, 1631, with 11 maps signed by Nicolas Geilkeckio.
See also under BLAEU, G.; HONDIUS, H.; HORN, G.; JANSSON, J.; MERCATOR, G.

Cobbett, William 1762-1835 Pub., Geog. Brit. M
Geographical dictionary of England and Wales 1832 (52 crude maps), 1854.

Cocco, Jacomo fl. 1558-75 Cart. Ital. G (T)

Cock, Hendrik }
Coquus } fl. 1581/3 Cart. Dut. M

Cock, Hieronymous or Hiero 1510-70
Pub., Eng. Dut. M
Pub. a map of Palestine compiled by
Laicksteen-Sgrooten and eng. by J. and
L. à Doeticum.
See also under GUTIÉRREZ DIEGO.

Cockerill, Thomas} fl. 1674-1702 Pub.
Cockeril } Brit. M
See also under MORDEN R.

Codde, Capt. Pierre 17th cent. Cart.
Dut. M
Dunkirk and adjacent coast in G. and J.
Blaeu's *Theatrum orbis terrarum.*

Coeck, Gerard *See under* DE WIT, F.

Coello de Portugal y Quesada, Francisco
1820-98 Cart. Span. M
Atlas de España . . . Madrid, 1848-68.

Coffin, R. *See under* DONN, B.

Coignet, Michel 1549-1623 Mathematician, Cart. Fr. M

Colart, L. S. 19th cent. Cart. Fr. M
Histoires de France et d'Angleterre . . . (with 2
maps) Paris and Lond., 1841.

Colburn *See under* SUCHET, L. G.

Coldewey *See under* HOMANN, J. B.

Cole, Benjamin fl. 1695-1709 Cart.,
Eng. Brit. M
Map of Cambridgeshire 1710. Eng. by J.
Harris (Reduction pub. same year.)
Twenty miles round Oxford 1705.
See also under WOOD, B.

Cole, Charles Nalson 18th cent. Cart.
Brit. M
Reduced Fens [Cambs.] 1789. Eng. S. J.
Neele.

Cole, G. fl. 1810 Cart. Brit. M
Operated in partnership with J. Roper with
whom he pub. *The British atlas* in 1810.
See also under NIGHTINGALE, JOSEPH; WALLIS,
JAMES.

Cole, Humfray or Humfrey 1530?-91
Instrument-maker, Eng., Goldsmith, Cart.
Brit. M
Map of Canaan, 1572

Cole, James *See under* POKER, M.

Cole, William *See under* EBDEN, W.

Colin, Emanuel 17th cent. Pub. Dut. M
See also under DE HERRERA, ANTOINE.

Colin, Michiel} 17th cent. Pub. Dut. M
Colijn } *See also under* DE HERRERA Y
Colinium } TORDESILLAS, A.; LE MAIRE, J.

Colinaeus, Simon fl. 1525 Cart. Fr. M

Colins, Jacobus} 1563-1628 Cart. Dut. M
'Ortelianus' }
Nephew of A. ORTELIUS (q.v.)

Colles, Christopher 1738-1816 Surv.
Amer. M
*A survey of the roads of the United States of
America* . . . New York, 1789. Eng.
by C. Tiebout.

Collimitius *Same as* TANNSTETTER, G.

Collin, J. *See under* MORTIER, D.

Collin and Hannay *See under* MORSE, J.

Collins, B. C. *See under* TUNNICLIFF, W.

Collins, Christopher *See under* STEPHENSON,
JOHN

Collins, Capt. Greenville fl. 1669-96
Surv., Cart. Brit. M
Hydrographer to the King.
Great Britain's coasting pilot Lond., various
edns., 1693-1792; some pub. by Mount
and Page. Eng. by J. Clark, J. Harris,
F. Lamb, H. Moll, J. Moxon. This
publication included silhouettes of the
coastlines to assist recognition by pilots.
See also under LAURIE, R.

Collins, Henry George fl. 1850 Pub.
Brit. M
Travelling atlas of England and Wales 1850,
1852, 1868.
In 1848 he pub. an edn. of H. Teesdale's
New British atlas.
See also under CLARKE, BENJAMIN; MUDIE, R.

Collot, Victor [Georges Henri] 1751-1805
Cart. Fr. M
Voyage dans l'Amérique septentrionale . . .
Paris, A. Bertrand, 1826. Pub. simultan-
eously in English.

Collyer, I. *See under* FADEN, W.

Colom, Arnold fl. 1656-61 Cart. Dut.
M
Zee atlas Amster., 1656.
Atlas of werelts water deel Amster., 1660-1.
Zee atlas maritimo o mundo aquatico Amster.,
1669.
See also under COLOM, JACOB AERTZ.

Colom, Jacob Aertz 1599-1673 Cart.,
Bookseller, Pub. Dut. M
Da vyerighe colom 1632-3; Fr. edn., 1633;
English edns., 1640: *Fierie sea-colomne,*
1648: *Upright fyrie columne.*
De groote lichtende ofte vyerighe colom 1661,
1663.
Atlas maritimo . . . Amster., 1669. Maps by
A. Colom, J. A. Colom and Dirck
Davidsz.

Colom, Jacobus Arnoldus 1628-42 Cart.
Dut. M G

Coltman, Nathaniel fl. 1806-36 Cart.
Brit. M
See also under PLAYFAIR, J.

Colton 19th cent. Cart. Amer. M
Colton's map of the United States, etc. New
York, 1855.

Columbus, Christopher *See under*
VERARDUS, CAROLUS.

Comberford, Nicholas fl. 1646-66 Cart.
Brit. M

Combette, A. *See under* LEVASSEUR, V.

Comenio, Johann Amos} 1592-1670
Comenius } Cart. Czech. M
See also under BLAEU, G., HONDIUS, H.;
JANSSON, J.; MERCATOR, G.

Conclin, G. 19th cent. Pub. Amer. M
Conclin's new river guide . . . Cincinnati,
1849 (H. S. and J. Applegate), 1850
(ditto), 1851 (J. A. and U. P. James),
1855 (ditto).
See also under CUMINGS, S.

Conder, Thomas fl. 1780-1801 Eng.
Brit. M
A map of Yorkshire by T. Conder is in *A new
and complete abridgement* . . . *in the antiquities
of England and Wales* by Francis Grose,
Lond., 1798, and in *Historical descriptions of
antiquities of England and Wales* by Henry
Boswell, Lond., 1786.
See also under WILKINSON, R.

Conders ab Helpen, W. and F. fl. 1640
Carts. Dut.? M

Condet, I. 18th cent. Eng. Brit.? M
See under COVENS, J.; POCOCK, R.

Conrad and Co., J. *See under* ARROWSMITH,
AARON; LATOUR, A. L.; PINKERTON, J.

Constable 19th cent. Cart. Brit. M
Map of Edinburgh district, 1822.

Constable and Co., A. *See under*
ARROWSMITH, AARON; AINSLIE, J.

Contareni, Johannes Petrus fl. 1564
Cart. Ital. M

Contarini, Giovanni Matteo d. 1507
Cart. Ital. M
Compiled a world map which was the
first-known printed map of America
(1506). Eng. by Francesco Rosselli.

Cöntgen, G. T. *See under* THERBU, L.

Conyenberg, Jacobus *See under* JACOBSZ,
T.

Conzatti, B. *See under* SAVONAROLA, R.

Cook, A. and J. *See* BANKES, T.

Cook, Capt. James 1728-79 Navigator,
Cart. Brit. M
*A general chart exhibiting the discoveries made
by Capt. Cook* by Lieut. H. Roberts.
Eng. by W. Palmer, 1782
The North American pilot (with M. Lane)
Lond., Sayer and Bennett, 1779, 1783-4.
Three Voyages 8 vols. and folio atlas, 1773-
88.
A general chart of the island of Newfoundland
(with M. Lane) Lond., 1775.
Surveys for Thomas Jefferys' *American atlas,*
1776.
*Chart of the N.W. coast of America and N.E.
coast of Asia* Lond., Faden, 1784.
Le pilot de terre-neuve . . . (with M. Lane)
Paris, 1784.
Chart of the southern hemisphere . . . Lond.,
Strahan and Cadell, 1777. Eng. by
Guilielmus Whitchurch.

Cooke and Co. *See under* BLAKE, J. L.

Cooke, Charles fl. 1808 Cart. Brit. M
Map of Warwickshire, 1809 and other edns.

Cooke, George Alexander fl. 1802-10
Topg. Brit. M
Modern British traveller 25 vols.; with 46
maps, 1802-10, 1822.
Topography of Great Britain 26 vols. Lond.,
1822.
See also under GRAY, G. C.

Cooke, John *See under* EDWARDS, B.; REES,
A.

Cooke, O. D., and Co. *See under* WILLARD, E. H.; WOODBRIDGE, W. C.

Cooper, Richard 18th cent. Eng. Brit. M
See also under ADAIR, J.; EDGAR, W.

Cooper, H. fl. 1806-10 Eng. Brit. M
See also under PHILIPPS, SIR R.; PLAYFAIR, J.; REES, A.

Cooper, Mrs. Mary fl. 1740-61 Pub., Bookseller Brit. M

Cooper, T. *See under* WELLS, E.

Copernicus, Nicolaus} 1473-1543 Astron.
Kopernikus } Ger. M
Map of Prussia, etc., 1529.

Coppo, Pietro 1469-1555 Cart. Ital. M
Made a small map of British Isles (woodcut), *c.* 1525, based on Ptolemy, portolans and a general map of Istria (ditto), Venice, 1540.

Coqqus *Same as* COCK, HENDRIK (q.v.)

Coquart, A. *See under* DE FER, N.

Corbridge, James 18th cent. Surv. Brit. M
Map of Norfolk, 1730 (eng. E. Bowen), 1755.

Cordier, Ludovic and R. 17th cent. Engs. Fr. M
Worked for SANSON, N., PÈRE (q.v.).
See also under BLAEU, G.; DESNOS, L. C.; JAILLOT, C. H.

Corne, Louis *See under* AYROUARD, J.

Coronelli, P. Vincenzo Maria or Marco 1650-1718 Theologian, Cart., Geog. Ital. M G (C and T)
Atlante veneto 2 vols., Venice, 1690, 1691, 1695-7.
Eng. map of Ireland, based on Mercator's map of that country.
World map 1695.
Corso geografico universale Venice, 1692.
Gli argonauti Venice, 1693.
Isolario dell'Atlante veneto Venice, 1696-7.
An historical and geographical account of the Morea, Negropont . . . Lond., M. Gillyflower and W. Canning, 1687.
See also under BLAEU, G.; NOLIN, J. B.

Corsulensis, Vincentus 16th cent. Cart. Ital. M

Cossinet, Francis fl. 1659 Book- and Instrument-seller Brit. M G (C and T)

Costa, Gian Francesco fl. 1775 Cart. Ital. G (C and T)

Costansó, Miguel 18th cent. Cart. Span. M
Carta reducida del oceano Asiatico . . . Madrid, 1771.

Cotes, Thomas *See under* WOOD, W.

Coulier, P. J. 19th cent. Cart. Fr. M
Atlas général . . . Paris, 1850.

Courtalon, L'Abbé 18th cent. Cart. Fr. M
Atlas élémentaire . . . Paris, Julien; Boudet, 1774.

Coussin, H. *See under* AYROUARD, J.

Coutans, Guillaume b. 1724 Cart. Fr. M
Atlas topographique . . . Paris, C. Picquet, 1800.

Covens, Jean fl. 1740 Pub. Dut. M.
Worked with Corneille Mortier as Covens and Mortier.
Maps of Africa (1730), East Indies. Plans of The Hague.
L'Amerique septentrionale Amster., 1757.
Maps by G. Delisle, H. Jaillot, Müller, H. de Leth, F. de Wit, Sanson, N., Homann.
Atlas Nouveau . . . 9 vols. Amster., 1683-1761.
The maps are by a large number of carts. and engs., incl. J. Condet, Cassini, Delisle, van de Aa, C. Allard, Sanson, G. v. Gouwen, F. de Wit, H. Jaillot, N. Visscher, etc.
Mappe monde ou globe terrestre . . . Amster., 1703?.
Nieuwe atlas . . . Amster., 1730-9, 1740-1817.
Nouvel atlas . . . 1735?.
Veteris orbis tabulae geographicae. Amster., 1714?
See also under JULIEN, R. J.; OTTENS, R.; SCHENK, P.; VALCK, G.; VISSCHER, N., III.

Cowley, John fl. 1734-44 Astron., Geog. Brit. M X
See also under DODSLEY, R.

Cowperthwait, T., and Co. *See under* MITCHELL, S. A.

Cox, Thomas d. 1734 Topg. Brit. M

Cramoisy, S. *See under* TASSIN, N.

Craskell, Thomas, and Simpson, James 18th cent. Engineer (Craskell), Surv. (Simpson) Brit M
Map of the island of Jamaica . . . (3 maps) Lond., 1763.

Crato, Johann 16th cent. Printer Ger. M

Crawford, William 19th cent. Surv. Brit. M
Map of Dumfriesshire, 1804 (eng. J. Kirkwood), 1812, 1828 (eng. S. Hall).

Creech, William 18th cent. Pub. Brit. M

Creighton, R. fl. 1831 Cart. Brit. M
Map of Warwickshire for Lewis's *Topographical Dictionary*, eng. by T. Starling.
See under GREENWOOD, C. AND J.

Cremer *Same as* MERCATOR (q.v.)

Crepy *See under* DE FER, N.; LE ROUGE, G. L.

Crevier, Jean Baptiste Louis 1693-1765 Geog. Fr. M
Atlas de géographie ancienne . . . Paris, Ledoux et Tenré, 1819. Maps are by d'Anville; nos. 2 and 3 eng. by Ambroise Tardieu.

Crickenbourg *Same as* VAN CRICKENBOURG (q.v.)

Criginger, Johann 1521-71 Cart. Ger. M

Crissy, J. *See under* MARSHALL, J.

Crocker and Brewster *See under* PALMER, R.

Croisey *See under* DESNOS, L. C.

Cromberger, Jacobus *See under* MARTYR, P.

Crome *See under* SCHRAEMBL, F. A.

Crosley, W. *See under* LINDLEY.

Cross, J. *See under* EDGEWORTH, W.

Crozet *See under* DUFOUR, A. H.

Cruchley, George Frederick fl. 1822-75 Eng., Printer, Mapseller Brit. M G (C and T)

Pub. the 1862 edn. of J. Cary's *New and correct English atlas*.
Map of Lancashire, 1836.

Cruquius, N. S. fl. 1730 Hydrographer Dut. M

Cruttwell, Clement 1743-1808 Topg. Brit. M
Atlas to Cruttwell's Gazetteer . . . Lond., G. G. and J. Robinson, 1799; Longman, Hurst, Rees, and Orme, 1808.

Cruz, Alonso de Santa 1500-72 Cart. Span. G (T)

Cubitt, T. fl. 1795 Cart. Brit. M
Map of Jersey, 1795 (with W. Gardner; eng. J. Warner). Reprinted, 1863.

Cumings, Samuel 19th cent. Cart. Amer. M
The Western navigator . . . Philadelphia, E. Littell, 1822.
The Western pilot . . . Cincinnati, Morgan, Lodge and Fisher, 1825; N. and G. Guilford and Co., 1829, 1832, 1834; G. Conclin, 1838, 1840, 1843, 1848; J. A. and U. P. James, 1854.

Cumming, John fl. 1815 Pub. Irish. M

Cummings, Jacob Abbot 1772-1820 Geog. Amer. M
Cummings' Ancient and modern geography Boston, 1813 and subs. edns. to *c.* 1821. Maps eng. by T. Wightman.
School atlas to Cummings' Ancient and modern geography . . . Boston, Cummings and Hilliard, 1818. Maps eng. by W. B. Annin, M. Butler.

Cummings and Hilliard *See under* CUMMINGS, J. A.

Cundee, J. *See under* EVANS, REV. J.

Cuningham, William 1531-86 Cart. Brit. M
Plan of Norwich, 1559.

Curry, William, and Co. fl. 1844 Pub. Irish M

Cushee, E. fl. 1729-60 Cart. Brit. M G

Cushee, L. 18th cent. Globe-maker Brit. G

Cushee, R. fl. 1729-32 Globe-maker
Brit. G (T)

Cusinus, Hugo } 16th cent. Cart. Dut.
Cousin, Hugues} M

Cuspinianus, Johann} 1473-1529 Cart.,
Spiesshaimer } Humanist, Diplomat,
Historian Austrian
M

Custos, David and Raphael 17th cent.
Eng. Austrian M

Cysat, Johann Leopold 17th cent. Cart.
Ger.? M

Czerny, Josef fl. 1838 Cart. G (T)

d'Abbeville *See under* LUYTS, J.

d'Ablancourt, Nicolas Perrot 1606-64
Cart. Dut. M
Vervolg van de Neptunus of zee atlas . . .
[Africa] Amster., P. Mortier, 1700.

Dahlberg, Erik Jönsson, Graf 1625-1703
Soldier, Cart. Swed. M
Map of Denmark.

d'Albe *See under* BOSSI, L.

Dalla Rovere *Same as* ROVERE (q.v.)

Dalli Sonetti, Bartolommeo 15th cent.
Geog. Ital. M
Isolario . . . (with 49 woodcut maps)
Venice before 1485.

Dall Olmo, Rocco fl. 1542 Cart. Ital. M
Portolans.

Dalrymple, Alexander 1737-1808 Pub.,
Cart. Hydrographer of the Navy.
Brit. M
Many charts and plans of all parts of the
world, especially the East Indies.
See also under JEFFERYS, T.

Dalton, William Hugh fl. 1794 Topg.
Brit. M

Damianus, Petrus *See under* SCHOONEBECK,
H.

Damme *Same as* VAN DAMME (q.v.)

Danckertz, Cornelis *See under* CASSINI,
G. D.; TAVERNIER, M.

Danckertz, Justus}
Dankers } 1635-1701 Pub. Dut.
Danquerts } M G (T)
Danckerts }

Atlas Amster., 1703? 1710? The maps are
variously signed by Cornelis, Johannes,
Justinus, Justus or Theodorus Danckerts.
Other names (engs. and carts.) on the
maps are (according to the edn.): A. Van
Luchtenburg, Frederici de Wit, A. Schut,
I. de Bróen, Caroli Allard, Abraham
Allard, Remmet Tennisse Backer, P. du
Val, Nicolas Witsen.
See also under DE WIT, F.; OTTENS, R.;
TASSIN, N.; VISSCHER, N., III.

Danckwerth fl. 1652 Pub. Ger. M
Pub. in 1652 *Atlas of Schleswig-Holstein* by
John Mejer.

Danckworth, F. *See under* TANNER, H. S.

Dandeleux, H. *See under* LORRAIN, A.

Danet, G. *See under* DE FER, N.; DESNOS,
L. C.

Daniell, John fl. 1612-42 Cart. Irish M.

Dannheimer, J. M. *See under* VÖLTER, D.

Danti, Ignazio }
    Egnatio } 1536-86 Astron.,
Pellegrino Danti} Cosmographer Ital.
    de Rinaldi} G (T) M

d'Anville, Jean Baptiste Bourguignon
1697-1782 Cart. Fr. M
Sometimes his name appears in partnership
with others: d'Anville and Jefferies;
d'Anville and Pownall; d'Anville and
Robert.
*Nouvel atlas de la Chine, de la Tartarie
Chinoise, et du Thibet* le Haye, H.
Scheurleer, 1737.
Atlas général 1740 and subs. edns.
Géographie ancienne abrégée various Fr. and
English edns., 1769-1820.
Atlas antiquus Danvillanus minor . . . Nurem-
berg, Schneideri-Weigeliana, 1798-9.
A complete body of ancient geography Lond.,
Sayer (1771?), Sayer and Bennett (1775)
and subs. edns.
A new map of north America Lond., R.
Sayer, 1763.
The western coast of Africa . . . Lond., Sayer,
1789.
North America . . . Lond., 1775.
Carte des isles de l'Amérique . . . Nuremberg,
Homann, 1740.

Atlas and geography of the ancients . . . Lond., T. Chaplin for J. Davis, 1815.

Atlas to the ancient geography . . . New York, R. M'Dermut and D. D. Arden, 1814. Maps eng. by W. Sim and A. Anderson.

An atlas of antient geography . . . Lond., R. H. Laurie, 1820-1.

See also under Arrowsmith, A.; Bossi, L.; Crevier, J. B.; de la Harpe, J. F.; Faden, W.; Gendron, P.; Homann's Heirs; Jefferys, T.; Julien, R. J.; Kitchin, T.; Laurie, R.; Santini, P.; Sayer, R.; Schraembl, F. A.; von Reilly, F. J. J.; Wyld, J.

Danz, C. F. 19th cent. Cart. Ger. M
Acht tafeln zur physisch-medicinischen topographie des kreises schmalkalden . . . (with 3 maps) (with Caspar Friedrich Fuchs) Marburg, N. G. Elwert, 1848.

d'Après de Mannevillette, Jean} 1707-80
　　Baptiste Nicolas Denis} Cart. Fr.
Manneville　　　　　　　} M
The oriental pilot Lond., Laurie and Whittle, 1797.
Instructions sur la navigation des Indes orientales et de la Chine . . . Paris, Dezauche, 1775.
The East-India pilot . . . Lond., Laurie and Whittle, 1795.
Le neptune oriental . . . Paris, J. F. Robustel, 1745 and various subs. edns.
Supplément au neptune oriental . . . Paris, Demonville; Brest, Malassis, 1781.
See also under Laurie, R.

Darby, John fl. 1677 Cart., Pub. Brit. M
Worked on Part III of John Seller's *English pilot*, and on his *Atlas maritimus*.

d'Aristotile, Nicolo 16th cent. Printer Ital. M
Printed Bordone's *Isolario*.

d'Armendale *See under* Tavernier, M.

Darmentier, L. I. *See under* de Fer, N.

Darmet *See under* Charle, J. B. L.

Darton, William}
Darton, William} fl. 1810-37 Geog., Pub.
**　　and Sons} Brit. M**
Darton's new miniature atlas Lond., 1820, 1825.

A complete atlas of the English counties . . . (with Thomas Dix) Lond., 1822.
Union atlas Lond., 1812.
See also under Miller, R.

Darton and Clark fl. 1842 Pub. Brit. M

Darton and Harvey *See under* Walker, J.

Darwin, Charles 1809-82 Naturalist Brit. M

Dasauville, William *See under* Thomson, J., and Co.

Dasypodius, Conrad} *c.* 1532-1600 Cart.
Hasenfratz, K.　　} Swiss G (C and T)

Daubree *See under* Boutruche, A.

Dauthendeij *See under* de Wit, F.

Dauty et Roret *See under* Charle, J. B. L.

d'Avellar, Andrea 1546-*c.* 1622 Cosmographer Port. M

Da Verrazano, Giovanni fl. 1525 Cart. Ital. G (T)

Da Verrazono, Girolamo fl. 1522-9 Cart. Ital. M G

David, Johann, the Elder 1796-1846 Cart., Lithographer Austrian G
See also under Meyer, J.

Davidson *See* Mount

Davidsz, Dirck *See under* Colom, J. A.

Davies, Benjamin Rees fl. 1845 Eng., Cart. Brit. M
Post office map of Warwickshire, 1845.
See under Bryant, A.

Da Vincio *Same as* Vincio (q.v.)

Davis, B. *See under* Scott, J.

Davis, J. *See under* d'Anville, J. B.

Davis, John *See under* Fischer, J.

Davis, W. T. *See under* Wilkinson, R.

d'Avity, Pierre 1573-1635 Cart. Fr. M
Neuwe Archontologia cosmica Frankfurt am Main, M. Merian, 1646.

Dawson, E. B. *See under* Tanner, H. S.

Dawson, Robert Kearsley, Lieut. (R.E.) fl. 1832 Surv. Brit. M

Dawson, William *See under* Prior, J.

Day, W., and Masters, C. H. fl. 1782 Carts. Brit. M
Map of Somersetshire, 1782 and subs. edns. (eng. P. Begbie).

'D.B.' Signature sometimes used by Donato Bertelli.

de Afferden, Francisco 1653–1709 Cart. Span. M

El atlas abreviado . . . Amberes, 1709, H. and C. Verdussen (with many maps by I. Peeters and map of America by I. P. Harrewyn); Francisco Laso, 1711. A reprint was issued in 1725 by the widow of H. Verdussen.

de Alliaco, Petrus, Cardinal 1350–1420
of Cambrai Geog. Fr.
d'Ailly, Pierre } M

Ymago mundi . . . Louvain, Johann de Westphalia, 1483 and other edns. Contains a woodcut map showing climatic zones.

de Almeda, Romaô Eloij 19th cent. Cart. Port. M

Carte militar das principaes estradas de Portugal Lisbon, 1808.

Dean, James 1777–1849 Cart. Amer. M
An alphabetical atlas, or gazetteer of Vermont . . . Montpelier, S. Goss, 1808.

de Antillon y Marzo, Isidoro 1778–1820 Cart. Span. M
[Atlas of the world] Madrid, 1804. Maps eng. by J. Morata, Cardono, F. Selma, P. Gangoiti. They are dated 1801 and 1802.

de Araos, Juan 18th cent. Cart. Span. M
Map of Cuba Havana, 1788.

d'Armendale *See under* TAVERNIER, M.

de Arnoldi, Arnoldo fl. 1600 Cart. Ital. M

de Azara, Félix 1746–1821 Cart. Fr. M
Voyages dans l'Amérique méridionale . . . Paris, Dentu, 1809.

de Bailly, Robertus fl. 1525–30 Cart. Fr. G (T)

de Bains *See* DE BEINS

de Barbuda, Luiz Jorge *Same as* GEORGIUS, LUDOVICUS (q.v.)

de Beaulieu, Sébastian de Pontault, Sieur 1613–74 Cart. Fr. M
Plans et profils . . . *des principales villes et places fortes de France* . . . Paris, Beaurain, 1694?.

de Beaumont, Elie *See under* ANDRIVEAU-GOUJON.

de Beaurain, Jean 1696–1772 Cart. Fr. M
Carte d'Allemagne . . . Paris, 1765.
See also under DE BEAULIEU, S. DE P.; DE GRIMOARD, P. H.; JEFFERYS, T.; JULIEN, R. J.

de Beeldesnijder, Jan Jansz 16th cent. Cart. Dut. M

de Beins, Jean } 17th cent. Cart. Fr. M
de Bains }
Delphinatus (i.e. Dauphiné) Amster., G. and J. Blaeu, c. 1660.
See also under HONDIUS, H.; MERCATOR, G.; TAVERNIER, M.

de Belli *Same as* DOMINICO MACHANEUS (q.v.)

Debes, Luis Jacobson fl. 1673 Cart. Dan. M

de Bougainville, Hyacinthe Yves Philippe Polentin, Baron 1781–1846 Navigator Fr. M
Journal de la navigation autour du globe . . . *atlas* Paris, A. Bertrand, 1837.

de Bouillon, Gilles Boileau 16th cent. Cart. Fr. M

de Bourgoing, Jean François, Baron 1750–1811 Cart. Fr. M
Atlas . . . *de l'Espagne moderne* Paris, Tourneisen, 1807.

de Brahm, John Gerar William b. 1717 Cart. Brit. M
Atlantic pilot Lond., T. Spilsbury, 1772.
Plan of Carolina.
See also under BULL, GASCOIGN; JEFFERYS, T.; SAYER, R.

Debrett, J. *See under* BOUGARD, R.

de Brion, Martin } fl. 1540 Cart. Fr. M
Bironius, Gallus }

de Bróen, Johannes *See under* DANCKERTZ J.; DE WIT, F.; DE LA FEUILLE, J.; OTTENS, R.

de Brosses, Charles *See under* DE VAUGONDY, G. R.

de Bry, Theodore 1527 or 8–98 Goldsmith, Eng., Pub. Ger. M
See also under DE HERRERA Y TORDESILLAS, A.; WAGHENAER, L. J.; WHITE, J.

de Bylandt Palstercamp, Comte A. 19th cent. Geog. Fr. M
Théorie des volcans . . . (with 13 maps) Paris, F. G. Levrault, 1836.

de Caroly, Francis *See under* LAURIE, R.

de Cassine, Giovanni Battista fl. 1700 Cart., Friar Ital. G

de Castro, João 1500-48 Cart. Port. M
Rotiero de Goa a Dio.

de Caylus *See under* JEFFERYS, T.; JULIEN, R. J.

de Champlain, Samuel 1567-1635 Cart. Fr. M
Les voyages de sieur de Champlain . . . Paris, 1613 (with map eng. by David Pelletier).

de Chattilion *See under* TAVERNIER, M

de Chaves, Alonzo 16th cent. Cart. Span. M

de Chaves, Géronimo b. 1524 Cart. Span. M
Son of Alonzo de C.

de Chieze, Jacques 17th cent. Cart. Fr. M
La principalité d'Orange . . . Amster., Jansson, *c.* 1650.

Decker, G., and Co. *See under* PLATER, S.

de Classun *See under* TAVERNIER, M

de Coimbra, Heitor fl. 1524 Cart. Port. M

de Cristofano, Girolamo 16th cent. Cart. Ital. M

de Croock, Hubert} fl. 1529 Eng. Dut.
de Croc   } M

de Divinis, Eustachius fl. 1649 Astron. Ital. X

de Fer, Antoine fl. 1645-6 Mapseller Fr. M

de Fer, Nicolas 1646-1720 Pub., Geog., Eng. Fr. M
Very prolific.
Maps of Fr. Provinces.
La France triomphante sous la règne de Louis le grand 1693.
Les forces de l'Europe ou descriptions des principales villes. 8 parts with separate title pages. Various edns., 1693-7. Other issues by Mortier (1702) and van der Aa (1726).

Plusieurs cartes de France avec les routes et le plan des principales villes. Various edns., 1698-1763.
Atlas curieux 1700-5. Most maps are eng. by Carolus Inselin, but no. X is eng. by van Loon.
Cartes et descriptions générales et particulières au sujet de la succession de la corunne d'Espagne 1701-2.
Petit et nouveau atlas 1705.
Les postes de France et d'Italie 1700, 1728, 1760.
Le théâtre de la guerre dessus et aux environs du Rhein 1705. Maps eng. by H. van Loon, C. Inselin, A. Coquart, P. Starckman, L. G. Begule.
Introduction à la géographie . . . (with 6 maps) Paris, Denet, 1717-40?.
Atlas ou recüeil de cartes géographiques . . . Paris, 1709-28, 1746-53. Carts. and eng. incl., according to edn.: H. van Loon, N. Guerard, P. Starckman, Carolus Inselin, F. de la Pointe, Jacqueline Panouse, L. I. Darmentier, des Rosiers, A. Coquart, Guillaume Delahaye, P. de Rochefort. Some maps pub. after de Fer's death have also the imprint of J. F. Benard or Sr. G. Danet, and some have the imprint of J. B. Nolin, Dezauche or Crepy.
Atlas royal . . . Paris, 1699-1702. Carts. incl.: H. Jaillot, J. Goeree, Sr. Vaulthier, Sanson, de Vauban. Engs. incl. H. van Loon, K. Huyberts.
See also under BLAEU, G.; DESNOS, L. C.; HONDIUS, H.; JEFFERYS, T.; MERCATOR, G.; MORTIER, P.; SANSON, N., PÈRE; VALCK, G.

de Ferraris, Comte 18th cent. Cart. Dut. M
A map of the frontiers of the emperor and the Dutch . . . Lond., Faden, 1789.
Carte chorographique des Pays-Bas . . . Eng. by L. A. Dupuis. n.p., 1777.

de Fleurieu *See under* LAURIE, R.

de Freycinet, Louis Claude Desaules 1779-1842 Geog. Fr. M
Voyage autour du monde . . . atlas Paris, Pillet, 1826. The maps are by Mareau, Labiche et Bérard, L. I. Duperrey.

Voyages de découvertes aux terres australes . . . *atlas* Paris, 1807-16, 1812-15. Maps are by various hydrographers, according to the edn., and incl. Beautemps-Beaupré, Boullanger, Faure, Heirisson, Lesueur, Montbazin, Péron, Ransonnet and Ronsard.

de' Gianuzzi, Guilo di Pietro} 1499-1546
Pippi, Giulio } Architect,
Romano } Painter Ital.
G

Degli Rubertis, Luc' Antonio} fl. 1525
de Huberti, Luca Antonio } Cart. Ital.
M

de Gouvion Saint-Cyr, Laurent, Marquis
1764-1830 Soldier Fr. M
Atlas des cartes et plans relatifs aux campagnes du maréchal Gouvion St. Cyr . . . Paris, 1828. Printed by Sampier; eng. by Hacq et Warin.
Atlas des mémoires pour servir à l'histoire militaire sous le directoire, le consulat et l'empire Paris, Anselin, 1831.

de Graff, Isaac fl. 1705 Cart. Dut. M
Cartographer to the East India Co.

de Grimoard, Philippe Henri, Comte
1753-1815 Geog. Fr. M
Atlas to accompany Histoire des . . . *campagnes du maréchal de Turenne* Paris, Beaurain, 1782.

de Groot, J., and Warnars, G. *See under* TITION, T.

de Harme, L. F. 18th cent. Eng., Cart. Fr. M
Plan de la ville et fauxbourgs de Paris . . . Paris, 1766.

de Herrera y Tordesillas, Antoine 1559-1625 Cart. Span. M
Description des Indes occidentales Madrid, 1601; Amster., Emanuel Colin; Paris, Michel Soly, 1622; Madrid, N. R. Franco, 1726.
Historia general de los hechos de los castellanos en las islas i tierra firme del mar oceano . . . Madrid, N. R. Franco, 1726-7.
Nievve werelt, anders ghenaempt VVest-Indien Amster., M. Colijn, 1622.

Novvs orbis, sive descriptio Indiae occidentalis . . . Amster., M. Colinium, 1622.
Novi orbis pars duodecima . . . Frankfurt, Theodori de Bry, 1624.
See also under JEFFERYS, T.

de Hondt *Same as* HONDIUS (q.v.)

de Hooge, Cornelius} d. 1583 Eng. Dut.
Hogius } M
Worked for SAXTON (q.v.).

de Hooge, Romain} ?1646-1708 Cart.
Hogius } Dut. M
Atlas maritime. Amster., 1693.
Zee atlas tot het gebruik van de vlooten des konings van Groot Britanje. Amster., D. Mortier, 1694.
See also under BEEK, A.; VISSCHER, N., II AND III.

de Huberti, L. A. *Same as* DEGLI RUBERTI L. A. (q.v.)

de Jode, Cornelis 1568-1600
See under DE JODE, G.

de Jode, Gerard }
de Iudaeis } 1509-91 Cart.,
de Iudoeis, Gerardi} Pub. Dut. M
Judaeus }
Copy of Gastaldi's oval world map (1555).
Speculum orbis terrarum. Antwerp, 1578 (printed by Gerard Smits), 1593-1613 (with Cornelis de Jode). When reissued by Cornelis de J. in 1593, the title was *Speculum orbis terrae.*
See also under ALGOET, L.

de Jomini, Antoine Henri, Baron 1779-1869 Cart. Fr. M
Atlas portatif . . . Paris, Anselin, 1840?
Atlas pour servir l'intelligence de l'histoire critique et militaire des guerres de la révolution . . . Brussels, J. B. Petit, 1840, 1841.

de Jonghe, Clemendt 17th cent. Cart. Dut. M
Europae nova discriptio Amster., 1661.
Silesiae ducatus Amster., n.d.
XVII provinciarum inferioris Germaniae n.p., n.d. and many other maps.

de la Bella *Same as* DOMENICO MACHANEUS (q.v.)

**de Laborde, Alexandre Louis Joseph,
Comte** 1773-1842 Cart. Fr. M
Atlas de l'itinéraire descriptif de l'Espagne . . .
Paris, H. Nicolle, 1809.
Atlas del itinerario descriptivo de España
Valencia, J. Ferrer de Orga, 1826 (2nd
edn.).

**de Laborde, Léon Emmanuel Simon
Joseph, Marquis** 1807-69 Historian
Fr. M
*Commentaire géographique sur l'Exode et les
Nombres* . . . Paris, Leipzig, J. Renouard et
Cie, 1841 (with 19 maps).

de la Caille, L'Abbé⎫ 1713-62 Astron.,
Nicolas Louis⎭ Mathematician Fr.
de Lacaille ⎫X G
de Laet, Joannes 1593-1649 Geog. Dut.
M
Nieuvve Wereldt . . . Leyden, I. Elzevir,
1625. With 10 maps by Hessel Gerritsz.
Latin edn., 1633; French edn., 1640, each
with 14 maps.
Beschrijvinghe van West-Indien . . . Leyden,
Elzevir. With 14 maps by Gerritsz.
*Historie ofter Iaerlijck verhael van de verrichting-
hen der geoctroyeerde West-Indische com-
pagnie.* Leyden, Elzevir, 1644. With 14
maps.

de la Feuille, Daniel 18th cent. Cart. Fr.
M
Atlas portatif . . . Amster., 1706-8. Carts.
incl. Paul de la Feuille and N. Sanson.
See also under BEEK, A.

de la Feuille, Jacobo b. 1668 Cart. Fr.
M
Atlas . . . Amster., 171[?] Carts. incl.: F.
de Wit, N. Visscher III, Mr. Samson [*sic*],
Ioannem de Ram, Ioannem Blaeu, Henrici
Hondÿ, Gerardo à Schagen, Mosem Pitt
et Stephanum Swart, Iohannes Andreas
Rauchhen, Gerard Mercator, Carolus
Allard, Joh. et Cornel, Blaeu, Ioannem
Ianssonium. Engs. incl.: I. de Bróen, G. v.
Gouwen, A. Goos.
See also under DESNOS, L. C.; DE WIT, F.

de la Feuille, Paul 18th cent. Cart. Fr.
M
Geographisch-toneel . . . Amster., J. Ratel-
band, 1732.
Les tablettes guerrières Amster., 1708, 1717.

The Millitary [sic] *Tablettes* . . . Amster.,
1707. English edn. of the foregoing,
taken from Daniel de la Feuille's *Atlas
portatif.* There were many other edns.
in Fr. and Dut.
See also under DE LA FEUILLE, D.

de la Fosse, J. B. 18th cent. Cart. Fr. M
Carte du comte de Flandre . . . Paris
Mondhare, 1780.
See also under NOLIN, J. B.

Delagrive, Le Sieur *See under* DESNOS,
L. C.

de la Guillotière, François fl. 1612-13
Cart. Fr. M

de la Harpe, Jean François 1739-1803
Cart. Fr. M
Abrégé de l'histoire générale des voyages . . .
Paris, E. Ledoux, 1780, 1820. Carts.
incl. Lieut. J. Cook, Laurent et Bellin,
and maps by d'Anville, Delisle and others
are reproduced.

de la Haye, Jean fl. 1602 Geog., Cart.
Fr. M
See also under DE FER, N.; LANGENES, B.

de la Hire, Philippe 1640-1718 Editor,
Compiler Fr. M G

de la Houue, Paul fl. 1620 Cart. Dut. M

de Lalin *See under* BRION DE LA TOUR, L.

Delamarche, Alexandre 1815-84 Cart.
Fr. M
*Atlas de la géographie ancienne, du moyen âge,
et moderne* . . . Paris, A. Grosselin et Cie,
1850.

Delamarche, Charles François 1740-1817
Cart. Fr. M G (C and T) O
*Institutions géographiques, ou description
générale du globe terrestre* . . . Paris,
Bellin, 1795. With world map eng. by
P. J. Picquet.
See also under DE VAUGONDY, G. R.; FORTIN,
J.; JULIEN, R. J.; LATTRÉ, J.

Delamarche, Félix 19th cent. Pub. Fr.
M
*Atlas de la géographie ancienne, du moyen âge,
et moderne* . . . Paris, 1820 (with C. Dien,
1827 (cf. with atlas of same title under
Alexandre Delamarche). Some maps by
Arrowsmith. Engs.: Barrière Frères,
Blondeau, Pélicier, Semen Jeune.

de Lannoy, Ferdinand 1542-79 Engineer, Cart. Fr. M

de la Paz, Principe *See under* TOFIÑO DE SAN MIGUEL, V.

de Lapérouse, Jean François de Galaup 1741-88 Explorer Fr. M
Voyage de la Pérouse autour du monde . . . Atlas 1797. Eng. by Ph. Triere. English edn., Lond., 1799.

de la Pointe, F. D. *See under* BLAEU, G.; DE FER, N.; DESNOS, L. C.

de Laporte, Joseph 1713-79 Cart., Pub. Fr. M
Atlas moderne portatif . . . Paris, 1781.
Atlas ou collection de cartes géographiques . . . Paris, Moutard, 1787.
See also under MALTE-BRUN, C.

de la Rochette, Louis Stanislas d'Arcy fl. 1795 Cart. Fr. M
The Dutch colony of the Cape of Good Hope Lond., Faden, 1782. Altered and reissued by James Wyld, 1782.
See also under FADEN, W.; JEFFERYS, T.; PALAIRET, J.; WYLD, J.

de la Rue, Philippe 17th cent. Cart. Fr. M
See also under LE CLERC, J.

de las Cases, Emmanuel }
(Marie Joseph Auguste } 1766-1842 Cart.
Emmanuel Dieu-Donné, } Fr. M
Comte de las Cases)
Also known as A. LE SAGE.
Atlante storico, geografico . . . Florence, Molini, Landi e Co., 1813-14.
Many other edns. in Ital., Fr. (*Atlas historique . . .*), Ger. (*Historisch-genealogisch-geographischer atlas*), English (*Le Sage's Historical atlas*).

De Lat, Jan 18th cent. Cart. Dut. M
Atlas portatif très exact . . . Deventer; Almelo, J. Keyser; 1747. Maps eng. by de Lat and Keyser after Delisle, M. Hasius and P. Lucas.
See also under SCHENK, P.

de la Tour 18th cent. Cart. Fr.? M
Plan of New Orleans Lond., Jefferys, 1759.
See also under DESNOS, L. C.

Delaune, J. fl. 1813 Astron. Fr. X

del Cano, Juan Sebastian fl. 1520-6 Cart. Span. M G (T)
Portolans.

del Dolfinado, Nicollo *Same as* DE NICOLAY (q.v.)

de l'Escluse, Charles *Same as* CLUSIUS, CAROLUS (q.v.)

de Leth, Hendrik fl. 1730 Eng., Cart., Pub. Dut. M
Plan of Amsterdam. Covens and Mortier, Amster., c. 1720.
Nieuwe geographische Nederlandische reise- en zak-atlas . . . Amster., c. 1740; J. C. Sepp, 1773. Some maps signed: C. et I. C. Sepp.

de Leucho, Jacobus Pentius *See under* PTOLEMAEUS, C.

de l'Isle, Claude 1644-1720 Geog. Fr. M
Father of Guillaume D.

Delisle, Guillaume }
de L'Isle } 1675-1726 Cart. Geog.
del Isle } Fr. M G (C and T)
Son of Claude D.
Atlas de géographie Paris, 1700-12.
Atlas géographique des quatre parties du monde (with P. Buache) Paris, Dezauche, 1769-99, 1789?, 1831? and other edns. Maps by the following carts. and engs. according to the edn.: des Bruslins, C. Simmoneau, des Rosiers, Berey, M. F. du Val, Jaillot, Denis, de Vaugondy and others.
Atlas géographique et universel . . . 2 vols. Paris, 1700-62; 1781-4, Dezauche; 1789-90 (ditto), and other edns. Maps by various carts., incl.: Jaillot, Le Rouge, Placide, de Vaugondy, Sanson, etc.
Atlante novissimo . . . 2 vols. Venice, 1740-50.
Atlas nouveau . . . Amster., Covens and Mortier, 1730, 1733, 1741?.
See also under BLAEU, G.; BUACHE, PHILIPPE; DE LA HARPE, J. F.; DE LAT, J.; HOMANN'S HEIRS; HUSSON, P.; JEFFERYS, T.; JULIEN, R. J.; KÖHLER, J. D.; LE CLERC, J.; LOTTER, T. C.; OTTENS, R.; RENARD, L.; VALCK, G.; VISSCHER, N., III.

Delisle, Joseph Nicolas 1688-1768 Cart., Astron. Fr. M
Atlas russicus . . . St. Petersburg, 1745. Some maps are drawn by I. Grimel, some eng. by G. I. Unvertzagt. The atlas exists also with Russ. text (same date, etc.) and Ger. text (ditto).

Delitsch, Otto 19th cent. Globe-maker Ger. G (T)

Delkeskamp, F. W. 19th cent. Cart. Ger. M
Panorama of the Rhine Lond., Samuel Leigh, 1830.

della Gatta, Johannes Franciscus 16th cent. Cart. Ital. M

de Lovenhorn, P. 18th/19th cent. Cart. Dan. M
Sailing instructions for the Kattegat Copenhagen, J. F. Schultz, 1800.

de Marchais, R.}
de Mareechais } *See under* Jefferys, T.

de Marre, Jan 1696-1763 Cart. Dut. M
See also under Van Keulen, Johannes, III.

de Medina, Pedro fl. 1540 Navigator Span. M
Arte de navegar . . . (with map of the New World) 1545.

de Mendoza, Diego Huardo fl. 1544 Cart. Port. M

de Mirici, Gaspard fl. 1537 Cart. Flemish G

de Mongenet, François d. *c.* 1592 Cart. Physician, Mathematician Fr. G (C and T)
Worked with Chassignet, D. (q.v.).

Demonville *See* D'Après de Mannevillette, J.

de Moussy, V. Martin 18th cent. Cart. Fr. M
Description géographique . . . de la confédération Argentine 3 vols. Paris, 1860-73.

de Musis, Julius 16th cent. Eng. Ital. M

de Nezon *See under* Julien, R. J.

Dengelsted *See under* Schenk, P.

de Nicolay, Nicolas } 1517- *c.* 1583 Cart.
del Dolfinado, Nicollo} Fr. M

Denis, Louis fl. 1785 Cart. Fr. M
Atlas géographique . . . Paris, Des Vents de la Doué, 1764?.
Empire des solipses . . . Paris, 1764.
Plan topographique . . . *de Paris* (with J. J. Pasquier) Paris, 1758.
See also under Delisle, G.

de Nobilibus, Petrus fl. 1560-79 Eng. Ital. M

de Nodal, Bartolomé García and Gonçalo fl. early 17th cent. Navigators Span. M
Relacion del viage . . . des Indias Madrid 1621 (with map of Tierra del Fuego).

de Norroy, Sieur *Same as* Du Pinet (q.v.)

Dentu *See under* de Azara, Félix; Marshall, J.

Denys, William *See under* Stephenson, J.

de Palestrina, Salvat } fl. 1503-11 Cart.
de Pilestrina, Salvatore} Ital. M

de Plaes, A.}
Plaets } *See under* Sanson, N., père
Plaetz }

Depping, G. B. 19th cent. Cart. Fr. M
L'Angleterre . . . 6 vols. Paris, 1828. Drawn by A. M. Perrot; eng. by M. Migneret.

de Prony, Gaspard Clair François Marie Riche, Baron 1755-1839 Cart. Fr. M
Atlas des marais Pontins . . . Paris, Firmin Didot, 1823.

de Ram, Joannem} 17th cent. Pub. Dut.
    Joannes} M G (T)
See also under de la Feuille, J.; Ottens, R.; Seller, J.; van Keulen, Joannes, II; Wolfgang, A.

de Rapin-Thoyras, Paul 1661-1725 Historian Brit. M
Atlas to accompany Rapin's History of England . . . Lond., J. Harrison, 1784-9. Carts. incl. J. Haywood. Eng. incl.: G. Allen, T. Bowen, Simpkins, E. Sudlow, G. Terry.

de Rinaldi *Same as* Rinaldi (q.v.)

de Rocheford, Jouvin} fl. 1675 Cart. Fr.
de Rochefort } M
See also under de Fer, N.; Desnos, L. C.

de Rossi, Domenico} fl. 1695 Cart. Ital.
de Rubeis } M G (C and T)
See also under DE ROSSI, G. G.

de Rossi, Giovanni Giacomo fl. 1674
 Cart., Pub. Ital. M
Mercurio geografico . . . Rome, 1685?, 1692-
 4, 1692-1714.
 Incl. maps by Guglielmo Sanson, Agostino
 Lubin, Giacomo Ameti, Jacobius Cantel-
 lius, Domenico de Rossi, Campiglia,
 according to edn.

de Rossi, Giuseppe fl. 1615 Cart. Ital.
 G (C and T)

de Rubeis *Same as* DE ROSSI *and* ROSSI (qq.v.)

de Rueda, Manuel 18th cent. Cart.
 Span. M
Atlas Americano . . . Havana, 1766.

de Ruyter, Gerard *See under* LAURIE, R.

de Sanctis, Gabriello 19th cent. Cart.
 Ital. M
Atlante corografico del regno delle due Sicilie . .
 Naples, 1843.

de Sandrart *Same as* SANDRART (q.v.)

**de Santarem, Manuel Francisco de Barrios
 e Sousa, de Mesquita de Macedo
 Leitãoe, Visconte** 1791-1865 Geog.
 Port. M
*Essai sur l'histoire de la cosmographie et de la
 cartographie pendant le moyen-âge* . . . *atlas*
 Paris, Maulde et Renou.
 Contains reproductions of many ancient
 maps, 1849-52.

des Barres, J. F. W. 18th cent. Pub. Surv.
 Cart. Brit. M
The harbour of Halifax, Nova Scotia Lond.
 1781.
The Atlantic neptune 4 vols. Lond., 1779.
A chart of the harbour of Rhode island . . .
 Lond., 1776.
See also under BLUNT, E. AND G. W.; LE
 ROUGE, G. L.

des Bruslins *See under* DELISLE, G.; DESNOS,
 L. C.; GENDRON, P.

Desceliers, Pierre 1487-1553 Cart. Fr.
 M

Descharles, Claude François Milliet 1621-
 78 Philosopher, Mathematician Fr. G

de Ségur, Louis Philippe, Comte 1753-
 1830 Cart. Fr. M
Atlas pour l'histoire universelle . . . Paris, A.
 Eymery, 1822; Furne et Cie, 1840.
*Atlas pour servir à l'histoire ancienne, romaine et
 du Bas-Empire* . . . Paris, A. Eymery, 1827.

Desliens, Nicolas 16th cent. Cart. Fr.
 M

Desmarest, Nicolas 1725-1805 Cart. Fr.
 M
Atlas encyclopédique . . . (with J. B. G. M.
 Bory de St. Vincent) Paris, H. Agasse,
 1827.
See also under BONNE, R.

Desnos, L. C. fl. 1750 Cart. Dan. M
 X G (C and T)
Almanach géographique . . . Paris, 1770.
Atlas chorographique . . . Paris, Savoye;
 Despilly, 1763-6.
Nouvel itinéraire générale . . . Paris, 1768.
 Some maps are by Brion de la Tour,
 Michel et Rizzi-Zannoni. Some are eng.
 by P. Starckman.
Nouvel atlas d'Angleterre . . . Paris, 1767.
Tableau analytique de la France . . . Paris,
 1766. Incl. maps by Michel and Rizzi-
 Zannoni.
*Atlas général, contenant le detail des quatre
 parties du monde* . . . Paris, 1767-9,
 1790-2.
 Carts. incl.: N. de Fer, L'Abbé Delagrive,
 I. B. Nolin, Delafosse, Jouuin de Rochefort,
 Le Rouge, G. Danet, Gaspard Baillieu,
 Hertiers d'Homann, Brion, de Tillemont,
 Bonne, S. G. Longchamps, Hubert Jaillot,
 Sanson, Rizzi-Zannoni, Le Sieur Moithey.
 Engs. incl.: H. van Loon, Chambon,
 Herriset, C. Inselin P. Starckman, F. D.
 La Pointe, des Bruslins, L. Cordier, T.
 Rousseau, Croisey, Jenvilliers, J. B.
 Liébaux, Dussy, Vallet, Perrier, Niquet.
See also under BELLIN, J.; BRION DE LA TOUR,
 L.; BUY DE MORNAS, C.; GOURNÉ, P. M.;
 RIZZI-ZANNONI, G. A.; ROCQUE, J.;
 SENEX, J.

Despilly *See under* DESNOS, L. C.

Desray *See under* BOISTE, P. C. V.

des Rosiers *See under* BLAEU, G.; DE FER,
 N.; DELISLE, G.

Dessiou, Joseph Foss} early 19th cent. Pub.
    I. Foss   } Fr. M
   A general chart of the West Indies . . . Lond.,
    W. Faden, 1808.
   Chart of the Adriatic sea . . . Lond., Faden,
    1806.
   A new general chart of the Mediterranean sea . . .
    Lond., Faden, 1839.
   See also under HEATHER, W.

de Stuers, François Vincent-Henri
    Antoine, Ridder 1792-1881 Explorer
    Fr. M
   Mémoires sur la guerre de l'île de Java de 1825 à
    1830 . . . *atlas* Leyden, S. and J. Lucht-
    mans, 1833.

de Templeux, Damien, Sieur du Frestoy
   See under BLAEU, G.; JANSSON, J.; VISSCHER,
    N., III

de Tillemont *See* TILLEMONT

Deuchino, E. *See under* ROSACCIO, G.

Deuer, Abraham Jansz 1666-1714 Eng.
    Dut. G
   See also under DE WIT, F.

Deur, Johannes fl. 1725 Eng. Dut. G
    (C and T)

Deutecum *Same as* DOETECUM

Deuvez, Arnold fl. 1693 Painter Fr. G

de Vallemont, Pierre le Lorrain, Abbé
    1649-1721 Cart. Fr. M
   Atlante portatile . . . Venice, G. Albrizzi,
    1748. Eng. incl. F. Polanzani.

de Vauban *See under* DE FER, N.

de Vaugondy, Didier Robert 1723-86
    Cart. Fr. M G (C and T)
   Partie de l'Amérique septentrional . . . Lond.,
    1761.
   See also under DE VAUGONDY, G. R.

de Vaugondy, Gilles Robert 1688-1766
    Cart. Fr. M G (C and T)
   Father of Didier Robert de V.
   Also called le Sieur Robert or Monsieur
    Robert.
   Spanish edns. of maps sometimes signed Sr.
    Robert or Sr. Roberto.
   Atlas portatif . . . Paris, Durand; Pissot,
    1748-9. Maps signed: Robert de Vau-
    gondy *fils*.
   Nouvel atlas portatif . . . Paris, 1762; Fortin,
    1778; Delamarche, 1784, 1794-1806.
    Many of the maps eng. by E. Dussy.

   Tablettes parisiennes . . . (incl. 3 maps)
    Paris, 1760.
   Atlas universel . . . (with D. Robert de
    Vaugondy) Paris, Boudet, 1757-8, 1757-
    86, 1783-99; Delamarche, 1793?. Some
    edns. incl. maps by Delamarche (sometimes
    called 'Lamarche'), Josué Fry et Pierre
    Jefferson.
   Maps of Pacific Ocean in Charles de Brosses's
    Histoire des navigations aux terres australes
    2 vols. Paris, 1756.
   See also under DELISLE, G.; GENDRON, P.;
    JULIEN, R. J.; KITCHIN, T.; SCHRAEMBL,
    F. A.; VON REILLY, F. J. J.

Deventer *Same as* VAN DEVENTER (q.v.)

de Villaroel, Domingo fl. 1589 Cart.
    Span.? M
   Portolans.

Devine, Thomas 19th cent. Cart.
    Canadian M
   Government map of Canada Quebec and
    Ontario, 1859.

de Visscher *Same as* VISSCHER (q.v.)

de Vries, Claas} *See under* LOOTS, J.; VAN
    Nicolas} KEULEN, JOANNES, II; VOOGT,
    C. J.

de Wale, Peter fl. 1520 Printer Dut.
    M
   Printed P. Apian's world map.

de Westphalia, Johann *See under* DE
    ALLIACO, P.

de Winter, A. *See under* DU SAUZET, H.

de Wit, Frederick} 1616-98 Pub., Cart.
    Witt   } Dut. M
   Zee karten Amster., 1675.
   Plan of Amsterdam.
   Nieut kaert-boeck Amster., c. 1720.
   Ducatum Livoniae et Curlandia Amster., c.
    1690.
   Regni Norvegia Amster., Covens and Mor-
    tier, c. 1744.
   Accuratissima principatus Cataloniae Amster.,
    c. 1688.
   Regni Suecial . . . Amster., Covens and
    Mortier, c. 1730.
   Atlas & zee-karten Amster., c. 1660.
   Urbis Moskvae Amster., c. 1660.
   Atlas Amster., c. 1690.

Carts. incl.: Allard, Blaeu, Dauthendeij, Hogenboom, Hondius, Janssen, Labanna, Thulier, Sweerts, Ten-Have, N. Visscher, II and III.

Atlas major . . . Amster., 1688?, 1695?, 1706?, 1707?.

Carts. incl., according to edn.: Allard, Christopher Browne, Danckertz, Norden, Jaillot, Jansson, Loon, Moll, Sanson, Schotanus à Sterringa, Valk, Visscher.

Atlas minor . . . 2 vols. Amster., 1634–1708.

Carts. and engs. incl.: C. Allard, Petri Schenk, N. Visscher, Ph. Tideman, G. van Gouwen, Guil le Vasseur, du Beauplan, G. Blaeu, J. Blaeu, G. Valk, H. Jaillot, Gerhard Mercator, Jacobo de la Feuille, Gerard Coeck, Joh. Jac. Stetter, Sanson, Abram Deur, L. von Anse, Joannes de Bróen.

Theatrum iconographicum omnium urbium et praecipuorum oppidorum Belgicarum XVII provinciarum per accurate delineatarum Amster., 1680?.

Theatrum praecipuorum totius Europae urbium . . . Amster., 1695?.

See also under Beek, A.; Braakman, A.; Danckerts, J.; de la Feuille, J.; Husson, P.; Jansson, J; Julien, R. J.; Ottens, R.; Renard, L.; Seller, John; Valck, G.; Visscher, N., III; Visscher, N., II; Wolfgang, A.

Dezauche fl. 1805 Pub. Fr. M
Pub. D'Après de Mannevillette's *Instruction sur la navigation des Indes* . . . Paris, 1775.
See also under de Fer, N.; Delisle, G.

Dheulland, Guillaume 1700–70 Pub., Cart. Fr. M
Carte nouvelle du duché de Brabant . . . Paris, 1747.
See also under Cassini de Thury, C. F.; Julien, R. J.

di Arnoldi *See* de Arnoldi

Diaz, Emmanuel 1574–1659 Missionary Port. G

di Bindoni, Augustino and Francesco 16th cent. Printers Ital. M

Dicey, Cluer fl. 1764–70 Printer, Pub. Brit. M
Pub. an edn. of Speed's maps *c.* 1770 with imprint: Dicey & Co.

See also under Saxton, C.

Dick, Thomas fl. 1840 Astron. Brit. X

Dickinson, J. 18th cent. Cart. Brit. M
Map of the south part of Yorkshire, 1750 (eng. R. Parr).

Dickson, J., and Vietch, J. 18th cent. Pub. Brit. M

Didot (Firmin Didot) *See under* de Prony, G. C. F. M. R.; Lemau de la Jaisse, P.; Letronne, A. J.

Dien, Charles or Karolus fl. 1841 Astron., Pub. Fr. M X G (C)
See also under Delamarche, F.

Diesth *See under* Ortel, A.

di li Sonetti, Bartolome fl. 1547 Cart. Ital. M

Dilly, C. *See under* Guthrie, W.

di Re, Sebastianus } fl. 1556–61 Eng.
Sebastianus à Regibus } Ital. M
    Clodiensis }
Pub. an edn. of George Lily's map of the British Isles.

di Rossi, Giuseppe } fl. *c.* 1615 Globe-
Rubeis Mediolanensis } maker Ital. G

Dirwaldt, Joseph 19th cent. Cart. Austrian M
Allgemeiner hand-atlas . . . Vienna, T. Mollin, 1816. Maps have imprint: Wien, bey Tranquillo Mollo.

Discepoli *See under* Rosaccio, G.

di Schmettau *Same as* von Schmettau (q.v.)

di Stefano, Alberto fl. 1645–50 Cart. Ital. M

di Vavassore, Giovanni } 16th cent. Printer,
    Andrea } Pub., Cart. Ital.
Valvassor } M
Vadagnino }
Woodcut map of Italy, Spain and British Isles, *c.* 1530–50.
See also under Köhler, J. D.

Divinis *Same as* de Divinis (q.v.)

Dix, Thomas fl. 1799–1821 Surv., Geog. Brit. M
A complete atlas of the English counties . . . (with William Darton) Lond., 1822.
A new map of the county of York . . . 1820.
The county of York, divided into its ridings . . . 1835.

Dixon, Jeremiah *See under* MASON, CHARLES

Djurberg *See under* SCHRAEMBL, F. A.

Djurberg and Roberts *See under* VON REILLY, F. J. J.

Dobaldo, J. *See under* ASCENSIO, J.

Dobree, Capt. *See under* STEPHENSON, J.

Dodsley, Robert 1703-64 Pub. Brit. M

Collaborated with John Cowley in *Geography of England*, 1744, and *New sett of pocket maps of all the counties of England and Wales*, 1745, actually a reprint of the first-named. Also pub. with J. Dodsley, Thomas Kitchin's *England illustrated*, 1764.

Doedtszoon, Cornelis} fl. 1600-7 Cart.
Doedes van Edam   **}** Dut. M

Doesburgh, T. *See under* ALLARD, K.

Doet *Same as* VAN DOET (q.v.)

Doetecum, Baptista} fl. 1592 Eng. Dut.
Deutecum   **}** M

Doetecum, Johannes
   or Jan and Lucas} fl. 1562 Eng.
Deuticum   **}** Dut. M
See also under WAGHENAER, L. J.

Doetz, Cornelius fl. 1589 Cart. Dut. M

d'Ogerolles, I. *See under* DU PINET, A.

Dolfinado *Same as* DE NICOLAY (q.v.)

Domenico Machaneus}
de la Bella   **}** fl. 1546 Cart.
de Belli   **}** Ital. M
Bellius   **}**

Dommerich, Johann Christoph 1725-67
Scholar Ger. G

Donald, Thomas fl. 1774–1802 Cart. Brit. M
A topographical map of the county of Norfolk Lond., Faden, 1797 (with Thomas Milne).
Map of Cumberland, 1774 (eng. T. Hodgkinson) and subs. edns.

Donauer, Johann}
Tonaver   **}** *c.* 1521-96 Painter,
Thonawer   **}** Architect Ger. G

Doncker, Hendrick 17th cent. Cart. Dut. M
De zee-atlas ofte water-woereld . . . Amster., 1600-61, 1665, 1666.

The lightning columne, or sea-mirrour . . . (with T. Jacobsz and H. Goos) Amster., C. Loots-Man, 1689-92. Signed maps are by Theunis Jacobsz, Casparus Loots-Man, Jacob Theunisz, Jacobus and Casparus Loots-Man.
See also under ROBIJN, J.

Donn, Benjamin fl. 1765 Cart. Brit. M
Devon Lond., 1765 (also known on vellum). Eng. by Thomas Jefferys. A reduction was pub. 1799 by W. Faden; eng. B. Baker.
Gloucestershire 1769 Eng. by R. Coffin. Reduction pub., 1778.

Doolittle, Amos *See under* CAREY, M.; GUTHRIE, W.

Doppelmayr, Johann} 1671-1750 Cart.,
      Gabriel} Physicist Ger. X
Doppelmaier   **}** G (C and T)
Worked with J. G. PUSCHNER (q.v.)

Dorret, James fl. 1750-61 Cart. Brit. M
See also under PALAIRET, J.; SCHRAEMBL, F. A.; VON REILLY, F. J. J.

Doué, Martino *See under* BLAEU, G.

Dourado, Fernão Vaz} 16th cent. Cart.
Vaz Durado   **}** Port. M

Dower, John fl. 1838 Cart., Eng. Brit. M
A new general atlas of the world . . . Lond., H. Teesdale and Co., 1838, 1842.
See also under PETERMANN, A. H.; MOULE, T.

Downes *See under* TAYLOR, A.

Draeck, Peter 16th cent. Cart. Dut. M

Drake, Sir Francis, Bart. 17th cent. Brit. M
Nephew of the famous Admiral
The world encompassed . . . Lond., Nicholas Bourne, 1628 (with eng. map).

Draper, I. *See under* JONES, T. W.

Drayton, J. *See under* CAREY, H. C.

Drayton, Michael 1563-1631 Poet Brit. M
Poly-Olbion. First issued with complete set of 22 allegorical maps, 1622 (London), although 18 of the maps had been pub. in 1612 and 1613.

Drentwett, Abr. *See under* SEUTTER, G. M.

Dreykorn, R. *See under* VON SCHLIEBEN, W. E. A.

Drinkwater, John 19th cent. Cart. Brit. M
Map of Isle of Man, 1826 (engs. J. and A. Walker).

Drogenham, G. *See under* DU SAUZET, H.

Drury, Luke d. 1845 Cart. Amer. M
A geography for schools . . . Providence, R. I.; Miller and Hutchens, 1822.

Dryander, Johannes} 1500-60 Cart.,
Eichmann } Mathematician, Astron., Physician Ger. M G

Dubrovin, Mark fl. 1731 Seaman, Cart. Russ. M

Dubuisson *See under* MENTELLE, E.

du Caille, Louis Alexandre 18th cent. Cart. Fr. M
Etrennes géographiques Paris, Ballard, 1760, 1760-1. Maps adapted by J. A. B. Rizzi-Zannoni.

Ducheti, Claudius fl. 1570-92 Eng., Cart. Ital. M
Copy of Gastaldi's oval world map (1570).

Dücker, Franciscus fl. 1666 Cart. Ger. M
Map of Salzburg, 1666.

Dudith, Andreas fl. 1579 Astron. Hungarian X

Dudley, Sir Robert } 1573-1649
Duke of Northumberland ⎱ Cart., Geog.
and Earl of Warwick ⎰ Brit. M
Arcano del mare . . . 1661, Florence, Jacopo Bagnoni and Anton-francesco Lucini; 1646-7, Francesco Onofri.

Dufart, P. *See under* WALCKENAER, C. A.

du Fayen *Same as* FAYEN (q.v.)

Duflot de Mofras, Eugène 1810-84 Cart. Fr. M
Exploration du territoire de l'Oregon, des Californies et de la mer Vermeille . . . Paris, A. Bertrand, 1844.

Dufour, Auguste Henri 1798-1865 Cart. Fr. M
Atlas de géographie numismatique . . . Paris, Crozet, 1838; map drawn and eng. by Ch. Dyonnet and Ch. Simon.

Le globe; atlas classique . . . Paris, J. Renouard et Cie, 1835.
See also under ANDRIVEAU-GOUJON; LETRONNE, A. J.

Dufour, G., et Cie *See under* VON HUMBOLDT, F. W. H. A.

Dufrénoy *See under* ANDRIVEAU-GOUJON

du Frestoy *Same as* DE TEMPLEUX (q.v.)

Dugdale, James fl. 1819 Topg. Brit. M
The new British traveller . . . Lond., 1819, J. Robins and Co. Eng. by Neele.

Dugdale, Thomas fl. 1835-60 Antiquarian Brit. M
Curiosities of Great Britain . . . (assisted by William Burnett, civil engineer) Lond., L. Tallis, 1842, 1844, 1846, 1848. A bibliographically complicated work. The first edn. was in 3 vols.: Vol. I (dated 1835) was pub. by Tallis and Co.; Vols. II and III (n.d.) were pub. by John and L. Tallis. This was illustrated with reprints of maps by G. Cole. The work was later pub. in edns. ranging from 4 to 11 vols. Cole's maps were later replaced by others drawn and eng. by J. Archer. Railways and outline colouring were added to the later edns. See p. 88, Whitaker, H., *The Harold Whitaker Collection* (see *Bibliography* for full details).

Dugdale, Sir William 1605-86 Geog., Cart. Brit. M
Maps of the fens, 1662, 1772 (eng. W. Hollar); Romney Marsh, 1662, 1772; Warwickshire, 1656, 1765 (eng. R. Vaughan).
See also under ACERLEBOUT, J.; BEIGHTON, H.; NEWCOURT, R.

du Halde, J. B. 18th cent. Cart. Fr. M

Dulaure, Jacques Antoine 1755-1835 Cart. Fr. M
Histoire physique civile et morale de Paris . . . (with 5 maps) Paris, Guillaume et Cie, 1829.

Dumas, [Guillaume] Mathieu, Comte 1753-1837 Cart. Fr. M
Précis des événements militaires . . . Paris, Strasbourg, London, 1816-26. Maps are eng. by Ambroise Tardieu.

Dumas de Fores, Isaac *See under* Boisseau, J.

Duménil, P. *See under* Houzé, A. P.

Dumez *See under* Chainlaire, P. G.

Dumont d'Urville, Jules Sébastien César 1790-1842 Sailor, Cart. Fr. M

Voyage au pole sud dans l'Océanie sur les corvettes l'Astrolabe et Zélée . . . atlas Paris, Gide, 1842-8. Maps eng. by Hacq.

Voyage de la corvette l'Astrolabe . . . Paris, J. Tatsu, 1833. Maps are by Lottin, Gressien, Guilbert, E. Pâris and others.

Duncan, James fl. 1833 Pub. Brit. M

A complete county atlas of England and Wales . . . 1837 (reprints of Ebden's maps).

See under Ebden, W.

Duncan, W. 19th cent. Cart. Brit. M

Map of Co. Dublin, 1821.

Dunkin, Edwin fl. 1869 Astron. Brit. X

Dunn, Samuel d. 1794 Astron., Cart. Brit. X M

A map of the British empire in North America London, 1774.

A new atlas of the mundane system Lond., R. Sayer, 1774; Laurie and Whittle, 1786-9, 1796, 1800.

See also under Laurie, R.; Sayer, R.

du Perac, Stefano 1525-1604 Eng. Ital. M

Duperrey, Louis Isidore 1786-1865 Cart. Fr. M

Voyage autour du monde . . . Paris A. Bertrand, 1826-30.

See also under de Freycinet, L. C. D.

du Petit-Thouars, Abel Aubert 1793-1864 Cart. Fr. M

Voyage autour du monde . . . atlas Paris, Gide, 1840-55.

du Pinet, Antoine, Sieur de Noroy 1510?-1566? Cart. Fr. M

Plantz . . . de plusieurs villes . . . Lyons, I. d'Ogerolles, 1564.

Dupuis, L. A. *See under* de Ferraris, Comte.

Durand and Co. *See under* de Vaugondy, G. R.; Maverick, P.

Dürer, Albrecht, Artist (1471-1528), **and Stabius, Johann** Cart. (fl. 1497-1515) Ger. M

Woodcut map of the eastern hemisphere, 1515.

Durham, Thomas fl. 1595 Topg. Brit. M

Drew the map of the Isle of Man in Speed's *Theatre,* 1611. (*See* Speed, J.)

Durnford, Lieut. 18th cent. Cart., Brit. M

See also under Faden, W.

Dury, Andrew 18th cent. Pub., Cart. Brit. M

Collaborated with J. Andrews (q.v.) on maps of Herts., etc. Pub. maps of Capt. Montresor.

A new general and universal atlas . . . engraved by Mr. Kitchin and others Lond., A. Dury and R. Sayer, 1761; A. Dury, R. Sayer and C. Bowles, 1763?.

La porte-feuille nécessaire à tous les seigneurs quit font le tour d'Italie . . . Lond., 1774.

See also under Bell, P.; Haydon, W.; Williams, R.

Dury, Mrs. 18th cent. Pub. Brit. M

du Sauzet, Henri 18th cent. Pub. Fr. M

Worked in Amster.

Atlas de poche . . . Amster., 1734-8. Maps by N. Sanson; eng. by A. de Winter and G. Drogenham.

Atlas portatif . . . 2 vols. Amster., 1734, 1738. Carts.: P. M. Sanson le fils, N. Sanson, Petrus Kaerius, Jodocus Hondius, Gerard Mercator. Eng.: A. d'Winter.

Dussy, E. *See under* Desnos, L. C.; de Vaugondy, G. R.

du Temps, Jean *Same as* Temporal, J. (q.v.)

du Tralage *See under* Ottens, R.

Duval, Henri Louis Nicolas 1783-1854 Cart. Fr. M

Atlas universel des sciences . . . Paris, Terzuolo, 1837.

du Val, M. F. *See under* Delisle, G.

du Val, Pierre 1619-83 Geog., Cart.
Fr. M G
Cartes de géographie . . . Paris, A. de Fer, 1662, 1688-9 and other edns.
Cartes et tables de géographie . . . Paris, 1667.
Diverses cartes et tables pour la géographie ancienne . . . Paris, 1665, 1669?.
La géographie françoise Paris, 1677.
La monde ou la géographie universelle . . . Paris, N. Pepingue, 1670, du Val, 1682.
La monde chrestien . . . Paris, 1680?.
See also under BLAEU, G.; DANCKERTZ, J.; HORN, G.; JANSSON, J.; LE CLERC, J.; PLACIDE DE SAINT HÉLÈNE, PÈRE.

du Val d'Abbéville, Pierre 1619-1683
Cart. Fr. M
Nephew and pupil of N. SANSON (q.v.)

du Villard, Jean *c.* 1539-1610 Cart. Fr.
M

Duvotenay, Th. 19th cent. Cart. Fr. M
Atlas pour servir à l'intelligence des campagnes de la révolution française . . . Paris, Furne et Cie, 1846?. Eng. Ch. Dyonnet.
Duvotenay made several atlases after 1850.

Dworžak, Samuel fl. 1677 Eng. Czech.
M

Dyonnet, Ch. *See under* DUFOUR, A. H.; DUVOTENAY, TH.

Easburn 18th cent. Cart. Brit. M

East India House } *See under* JONES, FELIX;
East India Company} WALKER, J. AND C.

Ebden, William 19th cent. Cart. Brit.
M
An atlas of the English counties . . . Lond., S. Maunder(?), 1828. Eng. by Hoare and Reeves; some maps had been previously pub. by Hodgson and Co. and William Cole. They were reprinted *c.* 1833, and subsequently by James Duncan.
Collins's railway and telegraphic map of Yorkshire 1858.

Ebersperger, Johann Georg 1695-1760
Eng., Pub. Dut. M
Son-in-law of J. B. Homann. He managed the firm after Homann's death.

Eckebrecht, Philipp fl. 1627-30 Cart.
Ger. M

Eden, Richard *See under* BELLERO, I.

Edgar, William 18th cent. Surv. Brit.
M
Maps of Peeblesshire, 1741 (eng. R. Cooper); Stirlingshire, 1777.

Edge, Thomas fl. 1625 Cart., Seaman
Brit. M

Edgeworth, William fl. 1814-25 Cart.
Brit. M
Maps of Co. Longford, 1814 (eng. J. Thomson), and Co. Roscommon, 1817, 1825 (with R. Griffith; eng. J. Cross).

Edkins, S. S. 19th cent. Globeseller Brit.
G
Son-in-law of WILLIAM BARDIN (q.v.)

Edmands, Benjamin Franklin 19th cent.
Cart. Amer. M
The Boston school atlas . . . Boston, Lincoln and Edmands, 1832.
Maps eng. by Annin and Smith, G. Boynton, H. Morse.

Edward, Bryan 1743-1800 Geog. Brit.
M
History of the British West Indies . . . (with 5 maps) Lond., 1818. Some of the maps are signed by the engs., G. Allen and J. Cooke.
A new atlas of the West-India Islands . . . Philadelphia, I. Riley, 1818. The maps are by the following engs.: J. H. Seymour, Tanner and Marshall. The general map is 'Reduced by S. Lewis'.
An earlier edn. was issued at Charleston in 1810 by E. Norford, Willingdon and Co. The same engs. had worked on the maps, but the general map was 'Reduced by S. Louis' (obviously meant for Lewis).

Edwards, J. *See under* VANCOUVER, G.

Ehrict, C. *See under* MEYER, J.

Ehrmann fl. 1793 Cart. Ger. M

Eichmann, J. *Same as* DRYANDER, J. (q.v.)

Eimmart, George Christopher 1638-1705
Cart., Eng., Astron. Ger. G (C and T)
O

Eisenschmidt, Johann Caspar 18th cent.
Cart. Ger. M

Eitzing *Same as* VON AITZING, M. (q.v.)

Ellis, G. fl. 1819 Pub. Brit. M
Ellis's new and correct atlas of England and Wales . . . Lond., 1819. Maps are reprints of those in James Wallis's *New British atlas* (1812).
Ellis's general atlas of the world . . . Lond., 1823?.

Ellis, John fl. 1750–96 Eng. Brit. M
Ellis's English atlas . . . Lond., 1766, 1768, 1777.
See also under BOWLES, C.; FADEN, W.

Ellis, Thomas Joseph 19th cent. Cart. Brit. M
Maps of Huntingdonshire, 1824 and 1829 (eng. T. Foot), and Nottinghamshire, 1825 and 1827 (eng. T. Foot)

Elphinstone, John fl. 1745 Cart. Brit. M
Map of Lothians, 1744 (eng. T. Smith).

Elstrack, R. } 1571–1625? Eng. Brit. M
Elstracke } *See also under* PURCHAS, S.; SPEED, J.

Elwe, Jan Barend 18th cent. Cart., Pub. Dut. M
Atlas Amster., 1792.

Elzevir Press } *See under* CLUVER, P.;
Elzevir, Abraham} DE LAET, J.; HONDIUS, J.

Emmius, Ubbo 1547–1625 Cart., Humanist Dut. M
See also under HONDIUS, H.; JANSSON, J.; MERCATOR, G.; VAN DEN KEERE, P.

Emmoser, Gerhard fl. 1573 Astron. Austrian G (C)

Emslie, John fl. 1848 Eng., Cart. Brit. M
Map of Warwickshire and Worcestershire, London, J. Reynolds, 1848.

Ende *Same as* VAN DEN ENDE, J. (q.v.)

Enders, Johann }
   Friedrich } 1705–69 Mechanic, Eng.
Endersch, I. F. } Ger. M G (C)
See also under HOMANN'S HEIRS; SCHRAEMBL, F. A.; VON REILLY, F. J. J.

Enenckel von Hoheneck, Baron Georg Acacius fl. 1614 Cart. Ger. M

Engel, Le Bailly 18th cent. Cart. Fr. M
Map of Spitsbergen; Berne, F. Samuel Fetscherin, 1779.

Enouy, J. *See under* LAURIE, R.

Erhardt *See under* ANDRIVEAU-GOUJON.

Erich, Adolar fl. 1625 Cart. Ger. M

Erlinger, Georg 16th cent. Scholar, Cart. Ger. M

Erskine, R. 18th cent. Cart. Brit. M
Various plans of Iberian ports, etc., 1727–34.

España, Casildo *See under* MAESTRE, M. R.

Espinosa y Tello, José 1763–1815 Cart. Span. M
Relacion del viage por las goletas sutil y Mexicana . . . Madrid, 1802.

Etzlaub, Erhard 1460–1532 Cart. Ger. M
Das ist der Rom weg . . . Nuremberg, c. 1492 (numerous issues), woodcut.

Euler, Leonhard 1707–83 Cart., Physicist, Mathematician Ger. M
Atlas geographicus . . . Berlin, Michaelis, 1753 (with text in Latin and Ger.) and 1760 (with text in Latin, Ger. and Fr.).

E.V. Signature used by ENEA VICO (q.v.)

Evans, Rev. John 1767–1827 Cart. Brit. M
Map of North Wales (eng. R. Baugh), 1795. A reduction was pub. in 1797.
A new royal atlas . . . Lond., J. Cundee, 1810.

Evans, Lewis c. 1700–56 Surv. Brit. M
Maps for Thomas Jefferys's *American atlas* (1776).
Map of the middle British colonies in America, Philadelphia, 1755. Eng. by Jas. Turner. Known printed on silk. Chart of Van Diemen's Land (1822).
A new and general map of . . . *the United States of America* Lond., Laurie and Whittle, 1794 (unsigned).
See also under PALAIRET, J.

Evans, Thomas fl. 1801 Cart. Brit. M

Evert *See under* TAVERNIER, M.

Evreinov, Ivan fl. 1719–23 Cart. Russ. M
Worked with LUŽIN, FEDOR (q.v.)

Ewich *See under* HORN, G.

Ewing, Thomas 19th cent. Cart. Brit. M

Ewing's new general atlas . . . Edinburgh, Oliver and Boyd, 1828. Maps eng. by W. H. Lizars.

Ewoutszoon, Jan} fl. 1551-60 Eng., Pub.
Ewoutsz } Dut. M

Ewyk fl. 1750 Cart. Dut. M

Eymer, Fruger et Cie *See under* Lapié, A. É.

Eymery, A. *See under* Bullock, W.; de Ségur, L. P.

Eyre, Thomas fl. 1779-80 Cart. Brit. M

Map of Northamptonshire (with T. Jefferys), 1779 and subs. edns. eng. by W. Faden.

Eyzinger *Same as* von Aitzing, M. (q.v.)

Faber, Samuel 1657-1716 Cart. Ger. M G (C and T)

Samuelis Fabri . . . atlas scholastichodoeporicus... Nuremberg, C. Weigel, 171[?]. Weigel's name is on all the maps excepting nos. 2, 4-6, 12, 18, 35, 39, 43, 49. Many are engraved by Michael Kauffer (1673-1756), who sometimes signs: 'M.K.sc.'

Fabert, Ab. 17th cent. Cart. Dut. M

Territorium Metense . . . (i.e. Metz) Amster., Blaeu, *c.* 1660.

See also under Blaeu, G.; Hondius, H.; Jansson, J.

Fabri, Francis *See under* Wytfleet, C.

Fabricius, Antonius Bleynianus fl. 1624 Cart. Ger. M

Fabricius, David 1564-1617 Cart., Astron. Ger. M

Fabricius, John d. *c.* 1625 Astron. Ger. X

Fabricius, Paulus 1519-88 Cart., Mathematician Ger. M

Faden, William 18th/19th cent. Eng., Cart., Pub. Brit. M

Prolific. Successor to Thomas Jefferys (q.v.); traded as Faden and Jefferys.

A map of a part of Yucatan . . . within the bay of Honduras . . . Lond., 1787.

Plan of Quebec, Lond., 1776.

The North American atlas Lond., 1777.

Contains maps by C. J. Southier, P. Gerlach, Lieut. Durnford, W. C. Wilkinson, Thos. Jefferys, Lieut. Page, Charles Blaskowitz, Sauthier and Ratzer, W. Scull and Heap, Joshua Fisher, S. W. Werner, Thos. Hutchins, Fry and Jefferson, Henry Mouzon, W. Fuller, Lieut. C. Ross, Michael Lane, Capt. Jas. Cook, Capt. Holland, L. de la Rochette, etc. All of the names are not present on every copy of the atlas, as they were made up differently, according to requirements.

A correct map of France Lond., 1806.

Le petit neptune français Lond., 1793.

A topographical chart of . . . the river Tagus Lond., 1810.

Map of the Ottoman dominions in Asia . . . Lond., 1822.

The course of the Delaware river . . . Lond., 1778.

The United States of North America Lond., 1785.

Atlas minimus universalis . . . Lond., 1798.

Atlas of battles of the American revolution . . . New York, Bartlett and Welford, 1845?

General atlas . . . Lond., 1797, 1799, 1750-1836. Maps eng. (according to edn.) by B. Baker, I. A. Barnalet, I. Collyer, W. Faden, Thomas Foot, J. Ellis, S. Hall, Wm. Palmer, R. W. Seale, James Tyrer, Neele and Son, S. I. Neele, D. Wright, J. Walker.

Carts. incl. (according to edn.): Lieut. Henry Roberts, Daniel Paterson, Thos. Kitchin, Alexander Taylor, L. S. de la Rochette, d'Anville, Thomas Jefferys, Laurie and Whittle, Professor C. C. Lous and Admiral J. Nordenankar, Jasper Nantiat, James Rennell.

See also under Blackadder, J.; de Ferraris, Comte; Dessiou, J. F.; Donn, B.; Eyre, T.; Gream, T.; Hodgkinson, J.; Lapié, A. E.; Leard, J.; Nantiat, J.; Ratzer, B.; Sauthier, C. J.; Sayer, R.; Schraembl, F. A.; Tofiño de San Miguel, V.; von Reilly, F. J. J.

A reprint of W. Faden's *Catalogue* (1822) has recently (1963) been issued by the Map Collectors' Circle.

Fairman, D. *See under* Pinkerton, J.

Falda *See under* Köhler, J. D.

Faleleeff *See under* PIADYSHEV, V. P.

Falger, Anton *See under* GASPARI, A. C.

Falkenstein *See under* HOMANN, J. B.

Fannin, P. fl. 1789 Cart. Brit. M
Map of the Isle of Man. Eng. H. Ashby, 1789.

Faure *See under* DE FREYCINET, L. C. D.

Favoli, Ugo 1523-85 Cart. Ital. M
Theatri orbis terrarum . . . Antwerp, Plantin, 1585.

Fayanus, J. *See under* TAVERNIER, M.

Fayen, Jean *See under* HONDIUS, H.;
JANSSON, J.; MERCATOR, G.

F.B. Signature sometimes used by F.
BERTELLI (q.v.)

Felkl, L. 19th cent. Globe-maker. Ger.
G (T)

Fellweck, Joh. Georg fl. 1772 Globe-
maker Ger. G

Felton, J. 18th cent. Pub. Brit. M

Fer, A. de *Same as* DE FER, A. (q.v.)

Fer, F. *See under* WOLFGANG, A.

Fer, N. de *Same as* DE FER, N. (q.v.)

Ferguson, James 1710-76 Cart., Astron.
Brit. G (C and T) X

Fernandez, Pero} 16th cent. Pilot, Cart.
Fez   } Port. M

Ferrar, John 17th cent. Cart. Brit. M
A mapp of Virginia . . . Lond., Iohn
Overton, 1667. Eng. by John Goddard.

Ferrari, Gabriel Giolito 1536-78 Pub.,
Mapseller Ital. M

Ferraris *Same as* DE FERRARIS (q.v.)

Ferrer de Orga, J. *See under* DE LABORDE,
A. L. J.

Ferreri, Giovanni Paolo fl. 1600-24
Astron. Ital. G (C)

Feterus, Olasson 18th cent. Cart.
Scandinavian M

Fetscherin, F. Samuel 18th cent. Pub.
Swiss M
See also under ENGEL, LE BAILLY.

Fez *Same as* FERNANDEZ, P. (q.v.)

Fickler, S. B. fl. 1567 Printer Swiss M

Field, Barnum 19th cent. Cart. Amer.
M
*Atlas designed to accompany the American
school geography* . . . Boston, W. Hyde
and Co., 1832.

Figg, W. fl. 1861 Cart. Brit. M

Filiberto, Emanuele fl. 1570-5 Cart.
Ital. G (T)

Filson, John 18th cent. Cart. Brit. M
A map of Kentucky . . . Lond., John Stockdale, 1793.

Finaghenof *See under* PIADYSHEV, V. P.

Finckh, Georg Philipp fl. 1671 Cart.
Ger. M

Findlay, Alexander George 1812-75
Cart., Eng. Brit. M
General chart of the Mediterranean sea Lond.,
Laurie, 1839.
A classical atlas . . . Lond., W. Tegg and
Co., 1853.
A comparative atlas . . . Lond., W. Tegg
and Co., 1853.
A modern atlas . . . Lond., W. Tegg and Co.,
1850.
See also under KNOX, R.; REES, A.

Finé, Oronce   } 1494-1555 Cart.,
Finaeus, Orontius} Astron., Mathematician
Finaeus   } Fr. M
World map (woodcut), single-heart shape,
1519; double-heart shape in *Novvs orbis
regionum* . . . by Simon Grynaeus and John
Huttich (Paris, 1531); *De orbis situ libri
tres* by Pomponius Mela (Paris, 1540). An
eng. copy of this cordiform map was
made by the eng. Giovanni Cimerlino and
pub. at Venice in 1566. There are also
other edns., some single-, some double-
heart shape.

Finlayson, J. *See under* CAREY, H. C.

Finley, Anthony 19th cent. Cart., Pub.
Amer. M
Atlas classica . . . Philadelphia, 1829.
A new American atlas . . . Philadelphia, 1826.
A new general atlas . . . Philadelphia, 1824,
1829, 1830, 1831, 1833.
See also under HART, J. C.; MALTE-BRUN, C.

Fischer, Wilhelm 19th cent. Cart. Ger.
M
*Historischer und geographischer atlas von
Europa* . . . 2 vols. (with F. W. Streit,
Junr.) Berlin, W. Natoroff and Co.,
1834-7.

Fisher, Son and Co. fl. 1842-5 Pub. Brit.
M

Fisher, Joshua 18th cent. Cart. Brit. M
... *Chart of Delaware bay* ... (with John
Davis) Philadelphia, 1756. Eng. by Jas.
Turner.
See also under FADEN, W.; HAYDON, W.

Fisher, Ralph *See under* NORRIS, R.

Fisher, William 1656-91 Mapseller Brit.
M
See also under SELLER, J.

Flahaut *See under* ANDRIVEAU-GOUJON;
LETRONNE, A. J.

Flamsteed, John 1646-1719 Astron.
Brit. X G
Atlas Coelestis 1729 and subsequent edns.

Flandro *See under* JANSSON, J.; VALCK, G.

Flemming, C. *See under* SOHR, K.; VON
WEDELL, R.

Fletcher *See under* BAKER, T.

Fletcher, J. *See under* MATTHEWS, J.

Fleurieu *Same as* DE FLEURIEU (q.v.)

Flinders, Capt. Matthew 1774-1814 Surv.,
Cart. Brit. M
The maritime portion of South Australia (in
collaboration with Col. Light) Lond.,
J. Arrowsmith, 1839.
Voyage to terra Australis ... *atlas* Lond., G.
and W. Nichol, 1814.

Florenti fl. 1650 Astron. Ital. X

Florentius *Same as* VAN LANGREN (q.v.)

Florianus, Antonius} fl. 1550 Cart. Ital.
Floriano } M G (T)

Florianus, Johann d. 1585 Cart. Dut. M

Florini, Matteo fl. 1600 Eng. Ital. M

Folkes, Martin 1690-1754 Archaeologist
Brit. G

Fontana, Giovanni Battista *c.* 1524-87
Astron. Ital. X G (C)

Foot, Thomas *See under* ELLIS, T. J.; FADEN,
W.; GREAM, T.

Ford, J. *See under* CAHILL, D.

Forlani, Paulo fl. 1562-9 Eng., Cart.
Ital. M
Copies of Gastaldi's oval world map (1560,
1562).
See also under LAFRÉRY, A.

Forrest, William fl. 1816 Surv., Cart.
Brit. M
Maps of East Lothian, 1802 (eng. J. Kirkwood & Sons); Lanarkshire, 1816, 1818
(eng. J. and G. Menzies); West Lothian,
1818.

Forster, T. 18th cent. Cart. Brit. M
Plan of the city of Durham 1754.

Fortin, Jean 1750-1831 Pub. Fr. G (C
and T) O X
Celestial globe, Paris, 1780.
Terrestrial globe, Paris, Delamarche, 1786.
See also under DE VAUGONDY, G. R.

Foster, John 1648-81 Eng. Amer. M
Eng. a woodcut map, the first executed in
America, of New England for William
Hubbard's *A narrative of the troubles with the
Indians in New England* ... Boston, 1677.

Fowler, C. 19th cent. Cart. Brit. M
Map of Yorkshire, 1836, *c.* 1855 (eng. J.
Neele)

Fowler, W. fl. 1765 Eng. Brit. M
Maps of Berwickshire, 1826 (Sharp,
Greenwood and Fowler; eng. by J.
Dower), 1845; East Lothian, 1825 (Sharp,
Greenwood and Fowler; eng. by J.
Dower); Fifeshire and Kinross, 1828 and
subs. edns. (ditto, ditto); Midlothian, 1828
(ditto, ditto), 1845.

Fox, G. *See under* PINKERTON, J.

Fox, Luke}
Foxe } 1586-1635 Explorer Brit. M
North-VVest Fox ... Lond., 1635 (with 1
map, and eng. of armillary sphere).

Fox, Samuel 18th cent. Cart. Brit. M
Map of Derbyshire, 1760.

Fracan Montalboddo} 15th-16th cent.
Fracanzano } Cart. Ital. M
Itinerariū Portugallēsiū . . . (woodcut)
Milan? 1508

Fracastro, Girolamo 1478-1533 Philosopher Ital. G

Fráchus, Jacobus fl. 1600 Cart. Fr. M

Franco, N. R. *See under* DE HERRERA Y
TORDESILLAS, A.

Francus, Ysaacus}
François, Isaac } fl. 1592 Cart. Fr. M

Franks, J. H. *See under* SHERRIFF, J.

Franz, Johann Georg fl. 1804–7 Cart.
Ger   G (C and T)

Franz, Johann Michael 1700–61 Pub.
Ger.   M   G
Manager of the firm of HOMANN'S HEIRS
(q.v.)

Freducci Family fl. 1497–1556 Carts.
Ital.   M

Freitag *See under* HONDIUS, H.

Freloff *See under* PIADYSHEV, V. P.

French, G. R. *See under* GREENLEAF, J.

Frères Châtelain Libraries 18th cent.
Pub. Dut.   M
See also under GUEUDEVILLE.

Frese, Daniel fl. 1602 Cart. Ger.?   M

Freycinet *Same as* DE FREYCINET (q.v.)

Frezier *See under* VOOGT, C. J.

Fricx, Eugène Henri d. 1733 Cart.
Belgian   M
*Table des cartes des Pays-Bas et des frontières de
France . . .* Brussels, 1712. *See also
under* BEEK, A.; OTTENS, R.

Frietag, Gerardo *See under* BLAEU, G.;
JANSSON, J.; MERCATOR, G.

Frijlink, Hendrik 1800–86 Cart. Dut.
M
Frijlink's kleine school atlas Amster., 1847.

Frisch, P. J. fl. 1730 Eng. Ger.?   M

Frisius, Gemma} *See* GEMMA FRISIUS
Steen, C. V. D. }

Frisius, Laurent fl. 1522 Cart. Fr.   M

Froben, Hieronymus fl. 1533 Pub.
Swiss   M

Frobisher, Martin *See under* BESTE, G.

Froggett, I. W. 19th cent. Cart. Brit.
M
Map of 30 miles round London, 1831, 1833.

Froschouer, Christoffer 16th cent.
Pub. Swiss   M
See also under HONTER, J.

Frost, J. H. A. *See under* MORSE, J.

Fry, Joshua, and Jefferson, Peter 18th
cent. Carts. Brit.   M
A map of the inhabited part of Virginia . . .
Lond., Sayer and Jefferys, 1754?.
See also under DE VAUGONDY, G. R.; FADEN,
W.; JEFFERYS, T.

Fryer, J., and Sons 19th cent. Cart.
Brit.   M
Map of Northumberland, 1820 (eng. M.
Lambert).

Fuchs, Caspar Friedrich *See under* DANZ,
C. F.

Fullarton, Archibald 19th cent. Pub.,
Printer, Brit.   M
Eng. of maps in works by JOHN BELL (q.v.)
and others.
*The parliamentary gazetteer of England and
Wales . . .* Lond., Edinburgh and
Glasgow, 1843, 1845.

Fuller, William 18th cent. Cart. Brit.
M
Plan of Amelia island in east Florida Lond.,
Jefferys, 1770.
See also under FADEN, W.

Funk, Christlieb Benedikt 1734–1814
Physicist Ger.   G

Funk, David fl. 1555 Pub. Ger.   M

Funke, Karl Philipp 1752–1807 *See under*
VIETH, G. U. A.

Furlong, Lawrence 18th/19th cent.
Cart. Amer.   M
The American coast pilot . . . (with 15 maps)
Newburyport, E. M. Blunt, 1809. The
maps are eng. by Hooker.

Furne et Cie *See under* DUVOTENAY, T.;
TARDIEU, AMBROISE

Fürst, Pauls fl. 1600 Eng. Ger.   G (T)
M

Fürstaller, Josef J. d. 1775 Cart. Ger.
G. (T)

Furtenbach, Joseph d. 1667 Mathematician Ger.   G

Furtenbach, Martin fl. 1535 Astron.
Ger.   G (C)   O

G.A. Signature sometimes used by GASPARO
ARGARIA (q.v.)

Gadner, Georg} 16th cent. Cart. Ger.
Gadnerus   } M

Gahrceus, Johannes} 1530–74 Astron.,
Garcaeus   } Mathematician Ger.
Garz   } G

Gail, Jean Baptiste 1755-1829 Historian
Fr. M
*Atlas contenant, par ordre de temps, les cartes
relatives à la géographie d'Hérodote* . . .
Paris, 182[?].

Galignani, G. B., E. S., P. and F., Simon
16th/17th cent. Pub. Ital. M
See also under PORACCHI, T.

Gall, Rev. James fl. 1857 Astron. Brit.
M X

Gall and Inglis 19th cent. (still operating)
Pubs. Brit. M

Gallaher and White *See under* MORSE, S. E.

Galle, Giovanni fl. 1540 Cart. Dut. M

Galle, Philipp 1537-1612 Pub. Dut. M

Galle, Théodor 16th cent. Pub. Brit.
M
Son of Philipp G.; married the daughter of
PLANTIN (q.v.).

Galletti, Johann Georg August 1750-1828
Cart. Ger. M
Allgemeine weltkunde . . . Leipzig, J. F.
Gleditsch, 1807-10 and many subs. edns.
Maps. eng. by J. N. Champion.

Gangoiti, P. *See under* DE ANTILLON Y
MARZO, I.

Garcaeus, J. *Same as* GAHRCEUS, J. (q.v.)

Garden, William 18th cent. Surv. Brit.
M
Map of Kincardineshire, 1776 (eng. P.
Begbie), 1797 (pub. A. Arrowsmith)

Gardner, Thomas fl. 1719 Eng. Brit.
M
A pocket guide for the English traveller . . .
(road-maps) 1719.
See also under JEFFERYS, T.

Gardner, William fl. 1787-95 Cart.
Brit. M
Map of Guernsey, 1787 (eng. J. Warner).
See also under CUBITT, T.; LE ROUGE;
YEAKELL, T.

Garrett (or Garret), John fl. 1679-c.1718
Pub., Bookseller Brit. M
Pub. edns. of Mathew Simon's *Direction for
the English travitler.*
See also under HOLLAR, W.

Garz, J. *Same as* GAHRCEUS, J. (q.v.)

Gascoigne, Joel} fl. 1700-30 Cart. Brit.
Gascoyne } M
Map of Cornwall, 1700 (eng. J. Harris),
1730 (pub. Mount and Page).

Gaspari, Adam Christian 1752-1830 Cart.
Ger. M
Allgemeiner hand-atlas . . . Weimar, 1804-11,
1821. The following carts. and engs.
contributed maps to this atlas: C. F.
Weiland, Fried. Wilh. Streit, F. L.
Güssefeld, C. J. Maedel, A. Bürck, L.
Beyer, Jean Michel Mossner, Ferd. Götze,
I. G. L. Weidner, Adolph Stieler, August
Stieler, Kitchin, Jefferys and Beaufort,
I. C. M. Reinecke, D. F. Sotzmann,
G. R. von Schmidburg, Anton Falger,
C. G. Reichard. Maps have imprint:
'Weimar, Im Verlage des Geographischen
Instituts.'
. . . *Neuer methodischer schul-atlas* . . .
Weimar, 1793, 1799, and othér edns.
Maps drawn by F. L. Güssefeld. Engs.: C.
Müller, Friedr. Müller, I. C. Müller,
C. Westermaÿr.

Gassendi, Pierre 1592-1655 Astron. Fr.
X

Gastaldi, Giacomo} c. 1500-c. 1565 Cart.,
Gastaldo, Jacobo } Cosmographer Ital.
Castaldo } M
Oval world map, Venice, 1546. This was
copied by de Jode in 1555, Forlani in 1560
and 1562, F. Bertelli in 1562 and 1565,
L. and D. Bertelli in 1568, Camocio in
1569, Duchetti in 1570, and also by
Valgrisi and Ortelius.
See also under LAFRÉRY, A.

Gaulle, Francis *See under* MARTYR
D'ANGHIERA, P.

**Gaultier, Aloïsius Édouard Camille,
L'Abbé** 1746?-1818 Cart. Fr. M
Atlas de géographie . . . Paris, J. Renouard,
1810?.
A complete course of geography . . . (with 6
maps) Lond., 1792; W. and C.
Spilsbury, 1800.

Gavin and Son *See under* BROWN, THOMAS

Gavin, H. *See under* ARMSTRONG, A.

Gedda, Peter 1661-97 Pub. Swed. M

Geelkerck, Nicolaus }
Geilenkerlken } d. 1656 Cart.,
Geerkerken, Nicolaas} Eng. Ger.? M

Gefferys *Same as* JEFFERYS, T. (q.v.)

Geilkeckio, Nicolas *See under* CLUVERIUS,
PHILIP

Gellatly, J. 19th cent. Cart. Brit. M
Map of 12 miles round Edinburgh, 1840
(drawn by W. Johnston).

Geminus, Thomas fl. 1540-63 Pub., Eng.
Flemish M
Worked in London.

Gemma, Cornelis fl. 1578 Astron.
Flemish X

Gemma Frisius, Regnier } 1508-55 Cart.,
Phrisius Gemma } Mathematician
Regnier } Ger. M G
Reinerus } (C and T)
Steen, C. V. D. }
See also under À MYRICA, G.

Gendron, Pedro 18th cent. Cart. Span.
M
*Atlas ô compendio geographico del globo
terrestre* . . . Madrid, Barthelemi; Cadiz,
Bonardel, 1756-8. The following carts.
and engs. have maps in this work: Robert
de Vaugondy, Sr. Robert or Sr. Roberto,
Brion de la Tour (Sr. Brion), d'Anville,
des Bruslins, Laurent.

Genevensi *See under* JANSSON, J.

Genoi, D. *Same as* ZENOI, D. (q.v.)

Genovese, Battista fl. 1514 Cart. Ital.
M

Geoffroy, Lislet 19th cent. Cart. Fr. M
Chart of Madagascar Lond., 1819.

Geographisches Institut Weimar 19th
cent. Austrian G (C and T)

Georgius, Johannes fl. 1503 Cart. Ger.?
M

**Georgius, Gorgius or Georgio, Ludovico
or Ludovicus** fl. 1584 Cart. Ital. M
Pseudonym of Luig Jorge de Barbuda.
Chinae Antwerp, Plantin, c. 1584.
See also TEIXIERA, L.

Gephart *See under* OTTENS, R.

Gerard *See under* HONDIUS, H.; JANSSON,
J.; MERCATOR, G.

Gerardrem, Hesselum *See under* BLAEU, G.

Gerber, Johann Gustav d. 1734 Cart.
Ger. M

Gerhardus fl. 1715 Cart. Dut. G (C)

Gérin *See under* ANDRIVEAU-GOUJON

Gerlach, P. *See under* FADEN, W.

Gerrich, J. C. *See under* WEILAND, C. F.

Gerritsz, Adrian (c. 1525-79) **and Hessel,
Gerardus** (1581-1632) Carts. Dut. M
See also under DE LAET, J.; HUDSON, H.

Gessner, Abraham 1552-1613 Cart.,
Goldsmith Swiss G (A and T) O

G.H. Signature sometimes used by GEORGE
HUMBLE (q.v.)

Ghebellinus, Stephanus fl. 1584 Cart.
Dut. M

Gianelli, Giovanni fl. 1550 Astron.
Ital. G (C)

Gibson, John fl. 1750-87 Eng. Brit. M
*New and accurate maps of the counties of
England and Wales* . . . Lond., T. Carnan,
c. 1770.
*Atlas minimus . . . revised, corrected and improved
by Eman. Bowen, geographer to his majesty.*
Lond., J. Newberry, 1758; T. Carnan and
F. Newberry, Junr., 1774-9, 1792; Phila-
delphia, M. Carey, 1798.
See also under BOWEN, E.; PALAIRET, J.

Gide *See under* DUMONT D'URVILLE, J. S. C.;
DU PETIT-THOUARS, A. A.

Giftschütz, Karl 1753-1831 Priest
Austrian G

Gigante, Joanne}
Gigantus } *See under* BLAEU, G.

Gigas *See under* HONDIUS, H.; JANSSON, J.
MERCATOR, G.

Gilbert, Humphrey c. 1539-83 Navigator.
Brit. M
'A general map' in Gilbert's *Discourse of a
discoverie for a new passage to Cataia* Lond.,
Henry Middleton for Richard Ihones,
1576.

Gilbert, Joseph 18th cent. Cart. Brit.
M
See also under LE ROUGE, G. L.

Gill, Valentine fl. 1816 Surv., Cart. Irish M
Map of Wexford, eng. J. Bailey, E. Rumsey, W. Faden, 1811, 1816.

Gilleland, J. C. 19th cent. Cart. Amer. M
The Ohio and Mississippi pilot . . . (with 16 maps) Pittsburgh, R. Patterson and Lambdin, 1820.

Gillies, J. 18th cent. Cart. Brit. M
Map of Co. Down, 1755, 1767 (eng. J. Ridge).

Gillingham, E. *See under* Tanner, H. S.

Ginver, N. *See under* Stephenson, J.

Giolito 16th cent. Cart. Ital. M

Giordani, Vitale 1633–1711 Astron. Ital. O

Giovo of Como 16th cent. Cart. Ital. M

Giraldon fl. 1815 Globe-maker Fr. G

Girava, Hieronymous fl. 1556 Cosmographer, Cart. Span. M

Giustiniani, Agostino } 1470–1536 Cart.
Justinianus, Augustinus} Ital. M

Giustiniani, Francisco 18th cent. Pub., Cart. Ital. M
El atlas abreviado ò el nuevo compendio de la geografia universal . . . Leon de Francia, J. Certa, 1739.

GLA Signature sometimes used by George Lily (q.v.)

Glareanus, Henricus}
Loritus, H. } 1488–1563 Cart.,
Loritz } Geog. Ger. G (T)
von Mollis }
Plain gores.
Maps of East Asia and America.

Gleditsch, J. F. *See under* Galletti, J. G. A.

Glockendon, Georg Albrecht fl. 1500 Cart., Illuminator Ger. M G

Glynne (or **Glin**), **Richard** 1681–1755 Instrument maker. Brit. G

G. M. Junior Signature sometimes used by Gerhard Mercator the Younger (q.v.)

G. M. A. W. L. Signature sometimes used by G. M. Lodewycks (q.v.)

Goddard, J. 18th cent. Cart. Brit. M
Map of Norfolk, 1731 and subs. edns. (with W. Chase), 1740 and subs. edns. (with R. Goodman).

Goddard, John, Junr. fl. 1651–76 Eng. Brit. M
See also under Ferrar, J.

Goddard, Pierre François 1768–1838 Cart. Fr. G

Goddard and Goodman fl. 1730–40 Cart. Brit. M

Godfray, H. 19th cent. Cart. Brit. M
Map of Isle of Wight, 1849 (eng. J. Welland).

Godiche, A. H. and F. C. *See under* Pontoppidan, E.

Goeree, W.} *See under* De Fer, N.; Ottens, R.
Goerê } J. and J.; van Loon, J.

Göeschen, G. J. *See under* von Schlieben, W. E. A.

Goghe, John fl. 1567 Cart. Brit. M
Map of Ireland, 1567.

Goldsmith, Rev. J. Pseudonym of Sir Richard Philipps (q.v.)

Gonzaga, Gurzio c. 1536–99 Cart. Ital. G (T)

Goodman, R. *See under* Goddard, J.

Goodrich, Charles Augustus 1790–1862 Geog. Amer. M
Atlas accompanying Rev. C. A. Goodrich's outlines of modern geography . . . Boston, S. G. Goodrich; Philadelphia, M'Carty and Davis, 1826.

Goodrich, Samuel Griswold 1793–1860 Cart., Pub. Amer. M
Atlas designed to illustrate the Malte-Brun school geography Hartford, H. and F. J. Huntington, 1830, 1838.
A general atlas of the world . . . Boston, C. D. Strong, 1841. Engrs.: G. W. Boynton, Sherman and Smith, S. Stiles. No. 26 (Delaware) is drawn by F. Lucas, Junr.
See also under Bradford, T. G.; Goodrich, C. A.; Woodbridge, W. C.

Goos, Abraham fl. 1640 Eng. Dut. M G (C and T)
Speed's *Africae described.*
See also under De La Feuille, J.; Köhler, J. D.; Visscher, N., II.

Goos, Henry *See under* Doncker, H.

Goos, Pieter fl. 1654-66 Cart. Dut. M
Pas-kaarte van de zuyd-west-kust van Africa
   Amster., 1669.
Pascaerte van nova Hispania Chili . . .
   Amster., 1666.
*Paskaarte van het zuy de lijckste . . . van Rio
   de la Plata . . .* Amster., 1666.
*Paskaarte vertonende alle de zekusten van
   Europa . . .* Amster., *c.* 1665 (known on
   vellum).
Pascaertevan Groen-Landt . . . Amster., 1669.
*Noordoost cust van Asia van Japan tot Nova
   Zemla* Amster., 1669.
Paskaerte zynde t'oosterdeel van Oost Indien . . .
   Amster., 1669.
Paskaart van Brazil . . .   Amster., 1669.
De zee atlas, ofte water-weereld . . . Amster.,
   1666 and other edns.
The sea atlas of the water world . . . Amster.;
   1667.
*Atlas of portulan charts for sailing from Holland
   to the Mediterranean.* Amster., *c.* 1673.
L'atlas de la mer . . . Amster., 1670.
Le grand et nouveau miroir ou flambeau . . .
   Amster., 1671.
De lichtende colomne . . . Amster., 1656-7.
The lighting colomne or sea-mirrour . . .
   Amster., 1660-1.
De nieuwe groote zee-spiegel . . . Amster.,
   1676.
Pascaart van de noort zee Amster., 1669.
Pas caerte van Nieu Nederlandt . . . Amster.,
   1669.
*Paskaerte van de zuyt en noort . . . in Nieu
   Nederlandt* Amster., 1669.
Pascaert vande Caribes eylanden Amster., 1669.
Pascaerte van West Indien . . . Amster., 1669.
Nieuwe werelt kaert Amster., *c.* 1669.
   See also under HONDIUS, H.; HORN, G.;
   ROBIJN, J.; ROGGEVEEN, A.

Gordon, James 17th cent. Cart. Brit.
   M
Son of Robert G. (q.v.)

Gordon, Robert   } 1580-1661 Geog.
Gordonius, Robertus} Brit. M
Maps in Blaeu's *Atlas novus.*

Gordon, Thomas F. 1787-1860 Topg.
   Amer. M
Gazetteer of the state of New York . . . (with
   68 maps) Philadelphia, 1836.

Gordon, William fl. 1730-36 Surv.,
   Cart. Brit. M
Map of Bedfordshire, 1736. Eng. J.
   Carwitham.

Gorgius *Same as* GEORGIUS (q.v.).

Gorton, John fl. 1833. Topg. Brit. M
Operated with SIDNEY HALL (q.v.)
*A topographical dictionary of Gt. Britain and
   Ireland* 3 vols. Lond., Chapman and
   Hall, 1831-3.

Goss, S. *See under* DEAN, JAMES

Gosselin, C. *See under* GUIZOT, F. P. G.

Gotho, Olao Iohannis *See under* JANSSON, J.

Gottholdt fl. 1810 Cart. Ger. M

Götz, Andreas } 1698-1780 Cart. Ger.
Goetzio, Andrea} M
Brevis introductio as geographiam antiquam . . .
   Nuremberg, B. J. C. Weigel, 1729.
The maps are ornamented with illustrations
   of ancient mythology and history.

Götze, Ferd. *See under* GASPARI, A. C.

Gough *See under* STOCKDALE, J.

Goujon, J. *See under* ANDRIVEAU-GOUJON

Goulart, Jacob 17th cent. Cart. Dut. M
Lacus Lemmani locorum que . . . Amster.,
   G. Blaeu, *c.* 1660.
   See also under HONDIUS, H.; JANSSON, J.

Gould, Benjamin Apthorp fl. 1866
   Astron. Amer. X

Gountei (Ontagawa) Sadahide 1807-73?
   Artist Japanese M
Fuji ryodo ichiran no zu ('General map show-
   ing two ways to climb Mount Fuji')
   Edo, 1855 (woodcut).

Gourmet fl. 1763 Cart. Fr. M

Gourmont, Jerome}
                      }} fl. 1553 Pub. Fr. X
Gurmontius   }
Pub. of GUILAUME POSTELLUS (q.v.)

Gournay. *See under* BEEK, A.

Gourné, Pierre Mathias 1702-70? Cart.
   Fr. M
Atlas abrégé et portatif Paris, L. C. Desnos,
   1763.
Contains maps by Benjamin Brion and
   Janvier.

Gouvion Saint-Cyr *Same as* DE GOUVION
   SAINT CYR (q.v.)

Gouwen *Same as* VAN GOUWEN (q.v.)

Graff *Same as* DE GRAFF (q.v.)

Grandi, P. Francesco fl. 1750 Cart. Ital. G (T)

Grangé, Guillyn *See under* BRION DE LA TOUR, L.

Grant, E. S., and Co. *See under* BRADFORD, T. G.

Grante, J. A. fl. 1745 Cart. ? M

Grass, J. *Same as* HONTER, J. (q.v.)

Grassom, John 19th cent. Surv., Cart. Brit. M
Map of Stirlingshire, 1817, 1819, 1837.

Gravier, Y. *See under* ROUX, J.

Gray, George Carrington fl. 1824 Topg. Brit. M
Gray's new book of roads . . . Lond., 1824. Maps are reprints of those in Cooke's *Topography in Gt. Britain.*

Gray and Son *See under* BELL, J.

Gream, Thomas 18th cent. Cart. Brit. M
Map of Sussex, 1795, 1815 (eng. T. Foot; pub. W. Faden).
Reduction pub. 1799, 1819 (eng. I. Palmer; pub. W. Faden).
See also under YEAKELL, T.

Green, John *See under* LOTTER, T. C.; MEAD, B.

Green, J. L. fl. 1824 Astron. Amer. X

Green, William fl. 1804 Cart. Brit. M
The picture of England illustrated . . . 2 vols. Lond., I. Hatchard, 1804.

Greene (or **Green**), **Elizabeth** fl. 1688–89 Mapseller. Brit. M Daughter of Robert Greene (q.v.).

Greene (or **Green**), **Robert** d. 1688 Cart. Brit. M
A mapp of Ireland . . . Lond., Lea and Overton, 1686.
A new map of Scotland . . . Lond., Lea and Overton, 1686.

Greenleaf, Jeremiah 1791-1864 Cart. Amer. M
A new universal atlas . . . Brattleboro, Vt., G. R. French, 1842.

Greenleaf, Moses 1777-1834 Cart. Amer. M
Atlas accompanying Greenleaf's Map and statistical survey of Maine . . . Portland, Maine, Shirley and Hyde, 1829.

Greenwood, Charles Name sometimes erroneously given to CHRISTOPHER GREENWOOD (q.v.).

Greenwood, Christopher (1786-1855) **and John** fl. 1821-40 Carts. Brit. M
Atlas of the counties of England . . . Lond., J. and C. Walker, 1834.
Many separate county maps *c.* 1818-*c.* 1831.
Charles alone pub. *Map of the county palatine of Chester*, 1819; *Map of the county palatine of Lancaster*, 1818 (drawn by R. Creighton); *Map of the county of Middlesex*, 1819; *Map of the county palatine of Durham*, 1820.
See also under FOWLER, W.; KENTISH, N. L.; NEELE, J.; TEESDALE, H.

Greenwood, James fl. 1834 Cart. Brit. M

Gregorii, Johann Gottfried 1685-1770 Cart. Ger. M
Collaborated with Homann.

Gregory, I. 18th cent. Pub. Brit. M

Greischer, Matthias fl. 1683 Eng. Ger. M
Eng. J. A. Reiner's map of Hungary.

Grenier *See under* LAURIE, R.

Gressien *See under* DUMONT D'URVILLE, J. S. C.

Greuter, Matthäus 1556-1638 Astron. Ger. G (C and T)

Gridley, Richard 18th cent. Cart. Brit. M
A Plan of the city and fortifications of Louisburg Lond., T. Jefferys, 1757.

Grienbergerus, Christophoro fl. 1679 Astron. Dan. X

Grieninger, Johann 16th cent. Printer Ger. M

Grierson, George 18th cent. Cart., Pub. Irish M
Turkish empire in Europe, Asia and Africa Dublin, *c.* 1720.
See also under MOLL, H.

Griffin, W. 18th cent. Pub. Brit. M
See also under SPEER, CAPT. J. S.

Griffith, Richard fl. 1825 Surv., Cart.
Irish M
Maps of Leinster coal district, 1814.
See also under EDGEWORTH, W.

Grigg, J. *See under* HART, J. C.

Grigg and Elliot *See under* MALTE-BRUN,
C.

Grigny *See under* MALTE-BRUN, C.

Grijp, D. 17th cent. Eng. Dut. M
See also under BLAEU, J.

Grimel, I. *See under* DELISLE, J. N.

Grimm, Johann Ludwig 1816–48 Cart.
Ger. M

Grimoard *Same as* DE GRIMOARD (q.v.)

Grodecki, Waclaw d. 1591 Cart.
Polish M

Groot *Same as* DE GROOT (q.v.)

Gros, C. 18th/19th cent. Geog. Fr.? M
*A new genealogical, historical and chronological
atlas . . .* (with C. V. Lavoisne) Lond.,
J. Barfield, 1807.

Grose, Francis fl. 1777–87 Historian
Brit. M
See also under CONDER, T.; KITCHIN, T.;
and SELLER, J.

Grosselin et cie. *See under* DELAMARCHE, A.

Grosvenor, James *See under* STEPHENSON, J.

Grünewald, C. B. Ch. fl. 1847–50 Cart.
Ger. G
See also under VON SCHLIEBEN, W. E. A.

Gruninger, Joannes fl. 1522–5 Cart.
Ger. M

Grynaeus, Simon 16th cent. Geog. Ger.
M
See also under FINÉ, O.; HOLBEIN, H.

Gudbrandus Thorlacius *Same as*
THORLAKSSON (q.v.)

Guedeville, Nicholas} 1654?–1721 or 1722
Gueudeville } Cart. Fr. M
Atlas historique . . . 4 vols. Amster.,
chez les frères Châtelain Libraries, 1708–14.
Le nouveau théâtre du monde . . . Leyden,
P. van der Aa, 1713. Incl. maps by
Cassini and Vignola.

Guerard, N. *See under* DE FER, N.; PLACIDE
DE SAINT HÉLÈNE, PÈRE.

Guerreros, Juan Antonio 18th cent.
Cart. Span. M
Plano del rio de la Plata Buenos Ayres,
c. 1790.

Guicciardini, Johann Baptista fl. 1549
Cart. Ital. M

Guicciardini, Lodovico or Luigi 1521–89
Cart. Cosmographer, Historian Ital.
M
Descrittione de . . . tutti i paesi bassi . .
Antwerp, Plantin, 1588.

Guilbert *See under* DUMONT D'URVILLE,
J. S. C.

Guilford, N. and G. and Co. *See under*
CUMINGS, S.

Guillemin, Amedée fl. 1867 Astron.
Fr. X

Guilloterius, Fridericus fl. 1598 Cart.
Dut. M

Guizot, François Pierre Guillaume 1787–
1874 Historian Fr. M
*Vie, correspondence et écrits de Washington.
Atlas.* Paris, C. Gosselin, 1840.

Guler von Weineck, Graf Johansen
fl. 1616 Cart. Ger. M

Gumpp, Martin fl. 1674 Cart. Austrian
M

Gunman de Valle, Christopher d. 1682
Cart. Brit. M

Gurmontius *Same as* GOURMONT (q.v.)

Güssefeld, F. L. fl. 1797 Cart. Ger. M.
Charte von Nord America Nuremberg,
Homann's Heirs, 1797.
*Charte über die XIII vereinigte staaten von
Nort-America* Nuremberg, Héritières de
Homann, 1784.
See also under GASPARI, A. C.; HOMANN'S
HEIRS.

Guthrie, William 1708–70 Geog. Amer.
M
Atlas universel pour la géographie de Guthrie
Paris, H. Langlois, 1802.
General atlas for Guthrie's Geography . . .
Philadelphia, B. Warner, 1820.

The Atlas to Guthrie's System of Geography . . . Lond., C. Dilly and G. G. and J. Robinson, 1795. Eng.: J. and J. C. Russell. Incl. a plate of an armillary sphere, by T. Kitchin. Each map has this inscription in the upper margin: 'Engraved for Guthrie's new System of geography'; and in the lower margin: 'Published . . . by G. G. and J. Robinson, and J. Mawman, 1801.'

Another edn. was pub. by J. Johnson and F. and C. Rivington in 1808.

The general atlas for Carey's edition of Guthrie's Geography improved . . . Philadelphia, Mathew Carey, 1795. Carts.: Harding Harris, Samuel Lewis. Engs.: Thackara and Vallance; William Barker; S. Hill; Amos Doolittle; J. T. Scott; C. Tiebout; B. Tanner; J. Smither, W. Harrison, Junr.

Gutiérrez, Diego} 1485–1554 Cart.
Gutierus } Span. M
Americae sive quartae orbis partis nova et exactissima descriptio (with Hiero Cock) 1562.

Gutschóuen *See under* BEEK, A.

Gutteling, H. C. 18th cent. Pub. Dut. M
See also under LANGEWEG, D. I.

Guyet, L. } 16th cent. Cart. Fr.
Guyetus, Licinius} M

Guyeto, Licimo 17th cent. Cart. Ital. M
Anjou Blaeu, *c.* 1660.

Gyger, Hans Konrad 1599–1674 Cart. Swiss M

Gysbertsoon, Evert fl. 1601 Cart. Dut. M

Haan, Friedrich Gottlob 1771–1827 Mechanic, Cart. Ger. G

Haas *Same as* HASIO (q.v.)

Habrecht, Isaac 1544–1620 Cart., Astron. Ger. G (C and T) X

Hachette et Cie *See under* VIVIEN DE SAINT-MARTIN, L.

Hack, William fl. 1678–1700 Hydrographer, Cart. Brit. M

Hacq *See under* DE GOUVION ST. CYR, L.; DUMONT D'URVILLE, J. S. C.

Hadji Ahmad fl. 1534 Cart. Tunisian M

Haeyen, A. fl. 1585–1613 Cart. Dut. M

Hagelgans, Johann Georg d. 1705 Cart. Ger. M
Atlas historicus . . . (with 10 maps) n.p., 1751.

Hahn, Philipp Matthäus 1739–90 Astron., Theologian Ger. G (C)

Hailes, N. fl. 1820 Pub. Brit. X

Haipolt, Joh. fl. 1617 Astron. Ger.? G (C)

Hakluyt, Richard *See under* THORNE, R.

Haldane *See under* LAURIE, R.

Hall, J. *See under* JOHNSON, W.

Hall, Ralph 17th cent. Eng. Brit. M
Map of Virginia in the 1637 English edn. of Mercator's *Atlas.*

Hall, Sidney fl. 1818–60 Cart., Eng. Brit. M X
A new British atlas . . . Lond., Chapman and Hall, 1833, 1834, 1836.
A travelling county atlas . . . Lond., Chapman and Hall, 1842 and many subs. edns. The maps were issued again as *A new county atlas* in 1846.
Black's General Atlas . . . Edinburgh, A. and C. Black, Lond., Longman and Co., 1840, 1841 and subs. edns.
A new general atlas . . . Lond., Longman, Rees, Orme, Brown and Green, 1830; Longman, Brown, Green and Longmans, 1857. Various county maps.
See also under CRAWFORD, W.; FADEN, W.; HUGHES, W.; LENNY, J. G.; QUIN, E.; REES, A.; THOMSON, J.

Hall, W. *See under* SPEED, J.

Halley, Edmund } 1656–1742 Astron.
Halleius, Edmundus} Brit. X G
A new and correct chart . . . *in the western and southern oceans* . . . Lond., 1701. Eng. by I. Harris
Novum & accuratissimen totius terrarum orbis tabula . . . Amster., R. and I. Ottens, 1740.
See also under JULIEN, R. J.; OTTENS, R.; RENARD, L.

174

Halma, F. *See under* ALTING, M.; OTTENS, R.; SANSON, N., PÈRE; VISSCHER, N., III.

Halse fl. 1809 Cart. Brit. G

Ham, Thomas fl. 1856 Cart., Eng. Australian M
Also printed some of the famous 'half-length' stamps of the colony of Victoria.

Hamersvelt, Everard Symons *See under* BLAEU, G.; HONDIUS, H.; HORN, G.

Hammond, Capt. John fl. 1720 Cart. Brit. M
See also under STEPHENSON, J.

Hamond, John fl. 1592 Cart. Brit. M

Hand, J. *See under* ANDREWS, JOHN.

Handmann, Johann Jakob 1711-86
Goldsmith, Stamp-cutter Swiss G.

Handtke, F. *See under* SOHR, K.

Hansteen, Christopher 1784-1873 Cart. Norwegian M
Magnestischer atlas . . . Christiania, 1819.

Happel, Eberhard Werner 17th cent. Geog. Ger. M
Mundi mirabilis tripartiti Ulm, 1708 (with eng. chart)

Harding, C. L. fl. 1822 Astron. Ger. X

Harding, S. 18th cent. Pub. Brit. M
See also under LAWS, W.

Hardy fl. 1738 Eng. Brit.? G
See also under HONDIUS, H.; JANSSON, J.; MERCATOR, G.; TAVERNIER, M.; VISSCHER, N., III.

Hareius *See under* HORN, G.

Harenberg *See under* HOMANN'S HEIRS.

Harigoni *Same as* ARIGONI, FRA B. (q.v.)

Har(r)iot, Thomas 1560-1621 Geog., Astron. Mathematician Brit. M X
See under WHITE, J.

Harme *Same as* DE HARME (q.v.)

Harper and Bros. *See under* BREESE, S.

Harrewyn, I. P. *See under* DE AFFERDEN, F.

Harris *See under* MIDDLETON

Harris, Harding *See under* CAREY, M.; GUTHRIE, W.

Harris, John fl. 1686-1746 Eng. Cart. Brit. M
Plan of London 1700.
See also under GASCOIGNE, J.; HALLEY, E.; HOLMES, T.; MORDEN, R.; MORTIER, D.

Harris, John and James fl. 1842 Carts. Brit. M
County maps.

Harris, T. 19th cent. Globe-maker Brit. G

Harrison, John fl. 1791 Pub. Brit. M
See also under DE RAPIN-THOYRAS, P.; TINDAL N.

Harrison, Samuel *See under* MACPHERSON, D. AND A.

Harrison, W., Junr. *See under* GUTHRIE, W.; JONES, T. W.; MACPHERSON, D. AND A.; PINKERTON, J.; SAYER, R.; WILKINSON, R.

Hart, Joseph C. d. 1855 Cart. Amer. M
A modern atlas of fourteen maps . . . New York, R. Lockwood; Philadelphia, J. Grigg and A. Finley, 1828, 1830, and other edns.

Hartmann, Georgius 1489-1569 Instrument-maker Ger. G (C and T).

Harwar, George *See under* VAN KEULEN, J.

Has, Prof. J. M. 18th cent. Cart. Ger. M
La Guinée . . . Nuremberg, 1743.
Carte de l'Asie mineure . . . Nuremberg, Homann's Heirs, 1743.
See also under VON REILLY, F. J. J.

Hase, J. M. *Same as* HASIO (q.v.)

Haselberg a Reichenau, Johann fl. 1535 Cart. Dut. M

Hasenfratz, K. *Same as* DASYPODIUS, C.

Hasius, Johann Matthias}
Haas } 1684-1742 Cart.
Hase } Ger. M G
Americae mappa generalis, Nuremberg, Homann's Heirs, 1746.
Africa secundum legitimas projectionis stereographicae regulas Several edns., 1737-45, Nuremberg, Homann.

Vorstellung der grundrisse von den jenigen weltberüharten stalten . . . Nuremberg, Homann, 1745.

See also under ANDRIVEAU-GOUJON; HOMANN, J. B.; HOMANN'S HEIRS

Hasio, M. *See under* DE LAT, J.; HOMANN'S HEIRS.

Hatchard, I. *See under* GREEN, W.

Hauber, Eberhard David 1695-1795 Geog. Ger. M
Collaborated with the Homann establishment.

Hauer, Johann 16th cent. Cart., Eng. Ger. G (T)

Have *See under* OTTENS, R.

Haviland, W. *See under* TANNER, H. S.

Hawkesworth, John 1715?-73 Editor Brit. M
An account of the voyages undertaken . . . by commodore Byron, captain Wallis, captain Carteret and captain Cook . . . atlas Lond., W. Strahan and T. Cadell, 1773.

Hayashi Shihei} 1738-93 Cart. Japanese
Rin Se Fée } M

Haydon, W. 18th cent. Cart. Brit. M
A chart of the Delaware river taken from the chart pub. at Philadelphia by Joshua Fisher. Lond., Andrew Dury, 1776.

Hayes *See under* LAURIE, R.

Hayes, Walter fl. 1651-92 Instrument-maker Brit. G

Haywood, John 18th cent. Cart. Brit. M
Maps of Isle of Wight, 1781, and Warwickshire, 1788.
See also under DE RAPIN-THOYRAS, P.

Hazama Shigetomi fl. 1810 Astron., Cart. Japanese M

Heap *See* SCULL AND HEAP.

Heath, Thomas C. 1714-65 Globe- and Instrument-maker Brit. G

Heather, William fl. 1795-1801 Cart. Brit. M
Chart for '*The Officers in the honourable East India company's service*' Lond., 1797.
A new chart of the Atlantic or western ocean Lond., 1797.

A new and improved chart of the Cape of Good Hope . . . Lond., 1796.
The maritime atlas or seaman's complete pilot . . . Lond., 1804, 1808.
The new Mediterranean pilot . . . Lond. 1802; 1814 (pub. by J. W. Norie and Co. successors to W. Heather). The 1802 edn. is from 'surveys of Michelot, Bremond, & Ayrouard; and from those of . . . John Wilson and Joseph Foss Dessiou'.
The north sea and Baltic pilot. Lond., 1807., Maps are dated 1794-1807.

Hebb, Andrew fl. *c.* 1625-48 Bookseller, Pub. M
Pub. Camden's *Britannia*, 1637.

Heiden *Same as* HEYDEN (q.v.)

Heimburger, A. *See under* MEYER, J.

Heinfogel, Konrad 1470-1530 Mathematician, Philosopher Ger. M

Heirisson *See under* DE FREYCINET, L. C. D.

Heis, E. fl. 1872 Astron. Ger. X

Hellert, J. J. 19th cent. Cart. Fr. M
Nouvel atlas physique, politique et historique . . . Paris, Bellizard, Dufour et Cie; St. Petersburg, F. Bellizard et Cie, 1843.

Helweg, Martin } 1516-74 Cart. Ger.
Heilwig, Martino} M

Hemminga, Doco ab d. 1555 Cart. Dut. M

Hemmings, S. 18th cent. Cart. Brit. M
A map of that part of America which was the principal seat of war 1759.

Henecy and Fitzpatrick *See under* McCREA, W.

Hengsen, C. *See under* WEILAND, C. F.

Hennebergen, Kaspar} 1529-1600 Cart.
Henneberger } Ger. M
See also under HONDIUS, H.; JANSSON, J.; MERCATOR, G.

Hennet, G. fl. 1830 Cart. Brit. M
Map of the county palatine of Lancaster . . . H. Teesdale and Co., 1830.

Hennipin, R. P. Ludovico 18th cent. Cart. Ger. M
Franciscan missionary in N. Amer.
Amplissimae regionis Mississippi . . . Nuremberg, Homann, *c.* 1720.

Henrici, Alberti 16th/17th cent. Cart.
Dut. M.
See also under LINSCHÓTEN, J.

Henricpetrina *See under* STRABO

Hepe, Christoph}
Höppe   } d. *c.* 1634 Draughtsman,
Hoppe   } Painter Ger. G

Herault *See under* MENTELLE, E.

Herberstein *Same as* VON HERBERSTEIN
(q.v.)

Herbert, William fl. 1769 Cart. Brit.
M
Collaborated with JOHN ANDREWS (q.v.) on
map of Kent.

Herculis, M. *See under* NICOLOSI, G. B.

Herder, B. *See under* LÖWENBERG, J.; VON
KAUSLER, F. G. F.; WOERL, I. E.

Hermelin, S. G. 18th cent. Cart. Swed.
M
Geographiske chartor öfver Swerige . . .
Stockholm, 1797-1805.

Hermundt, Jacobus fl. 1697 Eng. Dut.
M

Herold, Adam} fl. 1649 Instrument-
Heroldt   } maker Ger. O

Herrenvelt *Same as* VAN HERRENVELDT (q.v.)

Herrera y Tordesillas *Same as* DE HERRERA
Y TORDESILLAS (q.v.)

Herrick, J. K. *See under* BELL, A., AND CO.

Herriset *See under* DESNOS, L. C.

Herschel, Sir William 1738-1822 Astron.
Brit. M

Herzberg, H. *See under* MEYER, JOSEPH

Hessel, Gerardus *See under* GERRITSZ, A.

Hevelius, Johannes}
Hevel   } 1611-87 Astron.
Hewelcke   } Danzig X G

Hevenesi, Gabriel fl. 1689 Cart.
Austrian M

Hewitt, N. R. *See under* JOHNSON, W.;
MACPHERSON, D. AND A.; REES, A.;
THOMSON, J., AND CO.

Heydans, Carl}
Heydanus   } 16th cent. Cart. Ger. M

Heyden, Christian} 1525-76 Astron.
Heiden   } Ger. G (C and T)

Heyden, G. and P. *Same as* À MYRICA, G.
AND P. (qq.v.).

Heyden, Jacobus ab } fl. 1620 Printer
Heyden, Jacob van der} Ger. M

Heydt, Johann Wolfgang fl. 1744 Cart.
Ger. M

Heylin, Peter 17th cent. Cosmographer
Brit. M

Heyns, Pieter b. 1537 Pub. Dut. M
Le miroir du monde . . . Antwerp, Plantin,
1579, 1583.

Heyns, Zacharias} 1570-1640 Pub.,
Hyens   } Mapseller Dut. M
Son of Pieter H.

Heywood, John fl. 1868 Pub. Brit. M

Hickling, C. *See under* JENKS, W.

Hilacomilus *Same as* WALDSEEMÜLLER
(q.v.)

Hill, John fl. 1754 Astron. Brit. X

Hill, Nathaniel fl. 1746-64 Eng. Brit.
M G (C and T)
See also under MORRIS, L.; WARBURTON, J.;
WING, J.

Hill, P., and Co. fl. 1780 Pub. Brit. M

Hill, S. *See under* GUTHRIE, W.; MALHAM, J.

Hillcock *See under* JEFFERYS, T.

Hillebrands, A. J. 19th cent. Cart. Dut.
M
*Atlas van de vereenigde staten van Noord-
Amerika . . .* Groningen, J. Oomkens,
1850?

Hilliard, Gray, Little and Wilkins *See
under* WORCESTER, J. E.

Hind, John Russell fl. 1855 Astron.
Brit. X

Hinrichs, J. C. 19th cent. Cart. Ger. M
Nouvelle carte de l'Allemagne Leipzig, 1806.
See also under STEIN, C. G. D.

Hinton, John d. 1781 Pub., Bookseller
Brit. M
Sold many of Emanuel Bowen's maps.

Hirschvogel, Augustin *c.* 1503-53 Eng.,
Cart. Ger. M

Hirtzgarter, Matthäus 17th cent. Cart.
M

Hitch, Charles d. 1764 Pub., Bookseller
Brit. M
Pub. Gibson's edn. of Camden's *Britannia*,
1753.

Hoare and Reeves *See under* EBDEN, W.;
MURRAY, T. L.

Hobson, William Colling fl. 1850
Cart. Brit. M
Hobson's fox-hunting atlas . . . Lond., J.
and C. Walker, various edns. (*See under*
WALKER, J. A. AND C.)

Hocker, Johann Ludwig 1670-1746
Mathematician Ger. G

Hodges, I. *See under* OSBORNE, T.

Hodgkinson, Joseph fl. 1783 Cart.
Brit. M
Map of Suffolk, 1783 (pub. W. Faden);
reduction pub. 1787 (ditto).
See also under DONALD, T.; JEFFERYS, T.

Hodgson, O. fl. 1820 Pub. Brit. M
Also operated as Hodgson and Co.
See under EBDEN, W.

Hodgson, Thomas 19th cent. Surv.
Brit. M
Map of Westmorland, 1828.

Hoeckner, Carl *See under* MEYER, JOSEPH

Hoefnagel, Joris } 1542-1600 Cart.
Hufnagel, Georg } Belgian M
Brightstowe (i.e. Bristol) Cologne, Braun
and Hogenberg, *c.* 1575.
Cantuarbury Cologne, Braun and Hogen-
berg, *c.* 1575.
Cestria (i.e. Chester) Cologne, Braun and
Hogenberg, *c.* 1575.
Civitas Exoniae (i.e. Exeter) Cologne, Braun
and Hogenberg, *c.* 1575.
Bird's-eye plans of Ireland Cologne, Braun
and Hogenberg, *c.* 1618
Nordovicum (i.e. Norwich) Cologne, Braun
and Hogenberg, *c.* 1575
Oxonium (i.e. Oxford) Cologne, Braun
and Hogenberg, *c.* 1575
Vindesorium (i.e. Windsor) Cologne, Braun
and Hogenberg, *c.* 1575
Edenburgum (i.e. Edinburgh) Cologne,
Braun and Hogenberg, *c.* 1575
Yorke, Shrowesbury, Lancaster and Richmont
Cologne, Braun and Hogenberg, *c.* 1618
See also under JANSSON, J.

Hoffman, Erhard fl. 1570 Astron. Ger.
X

Hoffman, Johannes 1629-98 Mapseller
Ger. M

Hogenberg, Franciscus } *c.* 1540-*c.* 1590
Hoghenberghe, Frans } Eng. Dut. M

Hogenberg, Remigius } *c.* 1536-*c.* 1588
Hoghenberghe } Flemish M
Worked in England, 1572-89.
See also under BRAUN AND HOGENBERG.

Hogenboom, A. *See under* DE WIT, F.;
OTTENS, R.

Hogg, Alexander *See under* WALPOOLE, G.
A.

Hogius *Same as* DE HOOGE (q.v.)

Hohem, Diogo fl. 1557-76 Cart. Port.
M
La carta del Navigare . . . Rome, Claude
Duchet, 1581-6

Hokusai 1760-1849 Artist Cart.
Japanese M

Holbein, Hans 1497-1543 Artist Ger. M
A map in *Novvs orbis regionum* . . . by Simon
Grynaeus and Johann Huttich (Basle,
1532) is sometimes attributed to him.

Hole, G. fl. 1697 Eng. Brit. M

Hole, William fl. 1600-46 Eng. Brit.
M
Engraved some of the maps in Camden's
Britannia.
See also under SMITH, CAPT. JOHN

Holland, Capt., later Major Samuel
18th cent. Cart., Surv. Brit. M
A map of the island of St. John Lond., 1775.
*A chorographical map of the country between
Albany, Oswego, Fort Fontenae, and, les
trois rivieres* Lond., 1775.
*A new and accurate chart of the north American
coast* . . . Lond., Laurie and Whittle,
1808.
*Chart for the navigation between Halifax and
Philadelphia* Lond., Laurie and Whittle,
1798.
See also under FADEN, W.; JEFFERYS, T.; LE
ROUGE.

Hollar, Wenceslaus or Wenzel   1607-77
Eng. Bohemian M
Worked in England.
The Kingdome of England and principality of Wales . . . Printed and sold by John Garrett. Generally known as 'Quarter-master's map'. Lond., Thomas Jenner, 1644; Lond., Garrett (who added roads to the maps), *c.* 1676.
Hollar also eng. some of the maps for Blome's *Speed's maps epitomiz'd.*
See also under BLOME, R.; DUGDALE, SIR W.; TAYLOR, T.

Holle, Lienhart}
Holl   } *See under* PTOLEMAEUS,
Hol   } CLAUDIUS.

Holmes, Thomas   17th cent. Surv. Brit. M
A map of ye improved part of Pennsylvania . . . Eng. by John Harris.

Holzwurm, Israel   17th cent. Engineer Austrian M
Map of Austria, 1628 (eng. H. Bahre).

Homann, Johann Baptist} 1663-1724
Homanno, Bapt.   } Cart., Pub.
Noribergae in Officina } Ger. M G
           Homanniana} (C and T)
The firm continued to flourish in Nuremberg after Homann's death (*see under* HOMANN'S HEIRS).
Hasio's *Africa.*
Totius Africae nova repraesentatio   c. 1720.
Totius Americae . . . Nuremberg *c.* 1720.
Insulae Danicae . . . Nuremberg, *c.* 1720.
Tabula Generalis Jutiae   Nuremberg, *c.* 1720.
Aegyptus hodierna Nuremberg, *c.* 1760.
Regni Sueciae . . . Nuremberg, *c.* 1740.
Agri Parisiensis Nuremberg, *c.* 1730.
Insularum Maltae et Gozae Nuremberg,*c.*1720.
Map of Martinique Nuremberg,   *c.* 1762.
Regni Norvegiae Nuremberg, *c.* 1740
Imperii Persici . . . Amster., *c.* 1660.
Mappa geographica regni Poloniae Nurem-berg, 1773.
Ingermanlandiae . . . Amster., Covens and Mortier, 1734.
Accurate carte der uplandischen Scheren Nuremberg, *c.* 1740.
Nova tabula Scaniae Nuremberg, *c.* 1740.
Canton Lucern Nuremberg, *c.* 1763.

Belgii pars septentrionalis . . . Nuremberg *c.* 1720.
Virginia, Marylandia, et Carolina . . . Nurem-berg *c.* 1759.
Map of St Christopher Nuremberg, *c.* 1750.
Planiglobii terrestris (various versions) Nuremberg, *c.* 1702.
Atlas Germaniae specialis . . . Nuremberg, 1753. Carts. incl.: Arenhold, Coldewey, Falkenstein, Hasius, Hüber, Majer, Müller, Nell, Schreiber, Zollman.
Atlas novus terrarum . . . Nuremberg, 1702-50. Carts. incl.: T. Danckertz, Tobias Meyer, J. M. Hasio, Matthaeus Seutter, A. F. Zürner.
Grosser atlas Nuremberg, 1716, 1737-70 (maps dated 1702-48). Carts. incl.: Kaempffer, J. M. Hasio, M. Seutter, J. N. Bellin, T. Jefferys, Le Rouge; also included is a reproduction of H. Popple's map of North America.
Kleiner atlas scholasticus . . . Nuremberg and Hamburg, Johann Hubern, 1732?
Neuer atlas . . . Nuremberg, 1710-31, 1712-30. Carts. incl.: Nell, Müller, Seutter, Tillemon and N. Visscher, III.
See also under D'ANVILLE, J. B.; EBERSPERGER, J.; GREGORII, J. G.; HENNEPIN, R. P. F.; JULIEN, R. J.; KÖHLER, J. B.; NOLIN, J. B.; OTTENS, R.; VALCK, G.

Homann, Johann Christoph   1701-30
Pub. Ger. M
Son of Johann Baptist H.

Homann's Heirs   } 18th/19th cent.
Héritières de Homann } Pubs. Ger. M
Homannianos Heredes }
Successors to J. B. HOMANN (q.v.)
Atlas compendiarius . . . various edns. 1752-90. Carts. incl.: Guillaume Delisle (various spellings), Philippe Buache, Bellin, J. B. Nolin, G. M. Lowizio, Tobia Majero, I. A. Rizzi-Zannoni, J. P. Nell, I. M. Hasio, M. Hasio, A. G. Bohemio, F. L. Güssefeld.
Atlas geographicus . . . Nuremberg, 1759-84. Carts. incl.: F. L. Güssefeld, T. C. Lotter, P. Schenk, N. Visscher, I. F. Endersch, M. Seutter, A. Vindelicor, C. Niebuhr, J. W. Zollmann, I. Majer, A. G. Bohemio, T. Lopez.

Atlas mapparum geographicarum . . . Nurem-
berg, 1728-93. Carts. incl.: Lowitz,
Hasius, Kitchin, Majer, Nolin, Visscher,
Lopez, Güssefeld, Delisle, Rizzi-Zannoni,
Jaillot, Tillemon, Nell, Cerruti, Harenberg,
Bohemius.

Atlas novus republicae Helveticae . . . Nurem-
berg, 1769.
Maps by T. Mayer, Rizzi-Zannoni, Gabriele
Walser.

Atlas regni Bohemiae . . . Nuremberg, 1776.

Atlas Silesiae . . . Nuremberg, 1750-1808.

Bequemer hand-atlas . . . Nuremberg, 1754.
Carts. incl.: Lowitz, Hasius, Güssefeld,
Lopez, T. Mayer, Rizzi-Zannoni.

Homannischer atlas . . . Nuremberg, 1747-
57. Carts. incl.: Lowitz, Visscher, III,
Delisle, Hasius, Nolin, Valk, d'Anville,
Mayer, Tillemon.

Kleiner atlas . . . Nuremberg, 1803. Maps
by C. M. Pronner and F. L. Güssefeld.

Major atlas scholasticus . . . Nuremberg
1752-73. Carts. incl.: Haas, Majer,
Lowitz, Zannoni, Tollmann, Hübner,
Harenberg, Zürner.

Schul-atlas . . . Nuremberg 1743, 1745-6.

Städt-atlas . . . Nuremberg 1762. Maps by
I. Baumeister, C. B. Bestehorn, Perizot,
F. B. Werner, F. W. Zollman.

See also under Desnos, L. C.; Ebersperger,
J.; Has, J. M.; Lotter, T. C.; Sauthier,
C. J.

Hondius, Henricus} 1587-1638 Cart.
Hondt   } Dut. M G (C and T)
de Hondt   } X

Africae nova tabula Amster., J. Jansson, *c.*
1650.

India quae orientalis dicitur . . . Amster.,
J. Jansson, *c.* 1652.

Nova totius terrarum orbis geographica . . .
Amster., 1630.

Nouveau théâtre du monde . . . Amster.,
J. Jansson, 1639-40. Carts. inc.: Lluyd,
Pont, Ortelius, Lübin, Mercator, Massa,
Henneberger, Keer, Mellinger, Lauren-
berg, Stella, Scultetus, Gigas, Emmius,
Westenberg, Comenius, Lazius, Fabert,
Hardy, Fayen, Jubrien, Beins, Goulart,
Damme, Sprecher von Bernegg, Lang-
ren, van der Burght, Pynacker, Metius,

Brockenrode, Martini, Freitag, Wicheringe,
de Fer, Gerard, Secco, Labanna, Adrichem,
Cluver.

Engs. incl.: Keere, Hamersvelt, Rogiers,
Goos and Ende.

See also under de la Feuille, J.; de Wit, F.;
Franz, J. M.; Jansson, J.; Langenes,
B.; Mercator, G.; Ojea, F. F.; Tassin,
N.; Tavernier, M.

Hondius, Jodocus, Jodok}
      Jud or Josse} 1563-1611 Eng.,
Hondt   } Pub. Dut. M
de Hondt   } G (C and T) X

Africae, nova tabula Amster., J. Jansson,
1632.

Nieuwe caerte van . . . *Guiana* Amster., *c.* 1605

Andaluziae nova descripttio Antwerp, *c.* 1607.

Nova totius Europae descripttio (with Petrus
Kaerius) Amster., 1613.

Nova et accurata Italiae hodiernae descriptio
Batavorum, Abraham Elzevir, 1627.

See also under Barläeus, C.; Bertius,
P.; du Sauzet, H.; Jansson, J.; Merca-
tor, G.; Molyneux, E.; Ptolemaeus, C.;
Speed J.; Tavernier, M.; van der Berg,
P. Waghenaer, L. J.

Hondt *Same as* Hondius (q.v.)

Hondt, P. *See under* Horn, G.

Honter, Johann   }
Honterus, Johannes} fl. 1540 Cart. Ger.
Grass, J.   } M G

Rudimentorium cosmographicorum (with 13
woodcut maps) Kronstadt, 1542;
Zurich, Froschouer, 1549, 1564, 1570.

Hooge *Same as* de Hooge (q.v.)

Hooker fl. 1587 Cart. Brit. M

Hooker (18th/19th cent.) *See under* Furlong,
L.

Hooker Eng. *See under* Pinkerton, J.

Hooper, Samuel fl. 1770-93 Bookseller,
Pub. Brit. M
See also under Speer, J. M.

Hoorenhout, Jacques fl. 1540 Geographer, Cart. Dut. M

Höppe, Hoppe *Same as* Hepe (q.v.)

Hopper, Joachim 17th cent. Cart. Dut.
M

Horenbault, Jacques fl. 1620 Cart. Dut.
M

Horman, Robert fl. 1580 Cart. Brit. M

Hormann 18th cent. Pub. Ger. M
Published maps of J. Sandrart (q.v.)

Horn, Georg}
Hornius **}** 1620–70 Cart. Dut. M
Accuratissima orbis antiqui delineato Amster., J. Jansson, 1654.
Accuratissima orbis delineato Amster., J. Jansson, 1660; The Hague, P. Hondt, 1740. The 1660 edn. contains maps by Stella, Jansson, Adrichem, du Val, Ortelius, Ewich, Cluver, Briet, Harieus and Lluyd. Engs. incl.: Rogier, Keer, Goos, Hamersvelt, Ende and Saevry. The 1740 edn. incl. maps by du Val, Laurenberg, Blancard and Ortelius.
A compleat body of ancient geography ... The Hague; Lond., T. Osborne, 1741. Carts. and engs. as under *Accuratissima orbis delineato*, of which this is an English translation.
Description of the earth or ancient geography ... Amster., J. Jansson, 1700. Maps eng. by Everard, Symons, Hammersveldt, Solomon Rogiers, Pieter van den Keere, Abraham Goos, S. Saevry, Josua van den Ende.

Horsburgh, James 1762–1836 Cart. Pub. Brit. M
Atlas of the East-Indies and China Sea Lond., 1806–21.
Steel and Horsburgh's New and complete East-India pilot ... (with Penelope Steel) Lond., Steel and Goddard, 1817. The chart of Goa and Murmagoa roads is signed 'David Inverarity'.
See also under WALKER, J. AND C.

Horsfield, Thomas 19th cent. Cart. Brit. M
Map of the island of Banka Lond., c. 1822.
Map of the island of Java ... Lond., W. H. Allen and Co., 1852.

Horsley, J., and Cay, R. 18th cent. Surv. Brit. M
Map of Northumberland, 1753. Eng. A. Bell.

Houzé, Antoine Philippe 19th cent. Cart. Fr. M
Atlas universel historique et géographique ... Paris, P. Duménil, 1848, 1849.

Howe and Spalding *See under* Morse, J.

Hoxton, Walter 18th cent. Cart. Brit. M
... *Mapp of the bay of Chesepeack* ... Lond., W. Betts and E. Baldwin, 1735.
See also under SMITH, A.

Hoyau *See under* TRUSCHET

Hubbard, William *See under* FOSTER, J.

Huber, J. H. fl. 1712 Eng. Ger.. M

Huberinus, Maurit. fl. 1615 Astron. Ger. X

Hubinger fl. 1824 Cart. Ger.? G (T)

Hūble *Same as* HUMBLE (q.v.)

Hübnern, Johannes}
Hübner **}** 1668–1731 geographer
Hubner **}** M
See also under HOMANN, J. B.; PALAIRET, J.

Hübschmann, Donat fl. 1566 Eng. Ger. M

Huddard *See under* LAURIE, R.

Hudson, Henry d. 1611 Explorer Brit. M
Descriptio ac delineatio geographica detectionis freti ... Amster., Hessel Gerard, 1612 With one map.

Hueber *See* ANICH AND HUEBER

Hues, Robert 1553–1632 Historian Dut. G

Huet, Pierre Daniel *See under* LE CLERC, J.

Hufnagel, G. *Same as* HOEFNAGEL, G. (q.v.)

Hughes, William 1817–76 Topg. Brit. M
Black's general atlas ... (with S. Hall and others) Edinburgh, A. and C. Black; Lond., Longmans and Co., 1844 and subs. edns.
See also under QUIN, E.

Hulsbergh, H. *See under* MORTIER, D.

Hulsius, Lieven fl. 1597 Cart. Dut. M

Humble, George} d. 1640 Bookseller,
Hūble **}** Pub. Brit. M
Pub. maps in various edns. of Speed's *Theatre*.
Sometimes signed G. H.

Humble, William} fl. 1640–59 Bookseller,
Hūble **}** Pub. Brit. M
Pub. edns. of Speed's *Theatre*.

Humboldt *Same as* VON HUMBOLDT (q.v.)

Hunt, F., and Co. *See under* BRADFORD, T. G.

Huntington, H. and F. J. *See under* GOODRICH, S. G.

Huntington, Nathaniel Gilbert 19th cent. Cart. Amer. M
Huntington's School atlas . . . Hartford, E. Huntington and Co., 1833.

Hurd, Capt. T., R.N. fl. 1811 Cart. Hydrographer of the Navy. Brit. M

Husson, Pieter 17th/18th cent. Cart. Dut. M
Variae tabulae geographicae . . . Gravenhage, 1709?. Carts. incl.: Delisle; H. Jaillot; P. Schenck; G. Valck; Sanson; M. Van Medtman; N. Visscher III; de Witt; I. Massa.
See also under BEEK, A.; OTTENS, R.

Hutchings, Thomas 18th cent. Cart. Brit. M
See also under CAREY, M.; FADEN, W.; JEFFERYS, T.

Hutchings, W. F. *See under* PHILIPS, J.; SWIRE, W.

Huttich, John *See under* FINÉ, O.; HOLBEIN, H.

Huyberts, K. *See under* DE FER, N.

Huys, Pieter fl. 1568 Eng. Dut. M

H.W. Signature of H. WEIGEL (q.v.)

Hyde, W., and Co. *See under* FIELD, B.; WARREN, W.

Hyens, Z. *Same as* HEYNS, Z. (q.v.)

Hylacomilus *Same as* WALDSEEMÜLLER (q.v.)

Ieremin *See under* PIADYSHEV, V. P.

Ihones, Richard *See under* GILBERT, H.

Ilacomilus *Same as* WALDSEEMÜLLER (q.v.)

Imperial Academy of St. Petersburg Pub. Russ. M
The Russian discoveries (i.e. N.W. coast of America) Lond., 1775.

Ino Tadutaka *See* CHUKEI

Insclin, Carolus} *See under* BLAEU, G.; DE
Inselin } FER, N.; DESNOS, L. C.; PLACIDE DE SAINT HÉLÈNE, PÈRE.

Inverarity, David *See under* HORSBURGH, J.

Ishikawa Rin Tan fl. 1730 Cart. Japanese M

Ishikawa Toshiyuki} 1688–1713 Cart.
Ryûsên } Japanese M

Isidore of Seville, Bishop 7th cent. Polymath. Span. M
Etymologiae Augsburg, Gunther Zainer, 1472, with woodcut map $2\frac{1}{2}$ in dia., the first map ever printed.

Iungmeister, I. A. *See under* BARONOVSKII, S. I.

Iwanoff *See under* PIADYSHEV, V. P.

Jackson, Capt. *See under* SAYER, R.

Jacob fl. 1809 Cart. Brit.? G

Jacobsz, Caspar *See under* JACOBSZ, T.

Jacobsz, Jacob *See under his father's entry,* JACOBSZ, T.

Jacobsz, Jan} fl. 1540–1 Pub. Dut. M
Anthonisz }

Jacobzs, Theunis ̄ or Anthonie⌉
Jacob Theunifz (i.e.⌉ fl. 1648–1707 Cart. surname and Christ-⌐ Dut. M ian name reversed)⌋
Teunissen }
The Jacobsz family frequently used the name LOOTS-MAN, which means 'sea pilot', They collaborated as follows. Carts.: Anthony or Theunis Jacobsz, and his son Jacob Theunis. Printer: Caspar Jacobsz (he also drew some of the maps).
The fifth part of the new great sea-mirrour . . . Amster., John Loots, 1717. Maps are signed I. Loots or A. Roggeveen.
Lighting colom of the midland-sea Amster., C. Loots-man, 1692. Map 1 is by Theunis Jacobsz. The remainder are unsigned.
The lightning columne . . . Amster. C. Loots-man, 1680, 1692. Carts.: Theunis Iacobsz, Casparus Lootsman, Jacobus and Casparus Lootsman, Iacob Theunisz.
t'nieuw groot straets-boeck . . . Amster, A. Jacobsz, 1648.
Nieuwe groote geoctroyeerde verbeterde en vermeerderde Loots-man zeespiegel . . . Amster., C. Loots-man, 1707. Carts.: Theunis Jacobsz, Casparus Lootsman, Jacobus Conynenberg, Jacob and Casparus Lootsman, Jacob Theunisz.
See also under DONCKER, H.

Jaeger, J. C. *See under* VON ROESCH, J. F.

Jaeger, Johann Wilhelm Abraham 1718-90 Cart., Soldier Ger. M

Jaillot, Alexis Hubert *Same as* JAILLOT, C. H. A. (q.v.)

Jaillot, B. A. 18th cent. Cart. Fr. M
Carte des postes de France Paris, 1766.

Jaillot, Charles Hubert Alexius, Alexis Hubert or Hubert 1640-1712 Cart., Eng. Fr. M G (T)

États de l'Empire des Turcqs en Europe Amster., Covens and Mortier, *c.* 1730.

Europe Amster., Covens and Mortier, *c.* 1730.

Le Canada Paris, 1696.

Atlas . . . Paris, 1696. Incl. maps by N. and G. Sanson, de la Rue, etc.

Atlas nouveau . . . Paris, 1689.

Atlas françois . . . Paris, 1695-7, 1695-1701. Most of the maps are by Jaillot or Sanson; many eng. by Cordier.

See also under DE FER, N.; DELISLE, G.; DESNOS, L. C.; DE WIT, F.; HOMANN'S HEIRS; HUSSON, P.; JULIEN, R. J.; OTTENS, R.; SANSON, N., PÈRE; SCHENK, P.; VALCK, G.; VISSCHER, N., III.

Jaillot, Hubert *Same as* JAILLOT, C. H. A. (q.v.)

James, J. A. and U. P. *See under* CONCLIN, G.; CUMINGS, S.

James, Thomas *c.* 1593-1635 Navigator Brit. M

The strange and dangerous voyage of Captaine Thomas Iames . . . Lond., 1633 (with one map).

Jamieson, Alexander fl. 1822 Astron. Brit. X
Celestial Atlas Lond., various edns.

Jansson, Jan or John }
Ieansson, Iean }
Janssonius, Joannes } 1596-1664 Printer,
Janssonium, Joannem } Pub. Dut. M G
Janssen }
Johnson, John }
Ianssony, Ioannis }

Name sometimes is coupled with Waesburg: Jansson-Waesburg.

See note under G. BLAEU regarding the danger of confusing his name with Jansson's.

Printed many maps by J. and H. Hondius.

Genehoa, Jaloffi et Sierraliones regna Amster., Jansson's Heirs, *c.* 1690.

America nova delineata Amster., 1652.

Nova et accurata poli arctici . . . Amster., *c.* 1652.

Insularum Indiae orientalis nova descriptio Amster., 1652.

Mar di India Amster., *c.* 1650.

Accuratissima Brasiliae tabula Amster., *c.* 1650.

Capitaniarum de Phernambuca Amster., 1657.

Capitaniae de Cirii et Parnambuco Amster., *c.* 1650.

Essexiae descriptio (i.e. Essex) Amster., 1652.

Vectis insula (i.e. Isle of Wight) Amster., 1646.

Insula Zeilan (i.e. Ceylon) Amster., 1652.

Chili Amster., *c.* 1652.

Junnan, Quicheu, Quangsi et Quantung Amster., *c.* 1650.

China Amster., 1652.

Guina sive Amazonum regio Amster., 1650.

Guinea Amster. *c.* 1650.

Nova Hispania et Nova Galicia Amster., 1657.

Insularum, Moluccarum nova descriptio Amster., 1652.

Persia, sive Sophorum regnum Amster., *c.* 1660.

Peru Amster., 1657.

Nova Belgica et Anglia Nova Amster., 1652.

Belgii novi . . . Amster., 1657.

Virginiae partis australis . . . Amster., 1649.

Le nouvel atlas ou théâtre du monde . . . Amster., 1647, 1639-40 (with Hondius). Carts. incl.: Ortelius, Pont, Cluver, and many others all included below.

Nuevo atlas . . . Amster., 1653. Carts. incl.: Massa, Hondius, Flandro, Henneberger, Lübin, Magnus, Emmius, Scultetus, Gigas, Lazius, Comenius, Mercator, Lavaña, Secco, Jubrien, Templeux, Fabert, Hardy, Sanson, Du Val, Sprecher von Berneck, Goulart, Martini, Langren, Surhon, Pynacker, Matius, Wicheringe, Frietag.

Nouveau théâtre du monde ou Nouvel atlas
6 vols. (Vols. 1 to 5, text in Fr.; Vol. 6
text in Latin), 1647-57.

Novus atlas . . . 6 vols., 1657-8.

Atlantis majoris . . . Amster., 1650, 1657.

Atlas minor . . . Amster., 1651. Some maps
eng. by Pieter van den Keere.

La guerre d'Italie . . . Amster., 1702.

Illustriorum Hispaniae urbium tabulae . . .
Amster., 1652?

*Illustriorum principumque urbim septentrion-
alium Europae tabulae* . . . Amster., 1657?

Joannis Janssonii atlas contractus . . . 2 vols.,
Amster., 1666. Carts. incl.: Henricus
Hondius, Henneberger, Massa, Emmius,
Flandro, Mellinger, Laurenberg, Magnus,
Palbitzke, Lazius, Comenius, Frietag,
Mercator, Pynacker, Metius, Wicheringe,
Langren, Martini, Surhon, Templeux,
Hardy, Lavaña, Sprecher à Berneck,
van Loon, Ortelius, Genevensi.

Die neuwen atlantis . . . Amster., 1639,
1644. There were also French (1639) and
German (1640) edns. Carts. incl.: Massa,
O. I. Gotho, H. Hondius, Sprotta,
Lazio, Fabert, van Damme.

Nieuwen atlas . . . 3 vols., 1642-4; 5 vols.,
Amster., 1652-3; 6 vols., 1657-8. Carts.
incl.: Massa, Henricus Hondius, Gotho,
Henneberger, Keer, Mellinger, Lauren-
berg, Lubin, Stella, Palbitzke, Emmius,
Gigas, Scultetus, Westenberg, Lazius,
Comenius, Langren, Mercator, Martini,
Surhon, van der Burght, Pynacker,
Metius, Berckenrode, Frietag, Jubrien,
Fabert, Wicheringe, Templeux, Hardy,
Sanson, Du Val, Damme, Fayen, Goulart,
Sprecher von Bernegg, Ortelius,
Adrichem, Gerard, Fer, Secco, Labanna.

*Tooneel der vermaarste koop-steden en handel-
plaatsen van de geheelde wereld* . . . Amster.
1682. Maps are by Frederick de Wit,
Jan Jansson, Jodocus Hondius, Nicolas
Visscher III and Gerard Mercator. Many
are eng. by Georg Hoefnagel.

See also under Allard, K.; Bertius, P.;
Blaeu, G.; Braakman, A.; Cluverius,
P.; de Chieze, Jacques; de la Feuille,
J.; de Wit, F.; Hondius, H.; Horn, G.;
Mercator, G.; Ottens, R.; Ptolemaeus,
C.; Tassin, N.; Tavernier, M.; Visscher,
N., III.

Jansson's Heirs *See under* Jansson, J.

Jansz, Harmen and Marten fl. 1604-10
Cart. Dut. M

Janszoon, Guilielmus *Same as* Blaeu, G.
(q.v.)

Janvier, Antide 1751-1835 Cart. Fr. M
G

See also under Gourné, P. M.; Longchamps
et Janvier.

Jeans, J. W. fl. 1846 Astron. Brit. X

Jefferis *Same as* Jefferys, T. (q.v.)

Jefferys, Thomas}
Jefferis } 1695?-1771 Pub., Eng.
Jefferies } Brit. M
Gefferys }

Succeeded by W. Faden (q.v.)

American atlas Lond., 1776, 1778, 1775
(Sayer and Bennett), and other edns.
With surveys by William Scull, Henry
Mouzon, Lewis Evans, W. Carver, Capt.
Cook, Major Holland, Lieut. Ross,
Michael Lane, Gardner, Hillcock and
others.

Bedfordshire 1765 and other edns. Surv.:
J. Ainslie, T. Donald. Eng.: J. Hodgkinson.

Buckinghamshire A. Dury, 1770 and other
edns. Surv.: J. Ainslie, T. Donald.

British Honduras c. 1770.

Florida from the latest authorities.

A plan of the town of Northampton . . . Surv.
by Noble and Butlin, 1746.

An exact chart of the river St Laurence . . .
Lond., 1775.

A new map of Nova Scotia . . . Lond., 1775.

*A description of the Spanish islands and settle-
ments* . . . *West Indies* Lond., 1762.

The West India atlas Lond., Sayer and
Bennett, 1775; French edn. (Paris, Julien),
1777.

Mappa ou carta geographica . . . Lond.,
Faden, 1790.

A map of South Carolina . . . Lond., 1757.
Based on surveys by William de Brahm
and others.

The west coast of Florida and Louisiana
Lond., R. Sayer, 1775.

184

A map of the most inhabited parts of New England . . . Lond., 1774.

A complete pilot for the West Indies . . . Lond., Laurie and Whittle, 1794-5 and subs. edns. Sometimes it is entitled *Laurie and Whittle's complete pilot for the West-Indies*.

A description of the maritime parts of France . . . Lond., Faden and Jefferys, 1774. Incl. maps by or after Cassini de Thury, Belidar, Buache, Beaurain, Le Rouge, Michelot and Bremond, Michelot, de Fer and Tassin.

The West-India islands . . . Lond., Laurie and Whittle, 1775 and subs. edns.

A general topography of North America and the West Indies . . . Lond., Sayer and Jefferies, 1762, 1768. Maps by or after Emanuel Bowen, d'Anville, Delisle, Charles Morris, Capt. Holland, George Heap, N. Scull, Thomas Hutchings, Joshua Fry and Peter Jefferson, John Dalrymple, Arthur Mackay, William de Brahm, le Sieur Bully, de Herrera, Fran. Math. Celi, L. Delarochette, de Caylus, R. de Marchais, D. Ross.

See also under ARMSTRONG, A.; D'ANVILLE, J. B.; DONN, B.; EYRE, T.; FADEN, W.; GRIDLEY, R.; HOMANN, J. B.; JULIEN, R. J.; KITCHIN, T.; LAURIE, R.; LE ROUGE, G. L.; MAYO, W.; OUTHETT, J.; RATZER, B.; SAXTON, C.; SAYER, R.; VON REILLY, F. J.

Jenichen, Balthasar fl. 1570 Cart. Ger. M

Jenkinson, Anthony } d. 1611 Cart.
Jenkinsonus, Antonius} Brit. M
Map of Russia for the Ortelius atlas, based on his *Russiae* (Lond., 1562).

Jenks, William 1778-1866 Cart. Amer. M
The explanatory bible atlas . . . Boston, C. Hickling, 1847 and other edns.

Jenner, Thomas d. 1673 Eng., Printseller, Bookseller Brit. M
A new booke of mapps exactly describing Europe Lond., 1645?.
See also under HOLLAR, W.; SIMONS, M.; VAN LANGREN, J.

Jenvilliers *See under* DESNOS, L. C.

Jhones, Richard *See under* GILBERT, H.

Joachimus, Albertus Lyttichius fl. 1569 Cart. Ger. M

Jobbins, J. R. fl. 1842 Eng., Pub. Brit. M
The environs of London . . . *with all the railways to 1842*.

Jobson, Francis fl. 1860 Surv. Brit. M

Jocelyn, N. and S. S. *See under* MORSE, S. E.

Jode *Same as* DE JODE (q.v.)

Johnson, Joseph fl. 1790 Cart. Brit. M
Map of Warwickshire, 1788 and other edns.

Johnson, T. *See under* MORTIER, D.

Johnson, Thomas fl. 1847 Pub. Brit. M

Johnson, William *Same as* BLAEU, G. (q.v.)

Johnson, William 19th cent. Cart. Brit. M
Maps of Lanarkshire, 1840 (eng. J. Hall) and West Lothian, 1820 (eng. N. R. Hewitt).

Johnston, Alexander Keith 1804-71 Pub. Brit. M G
Stanford's library map of Australia Lond., 1879.
There was also a younger Alexander Keith Johnston (1844-79).

Johnston, John fl. 1636-8 Printer Brit. M
Printed in partnership with Hondius.

Johnston, Sir William 1802-88 Pub. Brit. M

Johnstone, James *See under* ARROWSMITH, A.

Jolivet, Jean fl. 1545-60 Geog. Fr. M

Jollain, Cl. 17th cent. Cart., Pub. Fr. M
Trésor des cartes géographiques . . . Paris, 1667.

Joly, Joseph Romain 1715-1805 Cart. Fr. M
Atlas de l'ancienne géographie universelle . . . Paris, A. A. Lottin, A. Bertrand, 1801.

Jomini *Same as* DE JOMINI (q.v.)

Jones, Benjamin *See under* MARSHALL, J.

Jones, E. *See under* PLAYFAIR, J.; REES, A.; SMITH, CHARLES

Jones, Felix and Hyslop, J. M. 19th cent. Carts. Brit. M
Vestiges of Assyria Lond., John Walker for the East India Co., 1855.

Jones, John 18th cent. Globe- and Instrument-maker Brit. G

Jones, Samuel 19th cent. Globe- and Instrument-maker Brit. G

Jones, T. W. 18th/19th cent. Cart. Amer. M
The traveller's directory . . . (with S. S. Moore) Philadelphia, M. Carey, 1802, 1804. Maps eng. by I. Draper, W. Harrison, Junr., F. Shallus, James Smither.
See also under MACPHERSON, D. AND A.

Jones, William fl. 1820 Globe- and Instrument-maker Brit. G (C and T)

Jones and Smith *See under* LAURIE, R.

Jonghe *Same as* DE JONGHE (q.v.)

Jonssonius, Johann fl. 1620 Cart. Dut. G (T)

Jorden, Mark } c. 1531-95 Cart. Dan.
Jordanus, Marcus} M

J.S. Signature sometimes used by JOHN SUDBURY.
Also the supposed signature of JOHANN SIZLIN (qq.v.).

Jubrien, Jean 17th cent. Cart. Fr. M
Diocese de Rheims Amster., Blaeu, c. 1660.
See also under HONDIUS, H.; JANSSON, J.; TAVERNIER, M

Judaeus *Same as* DE JODE (q.v.)

Juettner, J. *Same as* JÜTTNER, J. (q.v.)

Jugge, Richard 16th cent. Eng. Brit. M

Julien, Roch-Joseph 18th cent. Pub. Fr. M
Atlas géographique et militaire de la France . . . Paris, 1751.
Atlas topographique et militaire Paris, 1758.
Noveau théâtre de la guerre Paris, 1758.
Le théâtre du monde 2 vols., Paris, 1768. Carts. incl.: Covens and Mortier, Bailleul, Boullanger, d'Anville, de Vaugondy, Delisle, Nolin, Cantel, Tillemont, Beaurain, Homann, Robert, Jaillot, S. Angelo, Sanson, Delamarche, Le Rouge, Caille, Rizzi-Zannoni, Palairet, Mitchel,
Coronelli, Buache, Jefferys, Halley, Caylus, de Wit, Scheuchzer, de Nezon, Lemare, N. J. Visscher, Lagrive, Dheulland.
See also under COURTALON, L'ABBÉ; JEFFERYS, T.; VON REILLY, F. J. J.

Jung, Georg Conrad and Johann Georg fl. 1641 Cart. Ger. M

Jung, J. A. *See under* MALLET, A. M.

Jungmann, Carl *See under* WEILAND, C. F.

Justinianus, A. *Same as* GIUSTINIANI, A. (q.v.)

Jüttner, Josef} 1775-1848 Instrument-
Juettner } maker Ger. G (C and T) O

Kaempffer 18th cent. Cart. Ger. M
See also under HOMANN, J. B.; OTTENS, R.

Kaerius *Same as* VAN DEN KEERE (q.v.)

Kaiser, J. F. fl. 1830 Globe-maker Ger. G (T)

Kammermeister, E. *Same as* CAMERARIUS, E. (q.v.)

Kartaro, Mario }
Karterus, Mario} *Same as* CARTARO, M.

Kästner, Abraham Gotthelf 1719-1800 Mathematician Ger. G

Kauffer, Michael 1673-1756
See under FABER, S.; KÖHLER, J. D.

Kawatsi Ya Gisuke fl. 1749 Cart. Japanese M

Keenan, James 18th cent. Surv. Irish M
Map of Co. Kildare, 1752 (with J. Noble; eng. D. Pomarede).
See also under NOBLE, J.

Keer, Keere *Same as* VAN DEN KEERE (q.v.)

Keerberg, Joannis *See* ORTEL, ABRAHAM

Kegel *Same as* PYRAMIUS, C. (q.v.)

Keill, John fl. 1718 Astron. Brit. X

Keller, Christoph } 1638-1707 Cart.
Cellarius, Christophorus} Ger. M
Geographica antiqua . . . Rome, 1774; English edn. pub. in Lond. for F. and C. Rivington, 1799 and subs. edns. Maps eng. by R. W. Seale, W. H. Toms.

Kelly, Colonel 19th cent. Surv. Brit. M

Kelly and Co. Directory pubs., still extant Brit. M

Keltenhofer, Stephan fl. 1549 Cart. Ger.? M

Keltzl, Abraham 16th cent. Cart. Ger. M

Kempen, Godefridus fl. 1584 Cart. Ger. M

Kennedy, Robert *See under* MASON, C.

Kentish, N. L. 19th cent. Cart. Brit. M
Map of Hampshire 1826, 1834 (with C. and and J. Greenwood).

Kepert, H. *See under* WEILAND, C. F.

Kepler, Johann 1571-1630 Astron. Ger. M

Kerius *Same as* VAN DEN KEERE (q.v.)

Keschedt, Petrus fl. 1597 Cart. Ger. M

Keulen *Same as* VAN KEULEN (q.v.)

Keymer, W. 18th cent. Pub. Brit. M

Keyser, Jacob *See under* DE LAT, J.; OTTENS, R.

Khegel *Same as* PYRAMIUS, C. (q.v.)

Kies, Eberhardt fl. 1628 Cart. Ger. M

Kilian, Georg Christoph 1709-81 Eng. Ger. M
Supplement oder zuzaz zu dem atlas curieux ... 2 vols., Augsburg, 1738?
(Supplement to the *Atlas curieux* of Gabriel Bodenehr, pub. *c.* 1704.) Maps are eng. by Bodenehr, Kilian and G. F. Riecke.
Kleiner atlas ... Augsburg, 1757? Carts. incl.; G. Bodenehr and T. Mayer.

King, G. B. *See under* SEAMAN, J. V.

King, Gregory 1648-1712 Eng. Mapseller. Surveyor. Brit. M
Clerk to Sir William Dugdale, Norroy King of Arms; in 1677 appointed Rouge Dragon Pursuivant of Arms.

King, John fl. 1714 Pub., Cart. Brit. M
See also under BOWLES, JOHN; MOLL, H.

King, W. 19th cent. Surv. Brit. M
Map of Leicestershire, 1806.

Kip, William fl. 1598-1610 Eng. Brit. M
Eng. some of the maps in Norden's works, and in Camden's *Britannia*.

Kiprianov, Vasilij d. 1723 Cart., Eng. Russ. M

Kirby, John and Joshua and William 1737-66 Carts. Brit. M
Among engs. employed by the Kirbys were BASIRE (William Blake's master) and RYLAND (qq.v.). They eng. two edns.
(1736/7 and 1766 respectively) of the Kirbys' map of Suffolk.

Kircher fl. 1600 Astron. Ger. X

Kirilov, Ivan 1689-1737 Surv., Editor Russ. M

Kirkwood, James } fl. 1774 Eng.
Kirkwood, J., and Sons} Brit. M
See also under CRAWFORD, W.; FORREST, W.; KIRKWOOD, R.

Kirkwood, Robert d. 1818 Eng. Brit. M
Operated with James Kirkwood as Kirkwood and Son.
See also under THOMSON, J., AND CO.

Kitchin, Jefferys and Beaufort
See under GASPARI, A. C.

Kitchin, Thomas 1718-84 Eng., Pub. Brit. M
A plan of the navigable canals ... Liverpool, *c.* 1770.
Map of South America in Thomas Falkner's *A description of Patagonia* (Hereford, 1774).
A general atlas ... Lond., R. Sayer, 1768-88; *c.* 1783; Sayer and Bennett, 1787; Laurie and Whittle, 1797; and many other edns. Carts.incl.: d'Anville, de Vaugondy, T. Jefferys.
The small English atlas ... (with T. Jefferys) Lond., 1749, 1751, 1775, 1785, 1787.
The large English atlas ... Lond., 1760, 1767.
The royal English atlas ... (with Emanuel Bowen) Lond., R. Wilkinson, 1762, 1778.
England illustrated Lond., 1764.
Kitchin's pocket atlas ... Lond., 1769.
Miniature atlas of Scotland from latest surveys ... 1770?
A new universal atlas ... Lond., Laurie and Whittle, 1789-96 and many other edns.
Carts. incl.: d'Anville, de Vaugondy, Rennell.
Maps by Kitchin appeared in *The London Magazine* from 1747 to 1760, and were used in *A new and complete abridgement ... in the antiquities of England and Wales* by Francis Grose (London, 1798), and in *Historical descriptions of ... antiquities of England and Wales* by Henry Boswell (Lond., 1786).

See also under EMANUEL BOWEN (with whom he collaborated) and under the following entries: ADAIR, J.; ARMSTRONG, A.; BURDETT, P. P.; DODSLEY, R.; FADEN, W.; GUTHRIE, W.; HOMANN'S HEIRS; MITCHELL, J.; PALAIRET, J.; RUSSELL, P.; SAYER, R.; VON REILLY, F. J. J.

Klaudian, Mikulas } fl. 1518 Pub.,
Claudianus, Nicolaus} Bookseller
Bohemian M

Kleinknecht, L. V. *See under* WEILAND, C. F.

Klinckowström, Axel Leonhard 1775-1837 Cart. Swed. M
Atlas . . . om de förente staterne Stockholm, 1824.
Some maps are signed: John Melish.

Klinger, Johann George b. 1764 Cart. Ger. G (C and T)

Klint, Gustaf 1771-1840 Hydrographer Swed. M
Sverijes sjö-atlas Stockholm, 1832-45 (hydrographic charts of European coastlines).

Klose, J. G. B. *See under* SOHR, K.

Knaplock, D. *See under* MOLL, H.

Knapton, John (d. 1770) **and Paul** (d.1755) Pubs. Brit. M
Geographica classica . . . Lond., 1747 (8th edn.).
See also under MOLL, H.

Kneass *See under* CAREY, H. C.; MACPHERSON, D. and A.

Knight, Capt. J. fl. 1795 Cart. Brit. M
See also under TANNER, H. S.

Knight, William 18th cent. Pub. Brit. M

Knox, George *See under* McCREA, W.

Knox, James 19th cent. Cart. Brit. M
Map of Midlothian, 1812 and subs. edns.

Knox, Robert 19th cent. Cart. Brit. M
Map of 15 miles around Scarborough, and 1821 (eng. A. Findlay), 1849.

Koch, Christophe Guillaume 1737-1813 Cart. ? M
Maps and tables of chronology and genealogy . . . Lond., Baldwin and Cradock, 1831 (with 7 maps).

Köhler, Johann David} 1684-1755 Cart.,
Koehler   } Scholar Ger. M
Koeleri, Io. Davidis}
Atlas manualis . . . Nuremberg, C. Weigelio, 1724? Carts. incl: Abraham Goos, J. B. Homann, Hermann Moll.
Bequemer schul- und reisen-atlas . . . Nuremberg, C. Weigeln, 1734? Carts. incl: Goos, Moll, Homann, Valvasor, Falda, Müller, Kauffer, Reland, Delisle and Weigel.
Descriptio orbis antiqui . . . Nuremberg, 1720? (The same as C. Weigel's atlas of the same title, q.v.)

Koler, Johann fl. 1572 Printer Ger. M
Printed a Ger. version of Ortelius's *Theatrum.*

Koops, M. fl. 1796 Cart. Ger. M

Kopernikus, N. *Same as* COPERNICUS, N. (q.v.)

Korabinsky, J. M., and Townson, R. 18th/19th cent. Carts. Dut.? M
Nieuwe kaart van Hongaryen . . . The Hague, J. C. Leeuwestijn, 1801.

Kotzebue *Same as* VON KOTZEBUE (q.v.)

Kraay *See under* VOOGT, C. J.

Krafft, Georg Wolfgang 1701-54 Mathematician Ger. G

Kratz, W. *See under* WEILAND, C. F.

Kraus *See under* VALCK, G.

Kremer *Same as* MERCATOR (q.v.)

Krevelt *Same as* VAN KREVELT (q.v.)

Kribber, Cornelius 18th cent. Cart. Dut. M
Belgii par septentrionalis . . . Utrecht, 1751.

Krueger, A. fl. 1855 Astron. Ger. X

Krug, Lenhart fr. 1589 Silversmith Ger.? G (T)

Krünitz, Johann Georg 1728-96 Physician, Lexicographer Ger. G

Kruse, Karsten Christian 1753-1827 Cart. Ger. M
Atlas zur übersicht der geschichte aller Europäischen länder und staaten Leipzig, 1822?
Tabellen und charten zur allgemeinen geschichte der drey letzen jahrhunderte . . . Leipzig, 1821 (with 5 maps).

Krusenstern, A. J. *Same as* VON KRUSENSTERN, A. J. (q.v.)

Kruzenshtern, Ivan Fedorovich 1770–
1846 Sailor Russ. M
*Atlas to Captain Kruzenshtern's voyage around
the world* . . . St. Petersburg, 1813.
Title, etc., are in Russian.

Kühnen, G. W. *See under* MÜLLER, J. U.

Kümmer, K. W. fl. *c.* 1830 Globe-
maker Ger.? G (T)

Küssell, Melchoir fl. 1669 Eng. Ger.?
M

Kyrilov, Ivan 18th cent. Surv. Russ. M
Imperii Russici tabula generalis . . . St.
Petersburg, 1734.
The first eng. and pub. general map of
Russia.

Labanna, J. B.}
Labaña   }
Lavaña   }
See under DE WIT, F.; HONDIUS, H.; JANSSON,
J.; MERCATOR, G.; WOLFGANG, A.

Labiche et Bérard *See under* DE FREYCINET,
L. C. D.

Laborde *Same as* DE LABORDE (q.v.)

Lacaille *Same as* DE LA CAILLE (q.v.)

Lacam *See under* LAURIE, R.

Lackington, Allen, and Co. *See under*
LUFFMAN, J.

Laet *Same as* DE LAET (q.v.)

La Feuille *Same as* DE LA FEUILLE (q.v.)

Lafont *See under* MONIN, V.

La Fosse *Same as* DE LA FOSSE (q.v.)

Lafréry, Antoine}
Lafreri, Antony } 1512–77 Cart. Ital. M
Geografia tavole moderne . . . Rome, *c.* 1575.
Carts., engs. incl.: J. Gastaldi; J. F.
Camotii; P. Forlani; F. Berteli; B.
Zolterii; P. Ligorio and others.
See also under ARGARIA, G.; MAGNUS, O.;
VAN AELST, N.; VAN BOS, J. B.

Lagalla fl. 1614 Astron. Ital. X

Lagrive *See under* JULIEN, R. J.

La Guillotière *Same as* DE LA GUILLOTIÈRE
(q.v.)

La Haye *Same as* DE LA HAYE

La Hire *Same as* DE LA HIRE (q.v.)

La Houue *Same as* DE LA HOUUE (q.v.)

Laicksteen-Sgrooten}
Laicksteen, Peter } 16th cent. Carts.
Sgrooten, Chr.   } Dut. M
Map of Palestine (*see under* COCK, H.).

Laing, Major fl. 1830 Cart. Brit. M

Lakeman *See under* VISSCHER, N., III

Lalande, Joseph Jérôme le Français
1732–1807 Astron. Fr. G (C)

Lale *See under* VUILLEMIN, A. A.

Lalin *Same as* DE LALIN (q.v.)

Lamarche *Same as* DELAMARCHE (q.v.)

Lamb, Francis fl. 1670–1700 Eng., Pub.
Brit. M
A new map of East India Lond., Bassett and
Chiswell, 1676.
See also under BLOME, R.; PETTY, SIR W.;
THORNTON, J.; VAN KEULEN, J.

Lambert, M. and M. W. *See under* BELL,
J. T. W.; FRYER, J., AND SONS

Lang, Mauritius fl. 1664 Eng. Brit.?
M

Lane, Michael fl. 1775 Cart., Brit. M
G (T)
Lane's pocket globe Lond., 1818.
See also under COOK, CAPT. JAMES; FADEN.
W.; JEFFERYS, T.; LE ROUGE.

Langdon, Thomas 16th cent. Cart.
Brit. M

Langenes, Barent fl. end of 16th cent.
Cart. Dut. M
Caerte-thresoor . . . Amster., C. Claesz,
1599. Carts. incl.: Hondius, Keer,
Pigafetta, Wright. Fr. edn. (trs. by Jean
de la Haye), 1602; 1610?
Hand-boeck . . . (with 172 maps) Amster.,
C. Claesz, 1609.

Langeren *Same as* VAN LANGREN (q.v.)

Langeweg, D. I. 18th cent. Cart. Dut.
M
Nouveau plan de la Haye . . . The Hague,
H. C. Gutteling, 1776.

Langlands, George, and Son 18th cent.
Cart. Brit. M
Maps of Argyllshire, 1793 and 1801 (eng.
S. J. Neele).

Langley, Edward fl. 1804–35 Eng.,
Bookseller, Pub., Printer Brit. M
Operated also as a firm: Langley and Belch,
and Langley and Phelps.
*Langley's new county atlas of England and
Wales . . .* Lond., 1817-18.

Langlois, Hyacinthe 19th cent. Cart.,
Pub. Fr. M G
Atlas portatif et itinéraire de l'Europe . . .
Paris, 1817.
Grand atlas français départmental . . . Paris,
1856.

Langlumé, J. *See under* VUILLEMIN, A. A.

Langren *Same as* VAN LANGREN (q.v.)

Langrenus fl. 1630? Astron. Span. X

Lannoy *Same as* DE LANNOY (q.v.)

La Paz *Same as* DE LA PAZ (q.v.)

Lapérouse *Same as* DE LAPÉROUSE (q.v.)

Lapié, Alexander Émile 19th cent.
Cart. Fr. M
The French empire 1811; 1813, Lond.,
Faden.
Atlas universel . . . (with P. Lapie) Paris,
Eymer, Fruger et Cie, 1829-33, and other
edns.
See also under BOSSI, L.

Lapié, Pierre 1777-1851 Cart. Fr. M
G
Atlas classique et universel . . . Paris,
Magimel, Picquet, 1812. Each map is
inscribed in lower left-hand corner:
'Adam et Giraldon dirext.'
See also under LAPIÉ, A. E.

Laplace, Cyrille Pierre Théodore 1793-
1875 Cart. Fr. M
*Voyage autour du monde . . . atlas hydro-
graphique* Paris, 1833-9.

Lapointe *Same as* DE LA POINTE (q.v.)

La Popelinière 16th cent. Cart. Fr. M

Laporte *Same as* DE LAPORTE (q.v.)

Larkin, William fl. 1812-19 Cart. Irish
M
Maps of Co. Galway, 1819 and 1820
(reduction), Co. Leitrim, 1819, Co. Meath,
1812-17, and others.

La Rochette *Same as* DE LA ROCHETTE (q.v.)

Las Cases *Same as* DE LAS CASES

Laso Francisco *See under* DE AFFERDEN, F.

Lat *Same as* DE LAT (q.v.)

La Tour *Same as* DE LA TOUR (q.v.)

Latour, Arsène Lacarrière d. 1839 Cart.
Amer. M
*Atlas to the historical memoir of the war in
West Florida and Louisiana . . .* Philadel-
phia, J. Conrad and Co., 1816.

Lattré, Jean fl. 1772-35 Pub., Eng. Fr.
M G
Pub. maps by R. Bonne.
Atlas tôpographique des environs de Paris . . .
Paris, 1762?.
Petit atlas moderne . . . Paris, 1783, 1793;
Delamarche, 1821?
See also under RIZZI-ZANNONI, G. A.

Latz, Latzen *Same as* LAZIUS (q.v.)

Laurenberg, Joanne} *See under* BLAEU, G.;
Laurenbergio } HONDIUS, H.;
Lauremburg } HORN, G.; JANSSEN, J.;
MERCATOR, G.;
VISSCHER, N., III

Laurent *See under* DE LA HARPE, J. F.;
GENDRON, P.

Laurentii, Francesco di Niccoli *See
under* PTOLEMY, C.

Laurie and Whittle *See under* LAURIE, R.

Laurie, John fl. 1763 Cart. Brit. M
Map of Midlothian, 1763 (eng. A. Baillie)
and subs. edns.

Laurie, Richard Holmes *See under*
D'ANVILLE, J. B. B.; LAURIE, R.

Laurie, Robert c. 1755-1836 Pub., Eng.
Brit. M
Also operated in partnership with James
Whittle as Laurie and Whittle.
Pub. d'Après de Mannevillette's *The oriental
pilot,* Lond., 1797.
*A new and correct map of the British colonies in
North America . . .* Lond., 1794.
*A new and general map of the southern
dominions . . . United States of America*
Lond., 1794.
*Laurie and Whittle's New and improved English
atlas* Lond., 1807.
Laurie and Whittle's New traveller's companion
Lond., 1811, 1813 (7th edn. 'James
Whittle and Richard Holmes Laurie').
*Laurie and Whittle's new map of the county
of York* 1806.

The African pilot . . . Lond., 1801, 1816.

The complete East India pilot . . . Lond., 1800, 1803, 1806, 1810. 2 vols. Maps by or after: Bordee, Butler, Collins, d'Après de Mannevillette, d'Anville, de Fleurieu, Gerard de Ruyter, Grenier, Haldane, Hayes, Huddard, Jefferys, Lacam, Lemprière, Lesley, Lewis, Lindsay, Magin, Maxwell, Moffat, Murillo, Nicholson, Owen, Price, Popham, Stephenson, van Keulen, Woodville, etc.

The country trade East-India pilot . . . Lond., 1799, 1803.

A new and elegant imperial sheet atlas . . . Lond., 1796, 1798, 1800, 1808, 1813-14. Carts. incl.: Samuel Dunn, J. Enouy, J. Lodge, Mostyn John Armstrong, Francis de Caroly, Robert Mylne, Thomas Jefferys, John Roberts, John Stephenson, James Rennell, d'Anville.

A new juvenile atlas . . . Philadelphia, for J. Melish, J. Vallance and H. S. Tanner, by G. Palmer, 1814.

Laurie and Whittle's New and elegant general atlas . . . Lond., 1804. Engs.: Jones and Smith, S. J. Neele, George Allen, V. Woodthorpe. Maps are dated 1 October 1801.

See also under ARMSTRONG, A.; ARMSTRONG, M. J.; BACKHOUSE, T.; BICKHAM, G.; D'APRÈS DE MANNEVILLETTE, J.; DUNN, S.; EVANS, L.; FADEN, W.; FINDLAY, A. G.; HOLLAND, CAPT.; JEFFERYS, T.; KITCHIN, T.; PURDY, J.; SCALÉ, B.; STEPHENSON, J.

Lauterbach, Joh. Balth. fl. 1683 Astron. Ger.? G (C)

Laval, J. *See under* MALTE-BRUN, C.

Lavaña *Same as* LABANNA (q.v.)

Lavoisne, C. V. 19th cent. Cart. Amer. M

A new historical, chronological and geographical atlas . . . Philadelphia, M. Carey and Son, 1820, 1821; Lond., J. Barfield, 1807 (with C. Gros). Carts. incl.: E. Paguenaud, J. Aspin, John Melish.

Laws, William 18th cent. Surv. Brit. M
Plan of the harbour and town of Cartagena [America] Lond., S. Harding, 1741.

See also under CHASSEREAU, PETER.

Lawson, A. *See under* PINKERTON, J.

Lawson, John Parker fl. 1842 Geog. Brit. M

Lazarus, Ungar fl. 1515-28 Cart. Dut.? M

Lazius, Wolfgang} 1514-65 Cart.,
Lazio   } Historian, Physician
Latz   } Austrian M
Latzen   } *See also under* BLAEU, G.; HONDIUS, H.; JANSSON, J.; MERCATOR, G.; ORTEL, A.; VISSCHER, N., III.

Lea and Blanchard *See under* BUTLER, S.

Lea, Anne d. 1730 Pub. Brit. M
Widow of Philip Lea (q.v.)

Lea, Isaac 1792-1886 Cart. Amer. M
See under CAREY, H. C.

Lea, Philip d. 1700 Cart., Pub. Brit. M G (C and T)
A new mapp of America Lond., c. 1686.

A new mapp of Asia Lond., Lea and Overton, c. 1686.

A new map of Carolina.

Reissue of Saxton's *Shropshire*, Lond., Willdey, 1690.

A new mapp of Europe Lea and Overton, Lond., c. 1686.

Hydrographia universalis . . . Lond., 1700? Incl. a map of Londonderry by Capt. T. Phillips.

See also under ADAMS, J.; BEEK, A.; GREENE, R.; OVERTON, I.; SAXTON, C.; SELLER, J.; VISSCHER, N., III.

Leard, John 18th cent. Pub. Brit. M
Charts and plans of Jamaica . . . Lond., 1793. Maps eng. by S. I. Neele; some inscribed: 'Sold by Mount & Davidson, & W. Faden.'

Leavitt, Lord and Co. *See under* PALMER, R.

Le Boucher, Odet Julien 1744-1826 Cart. Fr. M
Atlas pour servir à l'intelligence . . . de la guerre de l'indépendance des États-Unis Paris, Anselin, 1830.

Le Clerc, Jean   } 1657-1736 Pub. Dut.
Clericus, Johannes} M
Atlas antiquus . . . Amster., Covens and Mortier, 1705. Carts. incl.: N. Sanson, Petrum Mortier, Pierre Daniel Huet, Ph. de la Rue, I. van Luchtenburg, G. Del Isle [sic], Guilielmi Sanson, P. du Val.
See also under TAVERNIER, M.

L'Écuy, Jean Marie, Abbé fl. 1578 Cart.

Ledoux, Charles Nicolas} 1736–1806
Ledoux, E. } Architect, Pub.
Ledoux et Tenré } Fr. *See under*
CREVIER, J. B. L.;
DE LA HARPE, J. F.

Leech, Major R. *See under* WALKER, J.
AND C.

Leeuwestijn, J. C. 18th/19th cent. Pub.
Dut. M
See also under KORABINSKY, J. M.

Lefebrise fl. 1814 Astron. Fr. X

Legatt, John *See under* SPEED, J.

Legrand, Augustin 19th cent. Geog. Fr.
M
Exposition géographique . . . Paris, Bulla,
1839.

Legrand, P. fl. 1720 Cart., Eng. Fr. M
G (T)
See also under ANCELIN.

Leigh, Samuel fl. 1820–42 Pub. Brit.
M
*Leigh's New pocket atlas of England and
Wales* . . . Lond., 1820, 1831 (3rd edn.),
1839 (7th edn.), 1843 (10th edn.). Eng.
by Sidney Hall. It is sometimes bound
up with *Leigh's New pocket road-book of
England and Wales.*
See also under DELKESKAMP, F. W.

Lejeune, T. *See under* MALTE-BRUN, C.

Le Keux, J. fl. 1842 Eng. Brit. .M

Lelewel, Joachim 1786–1861 Cart.
Polish M
Atlas de J. Lelewela . . . Vilna and Warsaw,
1818.

Le Lorrain, Pierre *Same as* DE VALLEMONT.
(q.v.)

Le Maire, Jacob 17th cent. Explorer
Dut. M
Spieghel der Australische navigatie . . .
Amster., Michiel Colijn, 1622.

Lemare *See under* JULIEN, R. J.

Lemau de la Jaisse 18th cent. Cart. Fr.
M
*Plans des principales places de guerre et villes
maritimes frontières du royaume de France* . . .
Paris, Didot, 1736.

Lemonnier, Pierre Charles fl. 1776
Astron. Fr. X

Le Moyne de Morgues, Jacques 16th cent.
Eng. Fr. M
Eng. a map of Florida in de Bry's *Brevis
narratio corum quae in Florida* . . .
Frankfurt-am-Main, 1591.

Lemprière, Capt. Clement 18th cent.
Cart. Brit. M
*A new and accurate survey of the island of
Minorca* Lond., Rocque, 1753.
Plan of the harbour and town of Cartagena
[America], Lond., 1741 (eng. I. Basire).
See also under LAURIE, R.; SAYER, R.;
STEPHENSON, J.

Lendrick, John fl. 1782–1808 Cart.
Irish M
Map of Co. Antrim, 1780 (eng. S. Pyle)
and subs. edns.

Lenex, John 18th cent. Cart. Brit. M
Turkey in Europe Lond., c. 1725.

Lenglet Dufresnoy, Nicolas 1674–1755
Geog. Fr. M
Kurzverfassete kinder geographie . . . (with 8
maps) Nuremberg, G. P. Monath, 1764.

Lenny, J. G. 19th cent. Cart. Brit. M
Map of 10 miles round Bury (Suffolk), 1823
(eng. S. Hall).

Lens, B. *See under* MOLL, H.

Lenthall, John fl. 1720 Cart. Brit. M
Playing-card maps copied from Robert
Morden's.

Leo, Sibrandus fl. 1579 Cart. Dut. M

Le Rouge, George Louis fl. 1741–79
Cart., Pub. Fr. M
Atlas Ameriquain septentrional . . . Paris, 1778.
Carts.: Holland, Evans, Scull, Mouzon,
Ross, Cook, Lane, Gilbert, Gardner, Dr.
Mitchel, Tryon, Bull, Bryan and Brahm,
R. Pocock.
Atlas nouveau portatif . . . Paris, 1748,
1756–9, 1767–73 (pub. by Crepy).
Pilote Américain septentrional . . . Paris, 1778
and other edns. Carts. incl.: Jefferys
(Jefferis or Gefferys in places), Lane,
Morris, des Barres, Smith, Blaskowitz
and Scull.

Recueil des fortifications, forts et ports de mer de France . . . Paris, 1760?

Recueil des plans de l'Amérique septentrionale . . . Paris, 1755.

Recueil des villes, ports d'Angleterre Paris, 1759.

Atlas général . . . Paris, 1741-62.

See also under DELISLE, G.; DESNOS, L. C.; JEFFERYS, T.; JULIEN, R. J.; MONTRESOR, CAPT.

Leroux *See under* LORRAIN, A.

Le Roy, Jacques 1633-1719 Geog., Historian Belgian M

Le Sage, A. Pseudonym of DE LAS CASES, E. (q.v.)

Lescarbot, Marc *See under* MILLOT, I.

L'Escluse *Same as* CLUSIUS, CAROLUS (q.v.)

Lesley *See under* LAURIE, R.

Leslie, John, Bishop of Ross fl. 1578-96 Cart. Brit. M

Scotiae regni antiquissimi . . . Rouen? 1578.

Lesueur *See under* DE FREYCINET, L. C. D.

Leth *Same as* DE LETH (q.v.)

Letronne, Antoine Jean 1787-1848 Hist. Fr. M

Atlas de géographie ancienne . . . Paris, Didot, 1827. Maps by Auguste Henri Dufour. Eng. by Flahaut.

Lettany, Franz 1792/3-1863 Military cart. Fr. M G

Leucho *Same as* DE LEUCHO (q.v.)

Levanto, Francesco Maria 17th cent. Cart. Ital. M

Primo parte dello specchio del mare . . . Genoa, Gerolamo Marino e Benedetto Celle, 1664.

Levasseur, Guilleaume d. 1643 Hydrographer, Pilot Fr. M

Levasseur, V. fl. 1847 Cart. Fr. M

Atlas national . . . *de la France* . . . Paris, A. Combette, 1847 and subs. edns.

Levasseur de Beauplan, Guil. d. 1685 Cart., Engineer Fr. M
Son of G. LEVASSEUR (q.v.)
Noted for work in Poland.
See also under DE WIT, F.

Levrault, F. G. *See under* DE BYLANDT PALSTERCAMP, A.

Lewis, J. *See under* LAURIE, R.; WILLETTS, J.

Lewis, Samuel fl. 1831 Pub., Cart. Brit. M
Operated as S. Lewis and Co.
Issued an atlas to accompany his *Topographical dictionary of England* . . . Lond., 1831, 1835, c. 1837, 1840, 1842, 1844, 1845, 1848.
Map of Scotland (accompanying his *Topographical Dictionary of Scotland*, 1846).
Atlas of the counties of Ireland (accompanying his *Topographical dictionary of Ireland*, 1846).
See also under ARROWSMITH, A.; CAREY, M.; CREIGHTON, R.; EDWARDS, B.; GUTHRIE, W.; LUCAS, F.; MARSHALL, J.; PINKERTON, J.

Lewis, William fl. 1819-36 Pub. Brit. M
Lewis's new travellers' guide . . . Lond., 1819, 1836. Maps originally pub. by J. Wallis in 1810; afterwards they appeared in *Martin's sportsman's almanack*. In the 1st (1819) edn., 23 county maps bear Martin's imprint, 1 has Lewis's imprint, 16 have no imprint. By the time another edn. had appeared, later in 1819, all the maps have the imprint: 'Publish'd by W. Lewis, Finch Lane.'
See also under WALLIS, J.

L'Honoré *See under* CHÂTELAIN, H. A.

L'Hoste, Jean d. 1631 Mathematician Fr. G

Lhuyd, Humphrey} 1527-1565 Cart.
Lloyd   } Brit. M
Map of Wales in Ortelius's *Atlas*, 1594.
See also under HONDIUS, H.; HORN, G.; MERCATOR, G.

Libri, Francesco fl. 1529 Cart. Ital. G (T)

Licinius, Fabio fl. 1544-70 Eng., Printer Ital. M

Liébaux, J. B. *See under* DESNOS, L. C.; THEVENOT, M

Liefrinck, Johannes 1518-73 Eng., Mapseller Ger. M

Liefrink, Mynken 16th cent. Cart. Dut. M

Liesvelt, J. fl. 1549 Pub. Dut. M

Light, Col. 19th cent. Cart. Brit. M
The district of Adelaide, South Australia
Lond., J. Arrowsmith, 1839.
See also under FLINDERS, CAPT. M.

Ligorio, Pirro } 1496?–1580? Cart.,
Pyrrhus, Ligorius} Architect, Artist Ital.
                    } M
See also under LAFRÉRY, A.

Lilius, Zacharias 15th cent. Geog. Ital.
M
De origine et laudibus scientiarium . . . (with
one woodcut map) Florence, Francesco
Bonaccorsi for Pierro Pacini, 1496, and
other edns.

Lily, George } fl. 1528–59 Cart. Brit.
Lilius, Georgius} M
Britanniae insulae . . . Rome, 1546. Signed
in floral decoration of cartouche at lower
right-hand corner: GLA (= *Georgius
Lilius Anglorum*). There were 8 edns. or
derivatives.

Lincoln and Edmands *See under* EDMANDS,
B. F.

Lindley, Joseph and Crossley, W.
fl. 1790–3 Carts. Brit. M
Map of Surrey, 1790 and subs. edns. (eng. B.
Baker).

Lindner, Cornelius 1694–1740 Mathematician Ger. G

Lindsay *See under* LAURIE, R.

Linschóten, Jan Huygen 1563–1610 Cart.
Dut. M
Histoire de la navigation aux Indes Orientales
Amster., Jean Evertsz Cloppenburch,
1619.
*Navigatio ac itinerarium Johannis Hugonis
Linscotani . . .* The Hague, Alberti
Henrici, 1599.

Littrow *Same as* VON LITTROW (q.v.)

Lizars, Daniel d. 1812 Eng. Brit. M
*The Edinburgh geographical and historical
atlas . . .* Edinburgh, J. Hamilton,
successor to D. Lizars, 1831?; 1842;
William Home Lizars, 1842?

Lizars, William Home 1788–1859 Eng.,
Pub. Brit. M
Eng. Blackwood's *Atlas*, 1830.
See also under EWING, T.; LIZARS, D.

Ljatkoj, Ivan fl. 1555 Soldier, Cart.
Russ. M

Lloyd *Same as* LHUYD (q.v.)

Lobeck, Tobias 18th cent. Cart. Ger.
M
Atlas geographicus portatilis . . . Augsburg,
1762. Maps eng. by Tob. Conr. Lotter.
Kurzgefasste geographie . . . Augsburg, A.
Brinhauser, 1762?

Locard et Davi *See under* ROUSSET, P.

Lochom *Same as* VAN LOCHOM (q.v.)

Lockwood, Anthony d. 1855 Master,
Royal Navy Brit. M
A brief description of Nova Scotia . . . Lond.,
1818. Maps eng. by J. Walker.

Lockwood, R. *See under* HART, J. C.

Lodesano, Francesco fl. 1544 Cart. Ital.
M

Lodewicks, Willem} fl. 1595–8 Cart.
Lodewycks   } Dut. M
Sometimes signed with the initials:
G. M. A. W. L.

Lodge, John fl. 1754–94 Geog., Eng.
Brit. M
See also under LAURIE, R.; REES, A.

Loggan, D. *See under* MORTIER, D.

Lohrmann, W. G. fl. 1824 Astron.
Ger. X

Lok, Michael 16th cent. Cart., Surv. Brit.
M

Lomonosov, Michael 1711–65 Geog.
Russ. M

Long, Major S. H. 19th cent. Cart.
Amer. M
Map of Arkansas, etc., in Carey and Lea's
A complete . . . American atlas, 1827.
See also under CAREY, H. C.

Longchamps, S. G. *See under* DESNOS, L. C.

Longchamps et Janvier 18th cent. Pub.
Fr. M

Longman and Co. *See under* HALL, S.

Longman, Brown, Green and Longmans
See under BUTLER, S.; HALL, S.

Longman, Hurst, Rees and Orme *See
under* CRUTTWELL, C.; PIMENTEL, M.

**Longman, Hurst, Rees, Orme and
Browne** *See under* REES, A.

Longman, Rees and Co. *See under* WALKER, J. A. AND C.

Longman, Rees, Orme, Brown and Green *See under* HALL, S.

Longmans and Co. *See under* HUGHES, W.

Longobardi, Niccoló 1566–1654 Missionary Ital. G

Loon *Same as* VAN LOON (q.v.)

Loots, Joannes} 17th/18th cent. Cart.
Lootz } Dut. M
Atlas . . . Amster., 1705? Carts. incl.: Claes de Vries, C. J. Vooght.
See also under JACOBSZ T.

Loots-man *Same as* JACOBSZ (q.v.)

López, Bastian or Sebastião fl. 1558 Cart. Span. or Port. M

López, Thomas or Tomas 18th/19th cent. Cart. Port. M
Mapa generál del reyno de Portugal Madrid, 1778.
A new general military map of . . . Portugal Lond., Stockdale, 1811.
Various maps of Spain. Madrid, 1765–98.
Atlas elemental . . . Madrid, 1792.
Atlas geográfico de España . . . Madrid, 1810–18, 1830–5.
Atlas geográfico de la America septentrional y meridional . . . Madrid, A. Sanz, 1758.
Atlas geográphico del reyno de España . . . Madrid, 1757.
See also under HOMANN'S HEIRS; SCHAEMBL, F. A.; TOFIÑO DE SAN MIGUEL, V.; VON REILLY, F. J. J.

Loputzkij, Stanislav fl. 1663–68 Artist, Cart. Russ. M

Lorichs, Melchoir fl. 1568 Cart. Ger. M

Lorin *See under* ARROWSMITH, A.

Loritus, H.} *Same as* GLAREANUS, H. (q.v.)
Loritz }

Lorrain, A. 19th cent. Cart. Fr. M
La France et ses colonies . . . Paris, Michel fils ainé, 1836. Engs. incl.: P. Tardieu, Leroux, Piat, H. Dandeleux.

Löscher, Valentinus Ernest fl. 1703 Cart. Ger. M

Loss *Same as* VAN DER LOSS (q.v.)

Lothian, John fl. 1825–35 Geog., Pub. Brit. M
Bible atlas . . . Hartford, N. Case, 1832.
Lothian's Historical atlas of Scotland . . . Edinburgh, 1829.

Lotter, Georg Friedrich *See under* LOTTER, T. C.

Lotter, Matthew Albrecht *See under* LOTTER, T. C.

Lotter, Tobias Conrad 1717–77 Cart., Pub. Ger. M
Pub. maps by Delisle, A. C. Seutter.
A map of the most inhabited part of New England Augsburg, 1776 (copied by Probst, 1779).
Recens edita totius Novi Belgii . . . Augsburg, 1760.
Atlas géographique . . . Nuremberg, Homann's Heirs, 1778. Carts. incl.: M. Seutt[er], Gabriel Walser, Johan Michael Probst, de l'Isle, Matt. Albrecht Lotter, Gottfrid Rogg, Georg Friedrich Lotter, J. M. Seligmann, John Green.
Atlas minor . . . Augsburg, 1744? The same book as Matthew Seutter's *Atlas minor*, but with 11 extra maps and Lotter's name substituted for Seutter's on 'Imperium Romano-Germanicum'; the 11 extra maps are each signed by Lotter.
Atlas novus . . . Augsburg, 1722? Carts. incl. de l'Isle and other members of the Lotter family.
See also under HOMANN'S HEIRS; LOBECK, T.; PROBST, J. M.; SEUTTER, G. M.

Lottin, A. A. *See under* DUMONT D'URVILLE, J. S. C.; JOLY, J. R.

Lotzbeck, J. L. *See under* REICHARD, C. G.

Louis, S. *Same as* LEWIS, S. (q.v.). *See also under* EDWARD, B.

Lous, Professor C. C. *See under* FADEN, W.

Lovenhorn *Same as* DE LOVENHORN (q.v.)

Low and Wallis *See under* PAYNE, J.

Löwenberg, Julius 1800–93 Cart. Ger. M
Historisch-geographischer atlas . . . Freiburg, Herder, 1839.

Lowizio, Georg Moritz} 1722-74 Cart.
Lowitz } Dut.? M G
} (C and T)
See also under HOMANN'S HEIRS.

Lowry, Joseph Wilson *See under* REES, A.;
SHARPE, J.

Lübbin, E. *Same as* LUBINO, E. (q.v.)

Lubieniczski, Stanislas fl. 1681 Astron.
Polish X

Lübin, Augustin or Agostino 1624-95
Cart. Fr. M *Orbis augustianus* . . . Paris,
P. Baudouyn, 1659.
See also under DE ROSSI, G. G.; HONDIUS, H.;
JANSSON, J.; MERCATOR, G.

Lubino, Eichardo }
Lübbin, Eilert } 1565-1621 Scholar
Lubinus, Eilhardus} Ger. M
Lübben, Eilert }
See also under BLAEU, G.

Lucas, Fielding, Junr. 1781-1852 Cart.,
Pub. Amer. M
A general atlas . . . Baltimore, 1823.
A new and elegant general atlas . . . Baltimore,
1816? Carts. incl.: S. Lewis, Samuel
Harrison. Eng. H. S. Tanner.
A new general atlas of the West India islands . . .
Baltimore, 1824?
See also under CAREY, H. C.; GOODRICH,
S. G.; MILLS, R.

Lucas, P. *See under* DE LAT, J.

Luchtenburg *Same as* VAN LUCHTENBURG
(q.v.)

Luchtmans, S. and J. *See under* DE STUERS,
F. V. H. A.

Lucini, Antonio Francesco 17th cent.
Eng., Pub. Ital. M
See also under DUDLEY, SIR R.

Lucini, Vincenzo} 16th cent. Printer, Eng.
Luchini } Ital. M

Lud, S. fl. 1627 Cart. Ital. G

Ludolf, Christian fl. 1683 Cart. Dut.
M

Ludovicus, Georgius *Same as* GEORGIUS, L.
(q.v.)

Luffman, John fl. 1776-1820 Geog., Eng.
Pub. Brit. M
*A new map of the seat of war between Russia
and France* Lond., c. 1812.

A representation of the coast of England . . .
London., 1803.
A new pocket atlas . . . *of England and Wales*
Lond., Lackington, Allen and Co., 1806.
*Luffman's Geographical and topographical
atlas* . . . Lond., 1815-16.
Select plans of the principal cities . . . *in the
world* . . . 2 vols., Lond., 1801-3.
See also under ARMSTRONG, A.; PRIOR, J.

Lütké, Fedor Petrovich, Graf 1797-1882
Explorer Russ. M
Atlas de voyage autour de monde . . . *sous les
ordres de Fréderic Lütké* . . . St. Petersburg,
1835. Titles and lettering in Russ. and
Fr.

Lützenkirchen, Wilhelm fl. 1597 Bookseller Ger. M

Luyts, Jan 1655-1721 Cart. Dut. M
Introductio ad geographiam novam et veterem . . .
(with 66 maps) Trajecti ad Rhenum, F.
Hahna, 1692. Carts. incl.: Sanson,
d'Abbeville.

Lužin, Fedor *See under* EVREINOV, I.

Lycosthenes, Conrad ? Astron. Ger.?
X

Lyne, Richard fl. 1574 Cart. Brit. M
Plan of Cambridge, 1574.

Lynslager *See under* VOOGT, C. J.

Lythe, Robert fl. 1860 Surv., Cart. Brit.
M

Maas, Abraham fl. 1725-9 Cart. Dut.
M
Worked in Russia.

Maccari, Giovanni fl. 1685 Instrument-
maker Ital. O

M'Carty and Davis *See under* GOODRICH,
C. A.

McCrea, W.} fl. 1795-1813 Cart. Irish
McRea } M
Maps of Donegal (1801), Monaghan (1795;
eng. Henecy and Fitzpatrick), Tyrone
(1813; with G. Knox).

M'Dermut, R., and Arden, D. D. *See
under* D'ANVILLE, J. B. B.

MacDougall, P. L. 19th cent. Cart.
Brit. M
Map of Guernsey, 1848.

McIntyre   *See under* Brown, T.

Mackay, Arthur   *See under* Jefferys, T.

Mackay, William   19th cent.   Cart.   Brit.
M
Map of Nova Scotia   Lond., 1834.

Mackenzie, Murdoch} d. 1797   Cart.,
McKenzie   } Surv.   Brit.   M
Surveyor to the Admiralty.   Succeeded in
that post by his nephew, Murdoch
McKenzie the Younger, in 1771.
*A maritime survey of Ireland and the west of
Great Britain*   2 vols., Lond., 1776.
*Orcades; or, a geographical and hydrographical
survey of the Orkney and Shetland islands*
1750; Dutch edn., 1753 (Amster., van
Keulen).

McLeod, W.   fl. 1861   Cart.   Brit.   M
A hand-atlas for class-teaching   Lond., 1858.
Eng. E. Weller.

Maclot, Jean Charles   1728-1805   Cart.
Fr.   M
Atlas général méthodique et élémentaire . . .
Paris, Desnos, 1770, 1786.   Maps by
Brion.

Macpherson, D. and Alexander   18th/19th
cent.   Carts.   Amer.   M
Atlas of ancient geography . . .   Philadelphia
1806.   Maps after A. Rees.   Engs. incl.:
S. Harrison, W. Harrison, Hewitt, Jones,
Kneass, Young and Delleker.
See also under Smith, Charles.

Macpherson, W. W.   *See under* Playfair, J.

Macquet, Ph.   *See under* Mentelle, E.

Mädler   *Same as* von Mädler (q.v.)

Maedel, Karl Jos. or}
      C. J., Senr.} *See under* Gaspari,
Mädel med.   } A. C.; Weiland, C. F.
Maedel, J., Junr.   }

Maestre, M. Rivera   19th cent.   Cart.
Span.   M
Atlas Guatemalteco . . .   Guatemala, 1832.
Engs. Casildo España, Fran. Cabrera.

Magdeburgus, Hiobus   1518-95   Cart.
Ger.   M

Maggiolo, Baldassare,}
   Giovanni, Jacobus,} 16th cent.   Carts.
            Vesconte} Ital.   M
Maiolo   }

Magimel, Picquet   *See under* Lapié, P.

Magini, Giovanni Antonio} 1555-1617
Magin, Anthoine   } Geog.   Ital.
                                       } M
Italia   Bologna, Nic. Tebaldini, 1620; 1642.
The first map of Italy by an Italian.
See also under Laurie, R.; Wytfleete, C.

Magnus Olaus} 1490-1558   Cart.   Swed.   M
Magnusson   }
Carta marina . . .   Rome, 1539.   Copied and
republished in 1572 by Antonio Lafreri
(q.v.)
See also under Jansson, J.

Maguire, C. E.   *See under* Allen, W.

Maiolo   *Same as* Maggiolo (q.v.)

Mair, Alexander   fl. 1603   Eng.   Ger.   X

Mair, John} fl. 1683-9   Surv.   Brit.   M
Marr   }

Maire, Christopher   18th cent.   Cart.
Brit.   M
Map of Durham, 1711.

Majero, Tobia} *See under* Homann, J. B.;
Majer   } Homann's Heirs.

Makowski, Tomasz   1575-1620?   Cart.
Polish   M

Malassis   *See under* d'Après de Mannevillette J.

Malby, Thomas} 19th cent.   Globe-maker
Malby and Co.   } Brit.   G

Malham, Rev. John   1747-1821   Cart.
Amer.   M
A naval atlas . . .   Philadelphia, W.
Spotswood (with J. Nancrede, Boston),
1804.   Engs.: B. Callender, S. Hill,
I. Norman, Rollinson.   Maps have
imprint: 'Spotswood and Nancrede,
Boston.'   The maps also were used in
Malham's Naval gazetteer (various edns.,
1795-1812), but have different imprints.

Mallat de Bassilan, J.   1806-63   Cart.   Fr.
M
Les Phillipines . . . atlas   Paris, A. Bertrand,
1846.   Maps eng. by H. Bonvalet.

Mallet, Alain Manesson   1630-1706?   Cart.
Fr.   M
Description de l'univers . . .   Paris, D. Thierry,
1683.   A Ger. edn. with map from different plates was pub. at Frankfurt-on-Main
by J. A. Jung in 1719.

Malte-Brun, Conrad fl. 1775–86 Cart.
  Ger. M
  Formerly MALTHE, CONRAD BRUUN.
  Atlas complet . . . Paris, F. Buisson, 1812;
    Brussels, The Hague, T. Lejeune, 1837.
    Some maps eng. by Grigny.
  A new general atlas . . . Philadelphia, Grigg
    and Elliot, 1837.
  Universal geography . . . Philadelphia, A.
    Finley, 1827–9; J. Laval, 1832.
  Géographie mathématique . . . (with Mentelle)
    Paris, H. Tardieu; Laporte 1804.
  See also under BOSSI, L.; GOODRICH, S. G.

Mammatt, E. fl. 1834 Geologist Brit.
  M

Mannert, Conrad *See under* WEIGEL UND
  SCHNEIDERSCHEN HANDLUNG.

Manning, J. 19th cent. Globe-maker
  Brit. G

Månsson, Johann fl. 1644 Sailor, Pub.
  Swed. M

Marchais *Same as* DE MARCHAIS (q.v.)

Marcollino, Francesco} 16th cent. Cart.
Marcolini   } Ital. M
  See also under ZENO, N.

Mareau *See under* DE FREYCINET, L. C. D.

Marent, Philip *See under* CELLARIUS, A.

Maria, Pietro 18th cent. Monk Ital. G

Mariette, Pierre 17th cent. Pub. Fr. M
  Worked on the *Atlas* of B. J. Briot.
  See also under SANSON, N., PÈRE.

Marino, Gerolamo and Celle, Benedetto
  17th cent. Pub. Ital. M
  See also under LEVANTO, F. M.

Marius   }
Marius Viterbicensis} fl. 1577 Astron.
       Romae} Ital. X

Markham, Rev. G. 18th cent. Cart. M

Marr, J. *Same as* MAIR, J. (q.v.)

Marre *Same as* DE MARRE (q.v.)

Marshall, John fl. 1695–1725 Pub.,
  Bookseller Brit. M

Marshall, John 1755–1835 Historian
  Amer. M
  Life of George Washington . . . *maps* . . .
    Philadelphia, C. P. Wayne, 1807; J.
    Crissy, 1832 (*Atlas to Marshall's Life of*

Washington; same maps as 1807 edn., but
  smaller scale); J. Crissy (ditto; eng. by
  J. Yeager), 1850; Fr. edn. 1807, Dentu,
  Paris. The maps in the Philadelphia,
  1807, edn. drawn by S. Lewis and eng.
  by B. Jones, I. H. Seymour, F. Shallus,
  and Tanner.

Marshall, Thomas *See under* EDWARDS, B.;
  PINKERTON, J.; TANNER, H. S.

Marshall, W., and Co. *See under* SMITH,
  R. C.

Marsigli, Luigi   }
  Ferdinando, Conte} 1658–1730 Cart.
Von Marsigli, Aloysius}
    Ferdinand, Graf} Ital. M
  La Hongrie et le Danube . . . The Hague,
  1741.

Martin, Benjamin fl. 1759–63 Cart.
  Brit. M

Martin, C. *See under* VON SCHLIEBEN,
  W. E. A.

Martin, D. *See under* WINTERBOTHAM, W.

Martin, Henry fl. 1833 Cart., Eng. Brit.
  M

Martin, P. I. fl. 1828 Geologist Brit. M

Martin, S. D. 19th cent. Surv. Brit. M

Martin de Lopez, Pedro 19th cent. Cart.
  Span. G (C and T)
  Geographical and celestial globes made of
  paper and card and arranged in gores
  which are formed by tightening up a
  small metal ring at the top of each.
  Madrid, *c.* 1840.

Martines, Johannes} fl. 1564–86 Cart.
Martins, João   } Port. M

Martinez de la Torre, F. *See under*
  ASENCIO, J.

Martini, Aegidius fl. 1616 Cart. Ital.
  M

Martini, Martino 1614–61 Jesuit mission-
  ary, Cart. Ital. M
  Novus atlas Siensis . . . Amster., J. Blaeu,
    1655 (*n.b.* Vol. 6 of W. and J. Blaeu's
    Toonel des Aerdriicx) Fr. edn. pub. simul-
    taneously.
  See also under HONDIUS, H.; JANSSON, J.;
    VAN DEN KEERE, P.

Martin's Sportsmen's Almanack   *See under* LEWIS, W.; WALLIS, J.

Martins, João   *Same as* MARTINES, J. (q.v.)

Martyn, Thomas   fl. 1748-84   Cart.   Brit. M
Map of Cornwall, 1748 and subs. edns.

Martyr, Peter   } 1455-1526   Geog.
Martyr d'Anghiera,} Ital.   M   G
        Pietro }
Opera . . . (with woodcut map)   Seville, Jacobus Cromberger, 1511.
De orbe novo . . .   Paris, Gvillelmum Avvray, 1587 (with eng. map by Francis Gaulle).

Mascaradi, V.   *See under* NICOLOSI, G. B.

Maschop, Godfried}
Mascop } fl. 1568   Cart.   Ger.
Mascopius } M
See also under ORTEL, A.

Masi e Compagni, G. T.   late 18th cent. Pub.   Ital.   M
Atlante dell' America   Livorno, 1777.

Mason, Charles and Dixon, Jeremiah
18th cent.   Cart.   Amer.   M
A plan of the boundary lines between the province of Maryland and the three lower counties of Delaware . . .   Philadelphia, Robert Kennedy, 1768.

Mason, John   1586-1635   Cart.   Brit.   M
Governor of Newfoundland, 1615.
Map of Newfoundland in William Vaughan's *The Golden Fleece* . . . Lond., 1626.

Massa, Isaac   1587-1635   Cart.   Dut.   M
See also under BLAEU, G.; HONDIUS, H.; HUSSON, P.; JANSSON, J.; MERCATOR, G.; VISSCHER, N., III; WOLFGANG, A.

Masters, C. H.   *See under* DAY, W.

Matal, Jean } 1520-97   Cart.   Ger.   M
Metellus, Johann}
Insularium orbis aliquot insularum . . . Coloniae Agrippinae, I. Christophori, 1601.

Mathew, William   17th cent.   Cart.   Brit. M

Mathews and Leigh   19th cent.   Pub. Brit.   M
The scripture atlas . . .   Lond., 1812.

Mathonière, Denis and Nicolas   fl. 1586-1621   Printer   Fr.   M

Matius   *See under* JANSSON, J.

Matthews, John, R. N.   18th cent.   Sailor Brit.   M
Twenty-one plans . . . of different actions in the West Indies . . .   Chester, J. Fletcher, 1784.

Maugein, Charles   *See under* PHILIPPE DE PRÉTOT, E. A.

Maunder, S.   *See under* EBDEN, W.

Maurer   *Same as* MURER (q.v.)

Mauritius, Johannis}
Miricio, Giovanni } fl. 1590   Geog.   Ger.
Myritius } M   G

Maurolico, Francesco   1494-1575   Astron. Ital.   M   G (C)

Maverick, Peter   1780-1831   Pub.   Amer. M
General atlas . . . (with Durand and Co.) New York, 1816.

Mawman, J.   *See under* GUTHRIE, W.

Maxwell, George   *See under* NORRIS, R.

Maxwell, J.   fl. 1714   Cart.   Brit.   M
Associated with John Senex.
See also under LAURIE, R.

Mayer, Johann Tobias, the Elder   1723-62 Cart., Astron.   Ger.   M   X   G
See also under KILIAN, G. C.; PALAIRET, J.

Mayer, Tobias, the Younger   1752-1830 Cart., Mathematician   Ger.   G

Mayo, Robert   1784-1864   Cart.   Amer. M
The atlas of ten select maps of ancient geography . . .   Philadelphia, J. F. Watson, 1813, 1814; J. Melish, 1815. Engs.: H. S. Tanner, J. Thackara, J. Vallance.

Mayo, William   18th cent.   Cart.   Brit. M
Barbadoes surveyed . . .   Lond., 1774 ('engraved and improved by Thomas Jefferys')

Mead, Bradock} d. 1757   Globe-dealer
Green, John } Brit.   M   G
Mead was Green's real name.   Green was a pseudonym.
Chart of North and South America 1753

Medebach *Same as* VOPEL (q.v.)

Medina *Same as* DE MEDINA (q.v.)

Medtman *Same as* VAN MEDTMAN (q.v.)

Mee *See under* BEEK, A.

Meijer, A. *See under* ALLARD, K.

Mejer, Johannes 1606-74 Surv., Cart. Dan. M

Mela, Pomponius fl. A.D. 50 Geog. Rom. M
Pomponii Melae de situ orbis . . . Lond., S. Birt, 1739.
See also under FINÉ, O.

Melish, John 1771-1822 Cart. Amer. M
A military and topographical atlas of the United States . . . Philadelphia, G. Palmer, 1813, 1815. Engs.: H. S. Tanner, J. Vallance.
See also under KLINCKOWSTRÖM, A. L.; LAURIE, R.; LAVOISNE, C. V.; MAYO, R.

Mellinger, Johannes} fl. 1568-93 Cart.
Mellingero } Ger. M
See also under BLAEU, G.; HONDIUS, H.; JANSSON, J.; MERCATOR, G.

Ménard et Desenne *See under* VIVIEN DE SAINT-MARTIN, L.

Mendoza *Same as* DE MENDOZA (q.v.)

Mengden, Georg 1628-1702 Cart., Pub. Dut. M

Menk fl. 1810 Cart. Ger. G

Mentelle, Edme 1730-1815 Cart. Fr. M
New map of Spain and Portugal Lond., Stockdale, 1808.
Atlas de la monarchie Prussienne Lond., 1788.
Atlas universel . . . (with Pierre Gregoire) Paris, Chanlaire, 1797-1801, 1807. Maps drawn and eng. by: André, P. F. Tardieu, Pasquier, Jean Valet, Dubuisson, Ph. Macquet, Herault, L. Auber, Blondeau.
Atlas de tableaux et de cartes gravé par P. F. Tardieu . . . Paris, Bernard, 1804-5. 'P. F. Tardieu' is a mistake for 'A. F. Tardieu'.
See also under MALTE-BRUN, C.

Menzies, J. and G. *See under* BROWN, T.; FORREST, W.; THOMSON, J., AND CO.

Mercator, Arnold 1537-87 Cart. Dut. M
Son of GERHARD, M. (q.v.)

Mercator, Gerhard}
Merkator } 1512-94 Cart. Dut.
Kremer } M G (C and T)
Kremer was Mercator's real name, which he later changed.
Atlas minor . . . (with Jodocus Hondius) Amster., 1606; Arnhem, 1621 (Latin text), 1628 (ditto, revised version), 1631 (German text), 1636 (ditto), 1651 (ditto), 1630 (French text), 1633 (ditto), and many other edns. Some of the maps in some edns. were eng. by van den Keere.
Atlas ofte afbeeldinghe vande gantsche weerldt . . . Amster., J. Jansson, 1634. Carts. incl.: Hardy, Lavaña, Jansson, Secco, Surhon, Cluver, H. Hondius, Beins, Sprecher von Berneck, Berckenrode, N. J. Visscher, Sinck, Wicheringe, Frietag, Metius, Emmius, Gigas, Mellinger, Laurenberg, Lübin, Keer, Palbitzke, Comenius, Lazius, Adrichomius.
Atlas or a geographical description . . . Amster., H. Hondius and John Johnson (i.e. Jansson) 1636. Carts. incl., Fayen, Gerard, Secco, Fer, Sprecher von Berneck, Labaña, H. Hondius, Lhuyd, Massa, Pont, Jansson, Henneberger, Laurenberg, Lübin, Palbitzke, Mellinger, Gigas, Emmius, Comenius, Berckenrode, Surhon, Wicheringe. There was also a Fr. Edn. (1609).
Atlas sive cosmographicae . . . Dusseldorf, 1585, Amster. (Hondius), 1611, 1613 and other edns.
Galliae tabulae geographicae Amster., 1606?
Gerardi Mercatoris atlas Amster., Hondius, 1607, 1619, 1630, 1632, 1628 (Fr.), 1630 (ditto), 1636 (ditto, Johannis Cloppenburg) and other edns. Carts. incl. (according to edn.): Hondius, Beins, Martin, Surhon, Lhuyd, Secco, Fayen, Mellinger, Bomparius, Lübin, Berckenrode, Henneberger, Massa, Pont, Palbitzke, Hardy, Wicheringe, Sprecher von Berneck, Gerard, Labaña, Fer. Some maps were eng. by Petrus Keere.
Gerardi Mercatoris et I. Hondii atlas . . . Amster., Hondius and Jansson, 1633,

1683 (Fr.) and other edns. Carts. as in preceding title.

Historia mundi or Mercator's atlas . . . Lond., Michael Sparke, 1637.

Italiae, Sclavoniae et Graeciae tabulae geographicae . . . Duisburg, 1589.

See also under À MYRICA, G.; BLAEU, G.; DE LA FEUILLE J.; DE WIT, F.; DU SAUZET, H.; HALL, RALPH; HONDIUS, H.; JANSSON, J.; THORNTON, J.

Mercator, Gerhard, the Younger fl. 1595 Cart. Dut. M
Son of Arnold M.
Sometimes signed: 'G. M. Junior.'
Irlandiae regnum . . . Amster., 1638.
Anglia regnum . . . Amster., 1638.
Eboracum, Lincolnia . . . Amster., 1638.
Also various other maps of British districts and regions. Same date as above.

Mercator, John fl. 1575 Cart. Dut. M
Son of Arnold M.

Mercator, Michael d. 1600 Cart. Dut. M
Son of Arnold M.

Mercator, Rumold 16th cent. Cart. Dut. M
Son of Gerhard, M., the elder.

Merian, Matthäus 1593-1650 Pub., Eng., Cart. Swiss M
See also under D'AVITY, P.; ZEILLER, M.

Merklas 19th cent. Globe-maker Austrian G

Merridew, H. *See under* ASTON, J.

Merula, Paul fl. 1665 Cosmographer Dut. M

Messier, Charles 1730-1817 Astron. Fr. G (C)

Metellus, Johann *Same as* MATAL, J. (q.v.)

Metio, Adriano }
Metius, Adriaan } 1571-1635 Astron. Dut.
or Adriaansz } M G
See also under BLAEU, G.; HONDIUS, H.; JANSSON, J.; MERCATOR, G.

Meyer, Johannes fl. 1652 Cart., Mathematician Dan. M

Meyer, Joseph 1796-1856 Cart. Ger. M

Neuester grosser schulatlas . . . Hildburghausen and New York, 1830-8.

Neuester universal-atlas . . . Philadelphia, 1830-40. Carts.: Capt. Radefeld and Lieut. Renner. Engs.: Joh. David, Adolf Gottschalk C. Ehricht, A. Heimburger, H. Herzberg, C. Metzeroth, R. von Rothenburg, E. Sporer, J. Zipter, Gustav Mezeroth, Carl Hoeckner.

Meyer, J. R. *See under* WEISS, J. H.

Meyer, Tobias *Same as* MAYER, T. (q.v.)

Mezeroth, C. and Gustav *See under* MEYER, JOSEPH.

Michaelis *See under* EULER, L.

Michaelis, Laurentius d. 1584 Cart. Dut. M

Michal, Jacques fl. 1725 Cart. Fr. M
Map of Sweden, 1725.

Michalet, Estienne } *See under* SANSON, N.,
Michallet } PÈRE; THEVENOT, M.

Michault, R. *See under* SANSON, N., PÈRE

Michel 18th cent. Cart. Fr. M
L'indicateur fidèle ou guide des voyageurs . . . Paris, 1765.
See also under DESNOS, L. C.; LORRAIN, A.

Michelot et Bremond 18th cent. Pub., Cart. Fr. M
Portulan atlas of the Mediterranean . . . Marseilles, 1715-26.
See also under HEATHER, W.; JEFFERYS, T.

Michie, R. S. *See under* MORISON, S. N.

Middleton fl. 1829 Cart. Brit. M
Middleton's New geographical game of a tour through England and Wales n.p., Harris, 1829

Middleton, Henry *See under* GILBERT, H.

Middleton, J. fl. 1842 Astron. Brit. X

Midwinter, D. fl. 1714 Cart. Brit. M
See also under BOWLES, J.

Migneret, M. *See under* DEPPING, G. B.

Millar, Andrew *See under* ADAIR, J.; OSBORNE, T.

Miller, Robert fl. 1810-21 Pub. Brit. M
Miller's New miniature atlas 1810, 1820. Republished as *Darton's New miniature atlas* in 1825.
See also DARTON, W.

Miller and Hutchens *See under* DRURY, L.

Millo, Antonio fl. 1557–90 Cart. Ital. M

Millot, Iean 17th cent. Pub., Eng. Fr. M
Eng. 3 maps in Marc Lescarbot's (1590–1630) *Histoire de la nouvelle France* . . . , which he also pub. in 1609 at Paris.
See also under SWELINC, I.

Mills, Robert 1781–1855 Cart. Amer. M
Atlas of the state of South Carolina . . Baltimore, F. Lucas, Junr., 1825.

Millward, T. 18th cent. Pub. Brit. M

Milne, Thomas fl. 1791 Surv. Brit. M
See also under DONALD, T.

Milner, Rev. Thomas d. 1882 Astron., Geol. Brit. M X
Descriptive atlas of astronomy and of physical and political geometry (with Thomas Petermann) Lond., W. S. Orr and Co., 1850.
See also under PETERMANN, A. H.

Miot, Vincenzo fl. 1700–10 Astron. Ital. G (C)

Mirici *Same as* DE MIRICI (q.v.)

Miricio, G. *Same as* MAURITIUS, J. (q.v.)

Miritius, Johannes fl. 1590 Geog. Ger. M

Mitchell, John}
Mitchel } d. 1768 Cart. Brit. M
A map of the British and French dominions in North America . . . Lond., 1755. Eng. by Thomas Kitchin.
See also under JULIEN, R. J.; LE ROUGE

Mitchell, Samuel Augustus 1792–1868 Cart. Amer. M
Mitchell's ancient atlas Philadelphia, T. Cowperthwait and Co., 1844; E. H. Butler and Co., 1859, and many other edns.
Mitchell's Atlas of outline maps . . . Philadelphia, T. Cowperthwait and Co., 1839.
Maps of New Jersey, Pennsylvania . . . Philadelphia, 1846.
A new universal atlas Philadelphia, 1849, 1850 and many subs. edns.

Mitchell's school atlas . . . Philadelphia, Thomas Cowperthwait and Co., 1839, and many other edns. Maps dated 1839. Eng. by W. Williams, J. H. Young.
Mitchell pub. many atlases after 1850.
See also under TYSON, J. W.

M.K.Sc. *See under* FABER, S.

Moers, Justus fl. 1575 Cart. Ger. M
Map of Waldeck, 1575.

Moffat, John fl. 1795 Eng. Brit. M
See under LAURIE, R.; THOMSON, J., AND CO.

Mogami Tokunai b. 1755 Cart. Japanese M

Mogg, Edward fl. 1808–26 Cart., Pub. Brit. M

Mogiol, Frantzesco fl. 1564 Cart. ? M

Mohammed, Diemat Eddin fl. 1575 Astron. Arab G (C)

Moithey, Maurelle Antoine *See under* DESNOS, L. C.; PHILIPPE DE PRÉTOT, E. A.

Moletti 16th cent. Cart. Ital. M

Molijns, Jean}
Mollijns } fl. 1565–99 Printer Dut.
Molanus } M

Molinax }
Molineux} *Same as* MOLYNEUX, E. (q.v.)

Moll, Herman 1688–1745 Geog., Pub. Bookseller Dut. M G (T)
Worked in England.
Map of Africa c. 1714.
A new and exact map . . . *within ye limits of ye South sea* . . . Lond., c. 1713.
Map of North America . . . Lond., c. 1714.
Map of Europe Lond., 1708.
A new map of the north parts of America . . . Lond., 1720.
Map of South America (in collaboration with B. Lens) Lond., Herman Moll, Bowles, Overton and King, c. 1713. Eng. by G. Vertue.
Map of Asia Lond., c. 1714 and (re-eng. by George Grierson) Dublin, c. 1720.
A map of the East-Indies . . . Lond., c. 1715 and (probably re-eng. by George Grierson) Dublin, c. 1720.
A new map of Great Britain Lond., c. 1715.
The north part of Great Britain Lond., 1714.
The south part of Great Britain Lond., 1710.

A new and exact map of . . . the Netherlands Lond., *c.* 1715 It is also known printed for John Bowles.

A new map of Italy Lond., *c.* 1715.

A new map of the upper part of Italy Lond., printed and sold by T. Bowles, *c.* 1715.

Atlas minor Lond., 1729.

A system of geography Lond., Churchill, 1701.

The world described . . . Lond., John Bowles, *c.* 1727.

A new map of Denmark and Sweden . . . Lond., *c.* 1715.

A new and exact map of Spain and Portugal Lond., 1711.

Florida . . . Lond., 1728.

A new and exact map of the dominions of the king . . . on ye continent of North America Lond., Bowles, 1715, *c.* 1731.

Virginia and Maryland Lond., 1729.

A map of the West Indies . . . Lond., Moll and King, *c.* 1713.

A new and correct map of the world . . . Lond., *c.* 1728.

A new description of England and Wales . . . Lond., 1724. The end margins of the maps in this atlas illustrated 'remarkable antiquities'.

Atlas manuale . . . Lond., A. and J. Churchill, 1709; I. Knapton, P. Knaplock, 1723.

Atlas minor . . . Lond., 1729, 1732? (T. and J. Bowles), 1736? (ditto), and many other edns., the later ones of which were pub. as *Bowles Atlas minor* (1745?, 1763?).

Forty-two new maps of Asia, Africa and America . . . Lond., J. Nicholson, 1716.

Geographica antiqua . . . Lond., 1721.

A set of thirty-two new and correct maps of the principal parts of Europe . . . Lond., 1727?

A system of geography Lond., T. Childe, 1701.

The world described 1709–20, 1709–36, Lond., J. Bowles.

A new and compleat atlas . . . Lond., 1708–20. Carts. incl.: D. Winter, T. Bowles, P. Overton.

Thirty-two new and accurate maps of the geography of the ancients . . . (with Thomas Bowles) 1721, 1732, 1739.

See also under BOWLES, J.; DE WIT, F.; KÖHLER, J. D.; MOORE, SIR JONAS; OTTENS, R.; SIMPSON, S.

Mollero, Christiano *See under* BLAEU, G.

Mollijns, Joannem *Same as* MOLIJNS, J. (q.v.)

Mollin, T. }
Mollo, Tranquillo} *See under* DIRWALDT, J.

Mollo, Eduard 1797–1842 Cart. Austrian G (C and T)

Mollweide, Carl Brandan 1774–1825 Mathematician, Astron. Ger. G

Molyneux, Emery}
Mollyneux } fl. 1587–1605 Cart., In-
Molineux } strument maker. Brit.
Molinax } G (C and T)
Made, in 1592, the first English globes. They were eng. by Jodocus Hondius. Molyneux emigrated to the Netherlands in 1596–7.

Mon, J. F. fl. 1629 Cart. Fr. M

Monachus, Franciscus fl. 1525 Cart. Dut. G (T)

Mondhare } 18th cent. Pub. Fr.
Mondhare St. Jean} M
See also under CLOUET, J. B. L.; DE LA FOSSE, J. B.; NOLIN, J. B.

Mongenet *Same as* DE MONGENET (q.v.)

Monin, V. [Charles V.] 19th cent. Cart. Fr. M
Petit atlas national . . . Paris, Binet, 1835; Lafont, 1841.

Monnier, P. d. 1843 Cart. Fr. M
Atlas de la martinique Paris, 1827–31.

Monno, Giovanni Grancisco fl. 1613–33 Cart. Ital. M

Montanus, Petrus *Same as* VAN DER BERG, P. (q.v.)

Montarini, Geminiano fl. 1664–5 Astron. Ital. X

Montbazin *See under* DE FREYCINET, L. C. D.

Monte, Urbino 1544–1613 Cart. Ital. M

Montresor, Major 18th cent. Cart. Brit. M
Map of Nova Scotia, or Acadia . . .
Province de New York (unsigned) Paris, Le Rouge, 1777.

Moore, Sir Jonas 1627-79 Geog. Brit. M
A new geography . . . Lond., R. Scott, 1681 (with 57 maps, incl. one of America by H. Moll).

Moore, S. S. *See under* JONES, T. W.

Morata, J. *See under* DE ANTILLON Y MARZO, I.

Morden, Robert d. 1703 Cart., Bookseller, Pub. Brit. M G (C and T)
See under CAMDEN, W., for edns. of *Britannia* pub. by him.
A new map of the British empire in America eng. by I. Harris, c. 1695.
Geography rectified Lond., 1680 (with Thomas Cockeril) and 1688.
A newe description of the whole worlde in twoe hemispheres . . . (unsigned) Lond., c. 1688.
To Capt. John Wood this map of the world . . . *is humbly dedicated* (with William Berry). Lond., 'sold at ye Atlas in Cornhill & at ye Globe in ye Strand', c. 1688.
The 52 countries [sic] of England and Wales described in a pack of cards Lond., 1676. English county maps issued as a pack of playing-cards. Reissued, without suitmarks (with H. Turpin) in 1750 as *A brief description of England and Wales* . . .
Fifty-six new and accurate maps of Great Britain . . . Lond., 1708.
Atlas terrestris . . . 1695, and other edns.
Map of Essex (with J. Pask), c. 1700.
See also under BEEK, A.; LENTHALL, J.; PETTY, SIR W.

Moreau de St. Méry, M. L. E. *See under* PONCE, N.

Moret, Balthasar (d. 1641)} Printers Dut.
    and Johannes} M
Moretus }

Morgan, William d. 1690 King's cosmographer, Master of the Revels in Ireland. Pub. M *See also under* OGILBY, J.

Morgan, Lodge and Fisher *See under* CUMINGS, S.

Morison, S. N. 19th cent. Surv. Brit. M
Map of Clackmannan, 1848 (eng. R. S. Michie).

Moroncelli, Silvester Amantius 1652-1719 Cart. Ital. G (C and T)

Moronobu, Hishikawa d. c. 1694 Artist Japanese M

Tōkaidō Bunken Ezo ('A measured pictorial map of the Tōkaidō') Edo, 1690 (woodcut).

Morphew, John fl. 1706-20 Pub., Bookseller Brit. M
Was a publisher of an atlas by MOLL (q.v.)

Morris, Charles *See under* JEFFERYS, T.

Morris, Lewis 18th cent. Cart. Brit. M
Plans of harbours, bars, bays and roads in St. George's channel . . . Lond., 1748. Eng. by Nath. Hill.
Plans of the principal harbours, bars, bays and roads in St. George's channel Shrewsbury, Sandford and Maddocks, 1801.

Morse, H. *See under* EDMANDS, B. F.

Morse, Jedidiah 1761-1826 Geog. Amer. M
The American geography . . . Lond., J. Stockdale, 1794.
Modern atlas . . . (with S. E. Morse) Boston, J. H. A. Frost, 1822; New York, Collin and Hannay, 1828.
A new universal atlas of the world . . . (with S. E. Morse) New Haven, Howe and Spalding, 1822.

Morse, Sidney Edwards 1794-1871 Cart. Pub. Amer. M
Also traded as Sidney Edwards Morse and Co.
The cerographic bible atlas . . . New York, 1844 and 1845.
A new universal atlas of the world . . . New Haven, N. and S. S. Jocelyn, 1825.
Nuevo sistema de geografía . . . (with J. C. Brigham) New York, Gallaher and White, 1827-8.
The cerographic missionary atlas New York, 1848.
See also under BREESE, S.; MORSE, J.

Mortier, Corneille *See under* BEEK, A.; COVENS, J.; OTTENS, R.; SCHENK, P.

Mortier, David b. 1673 Cart., Pub.
Nouveau théâtre de la Grande Bretagne . . . 5 vols., Lond., 1715-28. Engs. incl.: Badeslade, J. Collin, I. Harris, H. Hulsbergh, T. Johnson, Sutton Nichols, D. Loggan, I. Sailmaker, J. Spilbergh, H. Winstanley.
See also under DE HOOGE, R.; SCHENK, P.

Mortier, Pieter, or Peter,} d. 1711 Pub.
　　or Petrum} Dut. M
Atlas nouveau de cartes géographiques choisies...
　　Amster., 1703.
Les forces de l'Europe, Asie, Afrique et
　　Amérique . . . 2 vols., Amster., 1702?
　　Plates copied by Pieter Mortier from de
　　Fer's original edn. of 1694-7. Mortier's
　　plates were later purchased by van der Aa,
　　who republished them in 1726.

Mosburger *See under* BEEK, A.

Mossner, A. G., and Jean Michel *See*
　　under GASPARI, A. C.; VON SCHLIEBEN,
　　W. E. A.

Mottershead, Jasper fl. 1827-31 Cart.
　　Brit. M

Moule, Thomas 1784-1851 Topg.,
　　Bookseller, Pub. Brit. M
The English counties delineated . . . Lond.,
　　George Virtue, 1836, 1838, 1839 (eng. by
　　J. Dower.)
See also under BARCLAY, J. (for later edns.)
　　and SCHMOLLINGER, W.

Mount, W. and Page, T.　　　}
Mount, Richard　　　　　　　}
Mount and Davidson　　　　　} 18th/19th
Mount, R., and Page, T.　　　} cent. Pubs.
Page, T., and Mount W. and F.} Brit. M
Thornton and Mount　　　　　}
　　The firm was founded by Richard Mount
　　(d. 1722).
English Pilot . . . 1671 and subsequent edns.
　　to 1803.
The sea coasts of France . . . Lond., *c.* 1715.
A chart of the island of Hispaniola c. 1737.
The island of Jamaica Lond., 1737.
A new mapp of Carolina Lond., *c.* 1737.
A chart of the Caribe islands (unsigned)
　　Lond., *c.* 1737.
Atlas maritimus novus . . . Lond., 1702.
See also under COLLINS, CAPT. G.; GASCOIGNE,
　　J.; LEARD, J.; NORRIS, J.; SELLER, JOHN I.;
　　TIDDEMAN, M.

Moussy *Same as* DE MOUSSY (q.v.)

Moutard *See under* DE LAPORTE, J.

Mouzon, Henry 18th cent. Cart. Brit.
　　M
An accurate map of North and South Carolina
　　Lond., Sayer and Bennett, 1775.

See also under FADEN, W.; JEFFERYS, T.; LE
　　ROUGE.

Moxon, James d. 1708 Eng., Cart. Brit. M
　　There were two James Moxons, Senr. and
　　Junr. Their exact relationship to one
　　another and to JOSEPH MOXON (q.v.) is
　　uncertain.
A new discription of Carolina Lond., John
　　Ogilby, 1671.
See also under ADAIR, J.

Moxon, Joseph 1627-1691 Polymath
　　Brit. G (C and T)
See also under MOXON, JAMES.

Mudge, William 19th cent. Cart. Brit. M
　　Ordnance survey maps of Essex, 1805, and
　　Kent, 1801 and subs. edns.

Mudie, Robert 1777-1842 Compiler,
　　Pub. Brit. M
Gilbert's modern atlas of the earth . . . Lond.,
　　H. G. Collins, 1841?

Müelech, Johannes} 1516-73 Artist Ger.
Mielich　　　　　　} G

Müller, C. *See under* GASPARI, A. C.

Müller, Friedrich *See under* GASPARI, A. C.

Müller, G. *See under* CLUVER, P.; HOMANN,
　　J. B.; KÖHLER, J. D.; VON REILLY,
　　F. J. J.

Müller, Gerhard Friedrich 18th cent.
　　Cart. Ger. M
Nouvelle carte des decouvertes faites par des
　　vaisseaux Russes aux côtes inconnues de
　　l'Amerique septentrionale . . . . St.
　　Petersburg, 1754, 1758.

Müller, I. C. *See under* GASPARI, A. C.

Müller, Johann Cristoph 1673-1721 Cart.
　　Ger. M
Le royaume de Boheme . . . Amster., Covens
　　and Mortier, 1744.
Schweizerischer atlas . . . n.p., 1712?

Müller, Johann Ulbrich 17th cent. Cart.
　　Ger. M
Kurtz-bündige abbild- und vorstellung der
　　gantzen welt . . . Ulm, G. W. Kühnen,
　　1692.
See also under VALCK, G.

Müller, Johann Wolfgang b. 1765
　　Mathematician Ger. G

Müller, Philipp fl. 1619 Astron. Ger. X

Münster, Sebastian 1489-1552 Cart.,
Theologian Swiss M
Cosmographiae universalis . . . Basle, H.
Petri, 1540 (with 14 woodcut maps).
See also under Ptolemaeus, C.

Münzer, Hieronymus 1437-1508 Cart.
Ger. M
Map of Germany in Schedel's *Liber Chroni-
carum,* 1493.

Murer, Christoph} 1558-1614 Eng. Ger.
Maurer } M
Son of Joseph M.

Murer, Joseph or Jost} 1530-80 Glass-
Maurer, Josias } painter, Cart. Swiss
} M

Murillo *See under* Laurie, R.

Murphy, William fl. 1864 Cart. Brit.
M

Murray, John 1745-1843 Pub. Brit. M
The family firm still publishes guides, etc.,
with maps.

Murray, T. L. fl. 1830-4 Surv. Brit. M
An atlas of the English counties . . . Lond.,
1830, 1831. Eng. by Hoare and Reeves.

Murray, Fairman and Co. *See under* Rees,
A.

Musinus, Bartholomeus fl. 1560 Cart.
? M

Musis *Same as* de Musis (q.v.)

Mustafa, Resmi fl. 1785 Cart. ? M

Muth, Heinrich Ludwig (1673-1754) **and
Johann Philipp** (fl. 1721) Astrons.
Ger. G (C)

Mylne, Robert fl. 1819 Astron., Cart.
Brit. M X

Myrica, Gaspar *Same as* à Myrica, G.
(q.v.)

Myritius, G. *Same as* Mauritius, G. (q.v.)

Nagakubo Genshu} fl. 1779 Cart.
Sekisui } Japanese M

Nagel, Henricus 16th cent. Cart. Brit.?
M
Map of Scotland, 1595.

Nakamura Kan 18th cent. Cart.
Japanese M

Nakovalnik und Solomein fl. 1686 Cart.
Ger. M

Nancrede, J. *See under* Malham, J.

Nantiat, Jasper 18th/19th cent. Cart.
Brit. M
A new map of Spain and Portugal Lond.,
Faden, 1810.
See also under Faden, W.; Wyld, J.

Nash, John fl. 1666-89 Instrument-
maker Brit. G

Natoroff, W., and Co. *See under* Fischer,
W.

N.B. Signature of Bonifacius, Natalis
(q.v.)

Neale, John fl. 1751 Globe-maker Brit.
G

Neele, Josiah 19th cent. Eng. Brit. M
Son of Samuel John N. Eng. Greenwood's
map of Monmouthshire.

Neele, Samuel John, and Son fl. 1798-
1820 Engs. Brit. M X
See also under Albin, J.; Beaufort, D. A.;
Bell, J.; Brown, T.; Cole, C. N.;
Dugdale, J.; Faden, W.; Langlands,
G., and Son; Laurie, R.; Leard, J.;
Philipps, Sir R.; Playfair, J.; Robinson, J.;
Stackhouse T.; Thomson, J., and Co.;
Thorp, J.; Whitaker, G. and W. B.

Nell, J. P. *See under* Homann, J. B.;
Homann's Heirs.

Nelli, Nicolo fl. 1564 Eng. Ital. M

Nelson, T., and Sons Still extant pub.
Brit. M

Neugebauer, Salomon fl. 1612 Geog.
Ger. M

Nevill, A. Richard} fl. 1798 Cart. Irish
Nevil } M
Neville }
Map of Co. Wexford, 1798. Eng. G. Byrne

Nevill, Jacob} 18th cent. Cart. Irish M
Nevil }
Map of Co. Wicklow, 1760, and subs. edns.
Eng. G. Byrne.

Newbery, Francis, Junr.} 1743-1818 Pub.,
Newberry } Bookseller Brit.
} M
See also under Gibson, J.

Newbery, John} 1713-67 Pub., Bookseller
Newberry } Brit. M
See also under Gibson, J.

Newcourt, Richard　17th cent.　Cart.
　　Brit.　M
　　Map of Sedgemoor, 1662 (pub. W. Dugdale)
　　and 1772.

Newton, George　fl. 1817–36　Cart.,
　　Mathematician　Brit.　G (C and T)
　　New and improved terrestrial pocket globe
　　Lond., 1817. About 3″ diameter, with
　　horizon and brass meridian in shagreen
　　case, on the inside of which is the celestial
　　globe.　Silver stand.

Newton, James　b. 1748　Eng.　Brit.
　　G (C and T)

Nezon　*Same as* DE NEZON (q.v.)

Nicholls, Sutton}　fl. 1689–*c*.1713　Eng.　Brit.
Nichols　　　　}　M
　　Eng. several maps in Camden's *Britannia*
　　(1695 edn. of R. Morden).
　　See also under MORTIER, D.

Nichols, Francis　18th/19th cent.　Pub.,
　　Cart.　Amer.　M
　　A new atlas . . .　Philadelphia, 1811.

Nichols, G.　19th cent.　Cart.　Brit.　M
　　Map of Isle of Wight, 1844.

Nichols, John Bowyer, and Son　fl. 1835
　　Pub.　Brit.　M

Nicholson, John　fl. 1686–1715　Bookseller,
　　Pub.　Brit.　M
　　See also under LAURIE, R.; MOLL, H.

Nicol, G. and W.　*See under* FLINDERS,
　　CAPT. M.; STAUNTON, SIR G. L.

Nicolai, Arnold　16th cent.　Eng.　Dut.
　　M

Nicolai, C.　*Same ds* CLAESZ, C.

Nicolay　*Same as* DE NICOLAY (q.v.)

Nicolle, H.　*See under* DE LABORDE, A. L. J.

Nicollet, H.　*See under* ANDRIVEAU-GOUJON.

Nicolosi, Giovanni Battista}　1610–70
Nicolosius　　　　　　}　Geog.　Ital.
　　　　　　　　　　　　　}　M　G
　　Dell'hercole e studio geografico . . .　2 vols.
　　Rome, V. Mascaradi, 1660.
　　Hercules siculus sive studium geographicum . . .
　　2 vols.　Rome, M. Herculis, 1670–1.

Nicolson, Bishop William　1655–1727
　　Geog.　Brit.　M
　　The English atlas . . .　4 vols.　Oxford,

Moses Pitt, 1680–2. Unsigned; Nicolson
collaborated with Richard Peers in the
historical and geographical descriptions.
Some maps were based on those in
Jansson's atlas.
　　*The description of part of the empire of
　　Germany* . . .　Oxford, Moses Pitt, 1681.
　　*The description of the remaining part of the
　　empire of Germany* . . .　Oxford, Moses
　　Pitt, 1683.

Niebuhr, C.　*See under* HOMANN'S HEIRS:
　　SCHRAEMBL, F. A.

Nightingale, Joseph　fl. 1816–20　Topg.
　　Brit.　M
　　English topography . . .　Lond., Baldwin,
　　Cradock and Joy, 1816. Reprints of
　　maps originally intended for *The Beauties
　　of England and Wales* and issued in *The
　　British atlas* (Cole and Roper, 1810).

Niquet　*See under* DESNOS, L. C.

Nitzsche　*See under* TAITBOUT DE MARIGNY,
　　E.

Noble, E.　*See under* STOCKDALE, J.

Noble, John and Keenan, J.　fl. 1752
　　Carts.　Irish　M
　　See also under KEENAN, J.

Noble and Butlin　*See under* JEFFERYS, T.

Nodal　*Same as* DE NODAL (q.v.)

Noel, Lawrence　*Same as* NOWELL, L. (q.v.)

Nolin, Jean Baptiste}　Pub., Cart.　Fr.　M
1. Senior　1648–1708}　G
2. Junior　1686–1762}
　　Regnum Portugalliae . . .　Nuremberg,
　　Homann, 1736.
　　*Atlas général à l'usage des collèges et maisons
　　d'education* . . .　Paris, Mondhare, 1783.
　　Maps by Buache, de la Fosse, Nolin.
　　*Nouvelle édition du théâtre de la guerre en
　　Italie* . . .　Paris, 1718.
　　Le théâtre du monde . . .　Paris, 1700?–44
　　Carts.: Tillemon, P. Coronelli. Eng.:
　　L. B. T. Rousseau, H. van Loon, Bourgoin.
　　See also under BLAEU, G; CASSINI, J. D.;
　　DE FER, N.; DESNOS, L. C.; HOMANN'S
　　HEIRS; JULIEN, R. J.; PHILIPPE DE PRÉTOT,
　　E. A.; VALCK, G.

Nollet, Jean Antoine　1700–70　Cart.　Fr.
　　G (C and T)

Norden, John 1548-1626 Topg., Surv., Cart. Brit. M

Made maps for Camden's *Britannia*, probably the following: Hampshire, Hertfordshire, Kent, Middlesex, Surrey, Sussex.

Speculum Britanniae: an historical and chorographicall description of Middlesex and Hartfordshire Lond., Part I, Middlesex, 1593; Part II, Hertfordshire, 1598. Daniel Browne and James Woodman, 1723. Part I had 1 map and 2 plans, Part II, 1 map. All are reprinted in the 1723 edn., eng. by Senex.

See also under DE WIT, F.; SCHWYTZER, C.; SPEED, J.

Nordenankar, Admiral J. *See under* FADEN, W.

Norford, E., Willingdon and Co. *See under* EDWARDS, B.

Noribergae in Officina Homannia *See under* HOMANN, J. B.

Norie, John William 1772-1843 Pub. Hydrographer Brit. M

Traded as Norie and Co.

The Complete East India pilot Lond., 1816.

See also under HEATHER, W.

Norman, John (1748-1817) **and William** Pubs., Carts. Amer. M

The American pilot . . . Boston, 1792, 1794. Carts. incl.: Paul Pinkham, Osgood Carleton. Reprinted by W. Norman at Boston in 1798 and 1803.

See also under MALHAM, J.

Norris, John 1660?-1749 Cart. Brit. M

A complete sett of new charts . . . Lond., T. Page and W. and F. Mount, 1723.

Norris, Robert 18th/19th cent. Cart. Brit. M

The African pilot . . . Lond., Laurie and Whittle, 1794-1804. Carts. incl.: Wm Woodville, Ralph Fisher, Geo. Maxwell.

Norroy *Same as* DU PINET (q.v.)

Norton, John *See under* ORTEL, A.

Norton, Richard fl. 1620 Gunner, Engineer Cart. Brit. M

Norton, William fl. 1591 Cart. Brit. M

Norwood, Richard 17th cent. Mathematician, Surv. Brit. M

Nourse, J., and Vaillant, P. *See under* PALAIRET, J.

Nowell, Laurence, } 16th cent. Cart.
Dean of Lichfield } Brit. M
Noel }

N. St. Signature of N. STOPIUS (q.v.)

Nyon *See under* PHILIPPE DE PRÉTOT, E. A.

O'Beirne fl. 1830 Cart. Irish M

Ode, H. *See under* VANDERMAELEN, P. M. G.

Oddy, S. A. *See under* WALLIS, J.

Oeschläger, Adam *Same as* OLEARIUS, A. (q.v.)

Oernehufvuds, O. J. G. *Same as* SVART, O. H. (q.v.)

Oertel, A. *Same as* ORTEL, A. (q.v.)

Oger, Mathieu } 16th cent. Cart. Fr. M
Ogierius }

Ogilby, John 1600-76 Bookseller, Printer, Dancing master, Theatre manager, Geog. Brit. M

Britannia . . . Lond., 1675.

Britannia depicta or Ogilby improv'd . . . Lond., T. Bowles, 1720, 1736, 1751; C. Bowles, 1764.

Map of London (with W. Morgan).

Map of Essex (with W. Morgan), 1678 and subs. edns.

See also under BOWEN, E.; MOXON, JAMES; SAXTON, C.; SENEX, J.

Ojea, F. Fer 17th cent. Cart. Span. M

Gallacia regnum . . . Antwerp, Hondius, c. 1607.

Olaus Magnus 1490-1558 Cart. Swed. M

Olearius, Adam } 1599-1671 Mathema-
Oelschläger } tician, Cart. Ger. M
} G

Oleatus, Hieronymus } fl. 1567-70 Eng.
Olgiato, Girolaro } Ital. M

Oliva *Same as* OLIVES (q.v.)

Oliver, John fl. 1680-96 Surv., Eng. Brit. M

Maps of Essex, 1696; Hertfordshire, 1695; also worked with SELLER (q.v.) and others.

See also under OVERTON, P.

Oliver and Boyd *See under* EWING, T.; REID, A.

Olives, Bartolomeo} fl. 1532–81   Cart.
Oliva   } Ital.   M

Olives, Domingo} fl. 1568   Cart.   Ital.   M
Oliva   }

Olives, Jaume} fl. 1550   Cart.   Ital   M
Oliva   }

Oliveti, S. M. M.   *Same as* BUONSIGNORI, S.
(q.v.)

Olney, Jesse   1798–1872   Geog., Cart.
Amer.   M
A new and improved school atlas . . .   Hartford,
D. F. Robinson and Co., 1829 and other
edns.
Olney's school atlas . . .   New York, Pratt,
Woodford and Co., 1844–7.

Onofri, Francesco   *See under* DUDLEY, SIR R.

Oomkens, J.   *See under* HILLEBRANDS, A. J.

Oporinus, Johannes   16th cent.   Cart.
Swiss   M

Orell, Gessner et Cie   *See under* WALSER, G.

Orlandi, Johannes   fl. 1602–4   Eng.   Ital.
M

Ornehufud   18th cent.   Cart.   Dan.?   M

Orr, William S., and Co.   fl. 1852   Pub.
Brit.   M
See under MILNER, REV. T.

Ortelianus   *Same as* J. COLINS (q.v.)

Ortelius, Abraham}
Ortel   } 1527–98   Cart.,
Oertel   } Cosmographer   Dut.
Ortell   } M
Wortels   }
Abissinorum sive Prestsiosi Joannis imperium
Antwerp, Plantin, 1584.
Copy of Gastaldi's oval world map.
Hispaniae novae sive magnae . . .   Antwerp,
Plantin, 1612.
Fezzae et Marocchi . . .   Amster., Blaeu,
Plantin, 1665.
Persia, sive sophorum regni typus   Antwerp,
1612.
Americae sive novi orbis . . .   Antwerp,
Plantin, 1586.
Asiae nova descriptio . . .   Antwerp, Plantin,
1592.

Eryn (i.e. Ireland)   Antwerp, Plantin, *c.*
1594.
Scotiae tabula   Antwerp   Plantin, 1592.
Aegypti recentior descriptio   Antwerp, Plantin,
1592.
Gallia vetus   Amster., Blaeu, *c.* 1660.
Typus Galliae veteris   Amster., Blaeu, *c.*
1660.
Arragonia et Catalonia   Antwerp, *c.* 1610.
Castilliae veteris . . .   Antwerp, *c.* 1610.
Valentiae regni . . .   Antwerp, Plantin,
1584.
Regni Valentiae typus   Antwerp, *c.* 1610.
Tartariae . . .   Antwerp, Plantin, 1592.
Turcici imperii descriptio   Antwerp, Plantin,
1592.
Natoliae quae olim Asia minor . . .   Antwerp,
Plantin, 1592.
Theatrum orbis terrarum . . .   Antwerp,
Coppenium Diesth, 1570, 1571, 1573,
1574, and many other edns. in Latin,
Ital., English, Span., Ger., Dut., Fr.
Carts. incl.: J. Surhonio, Iacobo à
Daventria, Godefrido Mascop, Wolfgang
Lazio, Iacobo Castaldo, Aegidio Tschudo,
Martino Heilwig, Antonio Jenkinsono,
Ludoico Teisiera, Philippo Pigafetta,
etc. Various supplements, etc., to the
Theatrum were issued, e.g. *Addidamentum
iii. Theatri orbis terrarum* (Antwerp, 1584),
Addidamentum iv (Antwerp, Plantin, 1590,
etc.). There were, too, various epitomes
and abridged edns., e.g. *Abrégé du Théâtre* . . .
(Antwerp, I. B. Vrients, 1602); *An epitome
of Ortelius* (Lond., John Norton, 1602?);
Epitome theatri orbis terrarum . . . (Antwerp,
Keerberg, 1601; Plantin, 1612 and other
edns.).
See also under BLAEU, G. BRAAKMAN, A.;
CASTALDO, J.; COLINS, J.; HONDIUS, H.;
HORN, G.; JANSSON, J.; KOLER, J.; LHUYD,
H.; PARISIUS, P.; VISSCHER, N., III.;
WOLFGANG, A.

Os   *Same as* VAN OS (q.v.)

Osborne, I.   *See under* OSBORNE, T.

Osborne, T.   fl. 1748   Pub.   Brit.   M
Geographia magnae Britanniae . . .   (with S.
Birt, D. Browne, I. Hodges, I. Osborne,
A. Millar, I. Robinson)   1748.

Ottens, Reiner, Joachim and Joshua 18th cent. Pubs., Carts. Dut. M
Atlas sive geographia compendiosa . . . Amster., 1756? Carts. and engs. incl.:, Covens, de Leth, Delisle, de Wit, Fricx, Halley, Halma, Homann, Jaillot, Loon, Mortier, Pynacker, Reland, Sanson, Schenk, Specht, Starkenburg, Valk, van Keulen, N. J. Visscher, Voogt, Zürner, Anse, Broeck, Bróen, Goerê, Gouwen, Hogenboom, Keyser, Moll, Ruyter.
Atlas minor sive geographia compendiosa . . . Amster., 1695-1756 and other edns. Carts. and engs. incl. many of those mentioned above, and Gephart, Jansson, Kaempfer, Persoy, Placide, de Ram, Scheuchzer, Schut.
Many maps from the above atlases were reissued in an atlas by Ottens heirs (1740?) and incl. maps by van de Aa, Allard, J. Blaeu, Cantelli da Vignola, various members of the Danckertz family, Du Tralage, Have, Husson, W. Petty, Tjarde, as well as many named in the foregoing atlases.
See also under Halley, E.

Ottens Heirs *See under* Ottens, R.

Otterschaden, Johann 1580-1613 Cart. Ger. G (C and T)

Outhett, John 19th cent. Cart. Brit. M
Map of Dorsetshire, 1826. Also worked with Jefferys, T. (q.v.)

Outhier, Réginald 1694-1774 Astron. Fr. G

Overton, Henry d. 1751 Pub., Bookseller Brit. M
Issued a reprint of Speed in 1743; he was also pub. of Bowen's *Royal English atlas* of *c.* 1764.
See also under Richardson, W.

Overton, John 1640-1713 Cart. Brit. M
Collaborated with Philip Lea and H. Moll.
A new mapp of America c. 1686.
See also under Ferrar, J.

Overton, Philip d. 1745 Cart. Brit. M
Map of Essex, 1726 (with T. Bowles); Oxfordshire, 1715 (eng. J. Oliver); Sussex, 1740 (with J. Bowles).
See also under Moll, H.

Overton and Hoole 18th cent. Pub. Brit. M
A new plan of the garrison of Gibraltar Lond., 1726?

Owen, Francis *See under* Stephenson, J.

Owen, George 17th cent. Cart. Brit. M
Map of Pembrokeshire in Camden's *Britannia.*

Owen, John 18th cent. Cart. Brit. M
Co-operated with Emanuel Bowen (q.v.)
See also under Laurie, R.

Packe, Christopher fl. 1743 Physician, Cart. Brit.' M
Map of East Kent, 1743.

Pagano, Mattheo fl. 1538-62 Pub. Ital. M

Page, Lieut. 18th cent. Cart. Brit. M
See also under Faden, W.

Page, Thomas *See under* Mount.

Pagnano, Carlo 16th cent. Engineer Ital. M
Decretum super flumine Abduae . . . (with 1 map) Milan, 1520. The original issue had a woodcut map; there was an 18th-cent. reissue with an eng. map.

Paguenaud, E. *See under* Carey, H. C.; Lavoisne, C. V.

Palairet, Jean 1697-1774 Cart. Fr. M
Carte des possessions Angloises & Françoises . . . de l'Amerique septentrionale (unsigned). Eng. by Thos. Kitchin. Lond., Amster., Berlin and The Hague, 1755.
Atlas méthodique . . . Lond., J. Nourse and P. Vaillant, etc., 1755. Maps eng. by J. Gibson and T. Kitchin.
Bowles's universal atlas . . . Lond., Carrington Bowles, 1775-80; Bowles and Carver, 1794-8. Carts.: de la Rochette, Dorret, Evans, Hubner, Mayer, Patterson, Rouvier, Sheffield.
See also under Julien, R. J.

Palbitzke, Frid *See under* Jansson, J.; Mercator, G.

Palestrina *Same as* de Palestrina (q.v.)

Pallavicinio, Leone and Lucio fl. 1590-1604 Eng. Ital. M

Palmer, G. *See under* Laurie, R.; Melish, J.

Palmer, I. *See under* GREAM, T.

Palmer, Richard fl. 1670–1700 Cart., Eng., Stationer Brit. M
The bible atlas . . . New York, Leavitt, Lord and Co., Boston, Crocker and Brewster, 1836; New York, W. M. Brownson, 1847; and other edns.
Eng. for SELLER (q.v.)
See also under TAYLOR, T.

Palmer, Roger *Same as* CASTLEMAIN, EARL OF (q.v.)

Palmer, William fl. 1766–94 Eng. Brit. M
See also under COOK, CAPT. J.; FADEN, W.; SAYER, R.; WILKINSON, R.

Palmquist, Eric fl. 1673 Cart. Swed. M

Panagathus *Same as* ALGOET, L. (q.v.)

Panouse, Jacqueline *See under* DE FER, N.

Papen, Augustus 19th cent. Cart. Ger. M
Topographischer atlas des königreichs Hannover und herzogthums Braunschweig . . . Hanover, 1832–47.

Parentius, Gellio} fl. 1597 Military
Parenzius } engineer, Cart. Ital. M

Parides, Ignatius Gaston}
Parides, R. P. Ignatio } 1636–75 Astron.
Gastone } Fr. X

Parijs *Same as* VON PARIJS (q.v.)

Pâris, E. *See under* DUMONT D'URVILLE, J. S. C.

Parish, H. W. *See under* STAUNTON, SIR G. L.

Parisius, Prosperus 16th cent. Cart. Ital.? M
Made map of Calabria for Ortel.

Parker, H. fl. 1762 Cart. Brit. M
Executed, among others, maps for Bowen and Kitchin.

Parker, Samuel fl. 1749 Eng. Brit. M
Eng. for WARBURTON, J. (q.v.) and others.

Parliamentary Gazetteer of England and Wales *See under* BELL, J.

Parmentier, Jean 1494–1530 Cart., Cosmographer Fr. M G

Parr, Richard *See under* DICKINSON, J.

Parry, W. E. fl. 1816 Cart. Brit. X

Parsons, Samuel 18th cent. Pub. Brit. M

Pask, Joseph fl. 1860 Pub., Stationer Brit. M

Pasquier, J. J. *See under* DENIS, L.; MENTELLE, E.

Pass, J. fl. 1828 Eng. Brit. M
Eng. map of Warwickshire for the *Encylopaedia Londinensis* (1828).

Paterson, Daniel 18th cent. Cart. Brit. M
Circular map of 24 miles around London, 1791 and subs. edns.
See also under FADEN, W.

Patterson, R., and Lambdin *See under* GILLELAND, J. C.; PALAIRET, J.

Patteson, Rev. Edward 18th/19th cent. Geog., Cart. Brit. M
A general and classical atlas . . . 2 vols., Richmond, Surrey, G. A. Wall, 1804–6.

Pauli, Joannem *See under* CLUVER, P.

Payne, John fl. 1800 Cart. Amer. M
A new and complete system of universal geography . . . [*atlas*] New York, Low and Wallis, 1798–1800. Engs.: Anderson, Barker, Rollinson, Tanner, Tiebout, Scoles.

Pedrezano, G. B. *See under* PTOLEMAEUS, C.

Peers, Richard 1645–90 Geog. Brit. M
The description of the seventeen provinces of the Low Countries . . . Oxford, M. Pitt 1682.
See also under NICOLSON, W.

Peeters, Jacques 1637–95 Cart., Pub. Dut. M
L'atlas en abregé . . . Antwerp, 1692.
See also under DE AFFERDEN, F.

Pelet, Jean Jacques Germain, Baron 1779–1858 Soldier, Cart. Fr. M
Atlas des mémoires militaires relatifs à la succession d'Espagne sous Louis XIV Paris, 1836–48.

Pelham, Henry fl. 1787 Cart. Irish M
Map of Co. Clare, 1787.

Pélicier, A. fl. *c.* 1810 Eng. Fr. G

Pélicier fils, Pélicier, T. *See under* DELAMARCHE, F.

Pellet, J. L. *See under* BONNE, R.; RAYNAL, G. T. F.

Pelletier, David *See under* DE CHAMPLAIN, S.

Pentius de Leuchio, Jacobus fl. 1511 Cart. Ital. M

Perizot *See under* HOMANN'S HEIRS.

Perkins and Bacon *See under* STARLING, T.

Péron *See under* DE FREYCINET, L. C. D.

Perrier *See under* DESNOS, L. C.

Perrot, Aristide Michel 1793–1879 Geog. Fr. M
Atlas de 59 cartes . . . Paris, E. et A. Picard, 1843 (with Madame Alexandrine Aragon [b. 1798]).
See also under DEPPING, G. B.

Perry, George *See under* YATES, W.

Person, N. fl. 1694 Pub. Ger. M

Persoy *See under* OTTENS, R.

Perthes, J. *See under* BERGHAUS, H. K. W.; SPURNER VON MERZ, K.; STIELER, A.; VON SYDOW, E.; WILTSCH, J. E. T.

Petermann, August Heinrich 19th cent. Cart. Brit. M
The atlas of physical geography . . . (with Thomas Milner) Lond., Ward and Lock, 1850. Eng.: John Dower.

Petermann, Thomas *See under* MILNER, REV. T.

Peters, C. H. F. fl. 1882 Astron. Amer. X

Petersen, Mathias and Nicolaus 17th cent. Eng. Dan. M

Petit, J. B. *See under* DE JOMINI, A. H.

Petit, Pierre 1598–1667 Mathematician, Geog. G

Petit Bourbon, P. *See under* BLAEU, G.; TAVERNIER, M.

Petri, Henricus fl. 1540–52 Cart., Pub. Swiss M
See also under MÜNSTER, S.

Petrucci, Guilo Cesaro fl. 1571 Cart. Ital. M

Petrus, Henricus *See under* PTOLEMAEUS, C.

Petrus ab Aggere fl. 1520 Cart. Dut.? M

Petty, Sir William} 1623–87 Physician,
Petty, Guilielmi } Statistician, Economist
} Brit. M
A geographicall description of ye kingdom of Ireland . . . Lond., F. Lamb, R. Morden, J. Seller, 1689.
See also under OTTENS, R.; VALCK, G.

Peyer, Heinrich fl. 1685 Cart. Ger.? M

Peyrouin, A.} 17th cent. Pub., Eng. Fr.
Peyrounin } M
Employed by N. SANSON, PÈRE (q.v.)
See also under BLAEU, G.

Pfeffel, Johann Andreas 1674–1750 Eng., Pub. Ger. M

Phelipeau *See under* PONCE, N.

Phelps, Joseph fl. 1809–39 Stationer, Bookseller, Pub. Brit. M
See also under LANGLEY.

Philip, George b. 1799 Pub. Brit. M G (C and T) X
Founder of the firm of George Philip and Son.
Map of the Continent (lithographed) Liverpool, 1852.
Travelling map of Australia Lond., *c.* 1855.

Philippe de Prétot, Étienne André 1708–87 Cart. Fr. M
Atlas universelle . . . Paris, Nyon l'ainé, 1787.
Cosmographie universelle . . . Paris?, 1768. Incl. maps by J. B. Nolin and Chas. Maugein. Maps eng. by Maurelle Antoine Moithey and J. E. J. Vallet.

Philipps, J. *See under* SAYER, R.

Philipps, Sir Richard 1767–1840 Cart. Brit. M
A geographical and astronomical atlas . . . by the Rev. J. Goldsmith (this was the pseudonym of Philipps) Lond., G. and W. B. Whitaker, 1823. Engrs.: Cooper, Neele and Son.

Philips, John, and Hutchings, W. E. fl. 1832 Cart. Brit. M
Map of Staffordshire, 1832.

Phillips, Capt. T. *See under* LEA, P.

Phrisia, Henricus fl. 1568 Cart. Dut.? M

Phrisius, Gemma   *Same as* GEMMA FRISIUS (q.v.)

Piadyshev, Vasilii Petrovich   1769–1836 Cart. Russ. M
Atlas géographique de l'empire de Russie, du royaume de Pologne, et du grand-duché de Finlande . . . St. Petersburg, 1823–6. Titles of maps given in Russ., Polish, Fr. Names on maps given in Russ. and Roman types. Engs.: Chkatoff, Faleleeff, Finaghenof, Freloff, Ieremin, Iwanoff.

Piat   *See under* LORRAIN, A.

Picard, Jean   1620–82 Cart. Fr. M

Picard, Peter   1670–1737 Eng. Dut. M

Piccoli, Gregorio   fl. 1720 Eng. Ital. M

Piccolomini, Alessandro   1508–78 Astron. Ital. X

Pickart, P.   *See under* VOOGT, C. J.

Pickel   *Same as* CELTES (q.v.)

Pickett, J.   *See under* REES, A.

Picquet, Charles   19th cent. Cart., Pub. Fr. M
Carte de l'empire Français . . . Paris, 1815.
See also under ANDRIVEAU-GOUJON; COUTANS, G.; LAPIÉ, P.

Pieterssoon, Claes   fl. 1607 Cart. Dut. M

Pigafetta, Philipp or Filippo   b. 1533 Cart. Ital. M
Map of eastern and southern Africa Rome, 1590.
See also under LANGENES, B.; ORTEL, A.

Pigeon, Jean   fl. 1756–70 Cart. Fr. G (C and T)

Pigot, James and Co.   19th cent. Topg. Eng. Brit. M
Collaborated with ISAAC SLATER (q.v.).
Pigot and Co's. British atlas . . . Lond., 1829, 1830, 1831, 1840, 1842?.

Pilestrina   *Same as* DE PALESTRINA (q.v.)

Pilizzoni   *Same as* BASSO, F. (q.v.)

Pimentel, Manoel   1750–1818 Cart. Port. M
The Brazil pilot . . . Lond., Longman, Hurst, Rees and Orme, and A. Arrowsmith, 1809.

Pinadello, Giovanni   fl. 1595 Cart. Dut. M

Pine, John   18th cent. Cart. Brit. M
Plan of the cities of London and Westminster and borough of Southwark Lond., 1747 (with John Tinney; based on Rocque).

Pingré, A.   fl. 1784 Astron. Fr. X

Pinkerton, John   1758–1826 Cart., Geog. Brit. M
Maps to accompany Pinkerton's Modern geography . . . Philadelphia, Conrad and Co., 1804? Maps drawn by S. Lewis, G. Fox. Eng. by Thomas Marshall, Hooker, Tanner, D. Fairman, A. Lawson, W. Harrison, Junr.
A modern atlas . . . Lond., T. Cadell and W. Davies; Longman, Hurst, Orme and Brown, 1815. Philadelphia, T. Dobson and Son, 1818.

Pinkham, Paul   *See under* NORMAN, J.

Pinnock, William   fl. 1825 Topg. Brit. M

Pippi, Giulio   *Same as* DE' GIANUZZI, G. DI P.

Piri Re'is   d. 1554 Sailor, Cart. Turkish M

Pisani, Ottavio   b. 1575 Cart. Ital. M G

Pisarri, A.   *See under* ROSACCIO, G.

Piscator   *Same as* VISSCHER (q.v.)

Pissot   *See under* DE VAUGONDY, G. R.

Pitt, Moses   1641–97 Pub. Brit. M
A description of places next to the north-pole . . . Oxford, 1650.
See also under DE LA FEUILLE, J.; NICOLSON, W.; PEERS, R.; VISSCHER, N., III.

Placide de Saint Hélène, Le Père   1649–1734 Geog. Fr. M
Cartes de géographie . . . Paris, du Val, 1714? Engs.: C. Rousset, N. Guerard, C. Inselin, Roussel, C. A. Berey.
See also under DELISLE, G.; OTTENS, R.

Plaes, Plaéts, Plaetz   *Same as* DE PLAES (q.v.)

Plamann, Johann Ernst   1771–1834 Pedagogue Ger. G

Plancius, Peter}   1552–1622 Cart. Dut.
Platevoet   **} M   G (C and T)**

Plantin, Christopher}   1520?–1589 Printer,
Plantijn   **} Bookseller Dut. M**
One of the more important cartographic printers.

See Denucé's book in the *Bibliography*.

Printed the *Theatrum* and other works of Ortelius, and the works of Jacobo Castaldo, Ludovico Georgius, H. Chiaves, etc. (qq.v.)

See also under FAVOLI, U.; GALLE, T.; GUICCIARDINI, L.; HEYNS, P.; WAGHENAER, L. J.

Plater, Stanislaw, Comte 1784–1851 Cart., Historian Polish M
Plans des sièges et batailles qui ont eu lieu en Pologne . . . Posen, G. Decker and Co. 1828.

Platevoet *Same as* PLANCIUS, P. (q.v)

Platus, Carolus fl. 1578–89 Astron. Ital. G (C) O

Playfair, James 1738–1819 Geog. Brit. M
Atlas to Playfair's Geography . . . Lond., 1814. Carts. incl.: W. W. Macpherson. Engs. incl.: Cooper and B. Smith.
A new general atlas . . . Lond., 1814. Carts. incl.; N. Coltman, W. W. Macpherson. Engs. incl.: B. Smith, J. Bye, E. Jones, H. Cooper, S. I. Neele.

Plot, Robert *See under* BROWNE, J.; BURGHERS, M.

Plouich *See under* BEEK, A.

Pocock or Poeck, Ebenezer fl. 1831 Cart. Brit.? G (T)

Pocock, Richard 18th cent. Cart. Brit. M
Aegypti . . . Amster., Covens and Mortier, 1746. Eng. by I. Condet.
See also under LE ROUGE.

Poeck *Same as* POCOCK (q.v.)

Pograbius, Andreas} d. 1602 Cart.
Pograbski   **}** Polish M

Poirson, J. B. *See under* ARROWSMITH, A., BOSSI, L.

Poker, Matthew 17th cent. Surv. Brit. M
Romney Marsh 1737 (from Poker's survey of 1617). Eng. J. Cole.

Polanzini, I. *See under* DE VALLEMONT, PIERRE LE LORRAIN

Pomarede, D. *See under* KEENAN, J.

Ponce, Nicolas 1746–1831 Eng. Fr. M
Recueil de vues des lieux principaux de la colonie Française de Saint-Dominique . . . *accompagnées de cartes et plans* . . . Paris, Moreau de St. Méry, etc., 1795. All but two of the maps are by Phelipeau. The exceptions are 'Carte de l'isle St. Dominique . . .' by I. Sonis and eng. by Vallance, and 'Carte de la partie françoise de St. Dominique . . .' by Bellin, 'Augmentée par P. C. Varlé et autres', and eng. by J. T. Scott of Philadelphia.

Pont, Timothy b. *c.* 1560 Surv. Brit. M
Cuninghamia Amster., Blaeu, 1664.
Evia et Escia scotis (i.e. Ewesdale and Eskdale) Amster., Blaeu, 1664.
Levinin (i.e. Lennox) Amster., Blaeu, 1664.
Nithia vicecomitatus (i.e. the Nith valley) Amster., Blaeu, 1664.
The Steuartrie of Kircudbright Amster., Blaeu, 1654.
Glottiana praefectura superior Amster., Blaeu, 1654.
See also under HONDIUS, H.; JANSSON, J.; MERCATOR, G.

Pontano, Giovanni Giovio 1421–1503 Historian, Cart. Ital. M

Pontoppidan, C. J. fl. 1795 Cart. Norwegian M
Det sydlige Norge, det nordlige Norge Christiana, 1785, 1795.

Pontoppidan, Erik 1698–1764 Cart. Dan. M
Den Danske atlas eller konge-riget Dannemark . . . Copenhagen, A. H. Godiche, 1763-9.
Supplement til den Danske atlas eller konge-riget Dannemark . . . Copenhagen, A. H. Godiche and F. C. Godiche, 1774-81.

Popham *See under* LAURIE, R.

Poppele, E. fl. 1840 Cart. Fr. M

Poppey, C. *See under* WEILAND, C. F.

Popple, Henry fl. 1733 Cart. Brit. M
A map of the British Empire in America Lond., Stephen Austen, 1732.
See also under HOMANN, J. B.

Poracchi da Castiglione, Tomaso or Thomaso 1530-85 Cart. Ital. M
L'isole piv famose del mondo . . . Venice, S. Galignani and G. Porro, 1572, 1576, 1590, 1604; heirs of Simon Galignani, 1605; Padova, P. and F. Galignani, 1620, and other edns.

Porebski, Stanislav fl. 1563 Cart. Polish M

Porro, Girolamo fl. 1596 Pub., Eng. Ital. M
See also under PORACCHI, T.

Portantius, Johann fl. 1573 Cart., Mathematician Dut. M

Porter, Thomas fl. 1655 Cart. Brit. M

Postellus, Guilaume} 1510-81 Astron.
Postel, Guillaume } Cart. Fr. M X

Potel, Felice 19th cent. Pub. Ital. M
Atlante universale di geografia antica e moderna . . . Naples, 1850.

Potocki, Jan, Hrabia 1761-1815 Cart. Russ. M
Atlas archéologique de la Russie Européenne . . . St. Petersburg, 1823. Text and titles in Russ. and Fr.

Potter, Paracelte } 19th cent. Cart.
Potter, P., and Co.} Amer. M
Atlas of the world . . . Poughkeepsie, 1820.
See also under WILLETTS, J.

Pound, Thomas fl. 1677-91 Surv. Brit. M

Poussin, Guillaume Tell 1794-1876 Cart. Fr. M
Travaux d'améliorations intérieures projetés ou exécutés par le gouvernement général des États-Unis, de 1824-1831 Paris, Anselin et Carilian-Goeury, 1834.

Powell, William *See under* BELLERO, I.

Pownall, Thomas 1722-1805 Geog. Brit. M
A topographical description . . . *of North America* . . . Lond., J. Almon, 1776. Contains a large 'Map of the middle British colonies in North America'.
See also under D'ANVILLE, J. B.; SCHRAEMBL, F. A.

Praetorius, Johannes 1537-1616 Cart. Ger. G (C and T)

Prahl, Arnold Friedrich fl. 1746 Cart. Dut.? M

Pratt fl. 1705 Cart. Brit. M

Pratt, Woodford and Co. *See under* OLNEY, J.

Price, Charles d. 1733 Instrument maker Brit. M G

Price, H. and B. *See under* SCOTT, JOSEPH.

Price, Owen fl. 1769 Geog. Brit. M
See also under LAURIE, R.; RUSSELL, P.

Price, Capt. W. *See under* STEPHENSON, J.

Prichard, James Cowles 1786-1848 Historian Brit. M
Six ethnographical maps . . . Lond., New York, H. Ballière, 1843, 1851.

Pride, Thomas, and Luckombe, Philip fl. 1789 Cart. Brit. M
Map of 10 miles round Reading, 1790.

Pringle, Capt. 18th/19th cent. Surv. Brit. M

Pringles, George, Senr. and Junr. 19th cent. Carts. Brit. M

Prior, John, and Dawson, W. fl. 1777-9 Pub., Cart. Brit. M
Map of Leicestershire, 1779 (Eng. J. Luffman; surv. J. Whyman.)

Probst, Johann Balthasar 1673-1743 Eng. Ger. M

Probst, Johann Michael 18th cent. Cart. Ger. M
A map of the inhabited part of New England . . . Augsburg, 1779 (copied from Lotter's map of 1776).
See also under LOTTER, T. C.

Prockter, J. fl. 1750-76 Eng. Brit. M G (T)
Worked on one of Lane's globes.

Proctor, Richard Anthony 1834-88 Astron. Brit. X

Pronner, C. M. *See under* HOMANN'S HEIRS.

Prony *Same as* DE PRONY (q.v.)

Prunes, Johann fl. 1532 Cart. Span. M

Prunes, Matheus fl. 1553-90 Cart. Span. M

Prunes, Pietro Giovanni fl. 1651 Cart. Span. M

Pruthenus, Johannes Somer *See under* SOMERS, J.

Ptolemaeus, Claudius} d. A.D. 147? Geog.,
Ptolemy } Astron. Egyptian
Tolemeo, Claudio } M
The following are a few of the edns. of his works:
Cosmographia. The first edn. with engraved maps was pub. at Bologna in 1477, but dated '1462'. The first edn. with woodcut maps was pub. by Lienhart Holle at Ulm, 1482; other edns. incl. that pub. in 1486 (Ulm, Johannes Reger).
Liber geographiae 1511, Venice, Jacobus Pentius de Leucho; with the first printed map of any portion of the North American continent.
Geographia . . . 1482, Florence, Nicolaus Laurentii, in Ital. verse by Francesco di Niccolo Berlinghieri (1440-1501), 1511; 1513, Strasbourg, Schott, containing the first map printed in colours (see p. 56) and the first map of Switzerland; 1535, first edn. by Servetus, Lyons, Melchoir et Gaspar Treschel; 1540, first edn. by Münster, Basle, Henricus Petrus; 1541, Lyons, Gaspar Treschel for Hugo à Porta; 1548, first Ital. trs., contains the first allusion to S. America as a continent, Venice, G. B. Pedrezano; 1552, Basle, Henrich Petrus; 1562, Venice, Vincent Valgrisi; 1605, first edn. with Greek and Latin texts, Frankfurt, Jodocus Hondius and Cornelio Nicolai; 1617, Arnhem, Joannes Jansson. There were many other edns. (see p. 360, Bagrow *Die Geschichte der kartographie*; details in Bibliography).

Puke, I. *See under* Wilkinson, R.

Purchas, Samuel Author Brit. M
c. 1575-1626
Purchas his pilgrimage . . . Lond., 1625 and other edns. With a map by R. Elstracke.

Purdy, John fl. 1809-23 Cart. Brit. M
General chart of the West-India islands . . . Lond., Laurie, 1823.

Puschner, Johann George fl. 1720-8
Cart., Eng. Ger. G (C and T)
Worked with J. G. Doppelmayr. (q.v.)

Putsch, Joannes} 1516-42 Cart. Ger. M
Bucius }

Putte *Same as* van der Putte (q.v.)

Pyle, Stephen fl. 1770-80 Eng. Brit. M
See also under Armstrong, A.; Lendrick, J.

Pynacker, Corneliu *See under* Blaeu, G.; Hondius, H.; Jansson, J.; Ottens, R

Pyramius, Christophorus} fl. 1547 Cart.
Kegel } Belgian M
Khegel }

Pyrrhus, Ligorius *Same as* Ligorio, P. (q.v.)

Quad, Matthias} 1557-1613 Cart. Ger.
Quadum } M G
Geographisch handtbuch Cologne, Johan Buxemacher, 1600.
Europae totius orbis terrarum partis praestantissimae . . . Cologne, J. Bussemechers, 1592, 1594.
Fasciculus geographicus . . . Cologne, J. Buxemacher, 1608 (with 86 maps; they are the same as those in *Geographisch handtbuch*, 1600).

Quin, Edward 1793/4-1828 Cart. Brit. M
An historical atlas . . . Lond., R. B. Seeley and W. Burnside, 1836. Maps eng. by Sidney Hall. A new edn. was issued in 1846 by Seeley, Burnside and Seeley, but eng. by W. Hughes.

Rabus, Jacques fl. 1546 Globe-maker
Fr.? G (C)

Rad, Christoph d. 1710 Goldsmith Ger. G

Radefeld, Hauptmann (Captain), and
Renner, Prim. Lieut. 19th cent.
Carts. Ger. M
Atlas zum handgebrauche für die gesammte erdbeschreibung Hildburghausen, 1841.
See also under Meyer, Joseph.

Ram *Same as* de Ram (q.v.)

Ramble, Reuben fl. 1845 Topg. Brit. M
Juvenile maps, with borders of views. They are very rare in fine condition.

Ramusio fl. 1588 Cart. Ital.? M

Ransonnet *See under* de Freycinet, L. C. D.

Raspe, Gabriel Nicolaus 1712-85 Pub.
Ger. M
Schauplatz des gegenwaertigen kriegs . . . Nuremberg, 1757-64.

Ratelband, Johannes 1715-91 Pub. Dut. M

Kleyne en beknopte atlas, of tooneel des vorlogs in Europa Amster., 1735.

Ratzer, Bernard 18th cent. Cart. Brit. M

Plan of the city of New York ... Jefferys and Faden, 1776.

See also under FADEN, W.; SAUTHIER, C. J.

Rauch, Johann Andreas d. *c.* 1635 Painter, Cart. Ger. M

Rauchen, Johannes Andreas *See under* DE LA FEUILLE, J.

Rauchin, Mariae Magdalenae *See under* SCHERER, H.

Raus, Muhammed fl. 1590 Cart. Ger.? M

Rauw, Johann} fl. 1597 Cosmographer,
Ravis   } Cart. Ger. M

Raynal, Guillaume Thomas François 1713-96 Cart. Fr. M

Atlas portatif . . . Amster., E. van Herrenvelt, D. J. Changuion, 1773; later called *Atlas de toutes les parties connues du globe terrestre* . . . Geneva, J. L. Pellet, 1780; 1783-4 and other edns. Carts.: Rigobert Bonne, Nicolas Bellin, according to edn. Engs.: André, A. van Krevelt, J. van Schley.

Rea, Roger I and II fl. 1650 Booksellers, Pub. M

Pub. the 1650 edn. of Speed's *Theatre*.

Redmayne, M. 17th cent. Pub. Brit. M

Recreative pastime by card-play ... (Playing-cards with maps) Lond., 1676.

Rees, Abraham 1743-1825 Cart. Brit. M

The cyclopaedia; or universal dictionary of arts, sciences and literature . . . *atlas and modern atlas* S. F. Bradford; Murray, Fairman and Co., 1806. Another edn. was issued in 1820 in Lond. by Longman, Hurst, Rees, Orme and Brown; F. C. and J. Rivington. Engs. incl.: B. Smith, J. Bye, Hewitt, Sid. Hall, John Cooke, J. Pickett, Lowry, E. Jones, Alex. Findlay, J. Russell, Cooper, J. Lodge.

See also under MACPHERSON, D. AND A.

Reeve, James *See under* SMITH, CAPT. J.

Reger, Johannes *See under* PTOLEMAEUS, C.

Regnier *Same as* GEMMA, F. (q.v.)

Regrwill, Wolfgang fl. 1574 Cart. Ger. M

Reich, Erhard} fl. 1540 Cart. Ger. M
Reych   }

Reichard, Christian Gottlieb 1758-1837 Cart. Ger. M

Orbis terrarum antiquus . . . Nuremberg, F. Campii, 1824; J. L. Lotzbeck, 1861.

Orbis terrarum veteribus cognitus . . . Nuremberg, D. Campii, 1830, 1833.

See also under GASPARI, A. C.

Reichard, Heinrich August Ottokar 1752-1828 Cart. Ger. M

Atlas portatif et itinéraire de l'Europe Weimar, 1818-21. Carts. incl.: C. F. Weiland.

Reicherstorffer, Georg} fl. 1595 Cart.
Reicherstorff   } Ger. M

Reid, Alexander. 1802-60 Cart. Brit. M

An introductory atlas of modern geography ... Edinburgh, Oliver and Boyd; Lond., Simpkin, Marshall and Co., 1837. Maps drawn by A. Wright.

Reid, John *See under* WINTERBOTHAM, W.

Reid, W. H. fl. 1820 Pub. Brit. M
Operated with James Wallis, printer.

Reidig, M. *See under* STEIN, C. G. D.

Reilly *Same as* VON REILLY (q.v.)

Reimer, G. *See under* SCHOUW, J. F.

Reinecke fl. 1802 Cart. Ger. M
See also under GASPARI, A. C.

Reinel, Pedro fl. 1500-34 Cart. Port. M G

Reiner, Joannes Alexander fl. 1683 Cart. Ger.? M

Map of Hungary, 1683, eng. by M. GREISCHER (q.v.)

Reinerus *Same as* GEMMA, F. (q.v.)

Reinhold, Johannes 1511-53 Mathematician Ger. G

Reisch, George fl. 1508 Cart. Ger. M

Reland *See under* KÖHLER, J. D.; OTTENS, R.

Remondini, Giuseppe 18th/19th cent. Pub. Ital. M

Atlas géographique . . . Venice, 1801.

Renard, Louis fl. 1715 Cart., Pub. Fr.
M
Atlas de la navigation . . . Amster, 1715; R. J.
Ottens, 1739. Carts. incl. de Wit.
*Atlas van zeevaert en koophandel door de
geheele weereldt* . . . Amster., R. and J.
Ottens, 1745. Carts. incl. Jan Bruyst,
G. Delisle, E. Halley, N. Witsen.

Renat, Johan Gustav fl. 1715-33 Soldier,
Cart. Swed. M

Renaud, H. fl. 1835 Globe-maker
Belgian G

Rennell, Major James fl. 1773-84 Cart.
Brit. M
A Bengal atlas Lond., 1779, 1780, 1781.
See also under FADEN, W.; KITCHIN, T.;
LAURIE, R.; SAYER, R.

Renner, Lieut. *See under* RADEFELD,
HAUPTMANN.

Renouard, J. } *See under* DE LABORD,
Renouard et Cie} L. E. S. J.; DUFOUR, A. H.;
} GAULTIER, A. E. C.

Restrepo, José Manuel 1780-1864
Historian Span. M
*Historia de la revolucion de la república de
Colombia* . . . Paris, 1827.

Reuwick, Erhard *See under* VON BREYDEN-
BACH, B.

Reych *Same as* REICH (q.v.)

Reynolds, James fl. 1848-60 Pub. Brit.
M
See also under EMSLIE, J.

Reynolds, Nicholas fl. 1577 Eng. Brit.
M

Rheticus, Joachim} 1514-74 Cosmographer,
Georg} Cart. Ger. M
von Lauchen }

Rhind, William fl. 1842 Geog. Brit. M

Ribeiro, Diego fl. 1519 d. 1533 Cart.
Port. M

Ricci, Father Matteo 1552-1610 Cart.
Ital. M

Riccioli, John Baptista fl. 1651 Astron.
Ital. X

Richards and Scalé fl. 1764 Carts. Irish
M

Richardson, A. M. 19th cent. Cart.
Brit. M
Map of 10 miles round Newcastle, 1838.

Richardson, T. 18th cent. Cart. Brit. M
Map of New Forest, 1749, 1789.

Richardson, William 18th cent. Cart.
Brit. M
The harbour town and forts at Porto Bello . . .
Lond., Overton, 1740.

Richter, Johannes 1537-1616 Mathematician, Instrument-maker Ger. G

Ridge, J. *See under* GILLIES, J.

Ridgway and Sons, J. *See under* BANSEMER,
J. M.

Riecke, G. F. *See under* KILIAN, G. C.

Riedig, Christian Gottlieb 1768-1853
Mapseller Ger. M G

Riedinger, Johann Adam} 1680-1756
Reidiger } Cart. Swiss
Rüdinger } M G

Riedl, J. fl. 1840 Cart. Ger.? G (T)

Riedl, Mathias *See under* SCHERER, H.

Riedl von Leuenstern, J. 1811-40 Cart.,
Mathematician Austrian M G

Righettini, G. *See under* ROSACCIO, G.

Rihel, Wendelen *See under* ZIEGLER, J.

Riley, I. *See under* EDWARD, B.

Rinaldi, Pier Vincenzo} fl. 1571
Dante} Instrument-maker
De Rinaldi } Ital. O

Rios, Andrea fl. 1607 Cart. Ital.? M

Riss} *Same as* RÜST (q.v.)
Rist}

Riuj, Gerard *See under* WYTFLEET, C.

Rivière *See under* SANSON, N., PÈRE

Rivington, F., C. and J. *See under* GUTHRIE,
W.; KELLER, C.; REES, A.; SOCIETY
FOR THE PROPAGATION OF THE GOSPEL;
WALKER J.

Rizzi-Zannoni, Giovanni Antonio (1736-
1814) **and J. A. B.** Cart. Ital. M
Nuova carta della Lombardia . . . Naples,
1795.
Atlante marittimo delle due Sicilie . . . Naples,
1796.
Atlas géographique . . . Paris, Lattré, 1762.
*Atlas géographique et militaire ou théâtre de la
guerre présente en Allemagne* . . . Paris,
Lattré, 1763.

Atlas historique de la France ... Paris, Desnos, 1765.

Le petit neptune françois ... Paris, Desnos, 1765.

See also under DESNOS, L. C.; DU CAILLE, L. A.; HOMANN'S HEIRS; SCHRAEMBL, F. A.

Roades, William 18th cent. Eng. Brit. M

Robaert, Augustijn fl. 1600 Cart. Brit. M

Robert de Vaugondy, G. *Same as* DE VAUGONDY, G. R.

Roberts, Lieut. Henry *See under* COOK, Capt. J.; FADEN, W.

Roberts, John *See under* LAURIE, R.; SAYER, R.; SCHRAEMBL, F. A.; WINTERBOTHAM, W.

Robertson, G. fl. 1788 Cart. Brit. M

Robertson, James fl. 1822 Cart. Brit. M Map of Aberdeen, 1822.

Robijn, Jacobus} d. 1649 Cart. Dut. M
Robÿn }
Zee, zea-atlas-aquatique, del mar Amster., 1683. Carts. incl.: Pieter Goos, Hendrick Doncker.

Robins, James } d. 1836 Pub., Book-
Robins, J., and Co.} seller Brit. M
See also under DUGDALE, J.

Robinson, D. F., and Co. *See under* OLNEY, J.

Robinson, G. G. and J. *See under* CRUTT-WELL, C.; GUTHRIE, W.; VANCOUVER, G.

Robinson, John fl. 1819 Cart. Brit. M
See also under OSBORNE, T.
Map of Saffron Walden district, 1787 (eng. S. J. Neele).

Robinson, William 18th cent. Cart. Brit. M

Rocheford, Rochefort *Same as* DE ROCHEFORD (q.v.)

Rochette *Same as* DE LA ROCHETTE (q.v.)

Rocque, John d. 1762 Eng., Surv., Pub. Brit. (Huguenot extraction) M
A general map of N. America ... Lond., 1761.
Recüeil des villes ports d'Angleterre ... (with J. N. Bellin) Paris, Desnos, 1766.
An exact survey of the cities of London and Westminster, the borough of Southwark ... Lond., 1746.

The small British atlas ... Lond., 1753, 1764. In the 1753 edn. the maps have no plate numbers; these are provided in the 1764 edn. Maps had been previously issued in *The English Traveller*, 1746. In 1769, 33 of the maps were reissued in Russell and Price's *England Displayed*.

A set of plans and forts in America ... Lond. 1763. Maps eng. by P. Andrews.

See also under BELLIN, J.; LEMPRIÈRE, CAPT.; PINE, J.; RUSSELL, P.

Rocque, Mary Ann 18th cent. Pub. Brit. M
Wife of John R., she carried on his business after his death.

Rodrigues, Francis fl. 1524–30 Cart. Span.? M

Rodwell, M. M. fl. 1834 Cart. Brit. M

Roesch *Same as* VON ROESCH (q.v.)

Roger, Pierre fl. 1579 Cart. Fr. M
Worked on the *Theatrum* of Ortelius.

Rogg, Gottfrid *See under* LOTTER, T. C.; SEUTTER, G. M.

Roggeveen, Arent d. 1679 Cart. Dut. M
Het eerste deel van het brandende veen verlichtende alle de vaste kusten ende eylanden van geheel West-Indien ofte rio Amasones Amster., P. Goos, 1675.
La primera parte del monte ... Amster., P. Goos, 1680.
See also under JACOBSZ, T.

Rogier, Pierre 17th cent. Cart. Fr. M

Rogiers, Solomon *See under* BERTIUS, P.; BLAEU, G.; HONDIUS, H.; HORN, G.

Roijan *Same as* VAN ROIJAN (q.v.)

Roll, George (d. 1592) and⎫
   Reinhold, Johannes ⎬ Astron. Ger.
      (fl. 1586-9)⎭ G (C and T)
Rott }

Rollinson *See under* MALHAM, J.; PAYNE, J.

Rollos, G. fl. 1559–79 Eng., Geog., Mapseller Brit. M
See also under RUSSELL, P.

Roman, J. *See under* SCHENK, P.

Romano, Giulio *Same as* DE' GIANUZZI, G. DI P.

Romans, B. *See under* SAYER, R.

Romanus, Adrianus 1561-1615 Cart.
Ger. M
Parvum theatrum urbium . . . Frankfurt,
N. Bassaei, 1595, 1608.

Romstet, Chr. fl. 1692 Globe-maker
Ger.? G (T)

Ronsard *See under* DE FREYCINET, L. C. D.

Roper, John fl. 1801-10 Eng. Brit. M
See also under COLE, G.; WALLIS, JAMES;
WILKINSON, R.

Rosa, Vincenzo fl. 1790-93 Cart. Ital.
G (T)

Rosaccio, Giuseppe *c.* 1530-1620 Cart.
Ital. M X
Il mondo . . . (with 17 maps) Florence,
F. Tosi, 1595.
Mondo elementaire et celeste . . . (with 13
maps) Trevegi, E. Deuchino, 1604.
Teatro del cielo . . . (with 4 maps) Viterbo,
Discepoli, 1615. Trivegi, G. Righettini,
1642 (with 11 maps).
Teatro del mondo . . . (with 11 maps)
Bologna, A. Pisarri, 1688, 1724.

Rose *Same as* ROTZ (q.v.)

Roselli, Francesco *Same as* ROSSELLI, F. DI
L. (q.v.)

Rosiers *Same as* DES ROSIERS (q.v.)

Rosini, Pietro fl. 1760 Cart. Ital. G
(T)

Ross, Lieut. Charles fl. 1764-80 Surv.,
Cart. Brit. M
Maps of Dumbartonshire (1777), Lanarkshire
(1773; eng. by G. Cameron. 1775; eng.
by A. Baillie), Renfrewshire (1754),
Stirlingshire (1780).
Course of the river Mississippi Lond., Sayer,
1772, 1775.
See also under FADEN, W.; JEFFERYS, T.; LE
ROUGE.

Ross, D. *See under* JEFFERYS, T.

Ross, J. *See under* STEPHENSON, J.

Rosselli, Francesco di Lorenzo 1445-1520
Cart. Ital. M G (T)
See also under CONTARINI, M. G.

Rosser, W. H. fl. 1867 Astron. Brit.
X

Rossi, D., G. *Same as* DE ROSSI (q.v.)

Rossi, Luigi} 1764-1824 Cart. Ital. M
de Rubeis }
Nuovo atlante di geografiia universale . . .
Milan, Batelli and Fanfani, 1820-1.

Rossi, Tomasso} fl. 1539 Bookseller,
de Rubeis } Printer Ital. M

Rost, J. L. fl. 1726-43 Astron. Ger. X

Rostovcev, Alexij fl. 1727-34 Eng.
Russ. M

Rotenham, Sebastian 1474-1534 Humanist, Cart. Ger. M

Roth, Matheus fl. 1730 Pub. Dut. M

Rothenberg *Same as* VON ROTHENBERG (q.v.)

Rothmann, Christoph d. *c.* 1599 Mathematician, Astron. Ger. G

Rott *Same as* ROLL (q.v.)

Rotz, John} fl. 1542 Cart. Brit. M
Rose }
Geographer to Henry VIII

Rousseau, L. B. T. *See under* NOLIN, J. B.

Rousseau, T. *See under* DESNOS, L. C.

Roussel *See under* PLACIDE DE SAINT HÉLÈNE,
PÈRE

Roussell et Blottiere 18th/19th cent.
Carts. Fr. M
A map of the Pyrenees . . . Lond., A.
Arrowsmith, 1809.

Rousset, C. *See under* PLACIDE DE SAINTE
HÉLÈNE, PÈRE

Rousset, P. 19th cent. Cart. Fr. M
Atlas général . . . Paris, Locard et Davi,
1835.
See also under ANDRIVEAU-GOUJON.

Rousset de Missy, Jean 1686-1762
Astron., Cart. Fr. M X
Nieuwe astronomische geographische . . . *atlas*
Amster., H. de Leth, 1742?
Nouvel atlas géographique et historique . . .
Amster., H. de Leth, 1742?; H. de Leth
and S. J. Baalde, 1742?

Roussin, J. F. fl. 1663-9 Cart. Fr. M

Rouvier *See under* PALAIRET, J.

Rouville, Guillaume 19th Cent. Printer
Fr. M

Roux, Joseph 18th cent. Pub., Cart. Fr.
M
Nouveau recueil des plans, des ports . . . *de la
mer Mediterranée* . . . (with others).

Gênes, Y. Gravier, 1779, 1838; Marseilles, 1764.

Carte de la mer Mediterranée Marseilles, 1764.

Rover, Paolo fl. 1591 Cart. ? M

Rovere, Giuliano Della 1441–1513 Cart. Ital. G

Rovere, Giulio Feltrio Dalla fl. 1575 Cart. Ital. G (T)

Rowe, Robert *c.* 1775–1843 Eng., Geog., Pub. Brit. M
The English atlas 1816.
See also under TEESDALE, H.

Rowley, John d. 1728 Instrument-maker Brit. G

Roycroft, Thomas *See under* BLOME, R.

Royer, Augustine fl. 1679–1700 Astron. Fr. X

Rubeis Mediolanensis *Same as* DI ROSSI, G. (q.v.)

Rubertis, Luc'Antonio fl. 1525 Cart. Ital. M

Rubie, G. fl. 1830 Astron. Brit. X

Rüdinger, J. A. *Same as* RIEDINGER, J. A. (q.v.)

Rueda *Same as* DE RUEDA (q.v.)

Rueshammer, G.} fl. 1568–79 Scholar
Rueshaimer   } Ger. G

Ruff, E.   }
Ruff, E., and Co.} fl. 1837 Pub. Brit. M

Rughesi, Fausti fl. 1597 Printer Ital. M

Rumsey, E. *See under* GILL, V.

Ruphon fl. 1509 Cart. Ital.? M

Ruscelli 16th cent. Cart. Ital. M

Russell, John and J. C. 18th cent. Cart. Brit. M G
An American atlas . . . Lond., H. D. Symonds and J. Ridgway, 1795.
Russell's general atlas . . . Lond., Baldwin and Cradock.
Maps drawn and eng. by Russell and Sons.
See also under GUTHRIE, W.; REES, A.

Russell, P. fl. 1769 Topg. Brit. M
England displayed . . . (with Owen Price) Lond., 1769. Contains 23 maps by Rocque (1753); 10 by Kitchin (1747–60); 4 by G. Rollos; 1 by Thomas Bowen.
See also under ROCQUE, JOHN.

Russell and Sons *See* RUSSELL, J. AND J. C.

Russinus, Augustinus fl. 1590 Cart. ? M

Russus, Jacobus fl. 1520–88 Cart. ? M

Rüst, Hanns}
Rist   } 15th cent. Cart. Ger. M
Riss   }
Das ist die mapa mūdi . . . Augsburg, late 15th cent. (woodcut).

Rutlinger, Johannes or Jan d. 1609 Eng. Dut. M
See also under WAGHENAER, L. J.

Ruuolo, Franc 17th cent. Globe-maker ? G

Ruysch, Johannes fl. 1506 Cart. Dut. M

Ruyter, B. fl. 1790 Eng., Cart. Dut. M
Imperium Japonium Utrecht, G. Broedelet, 1715.
See also under OTTENS, R.

Ruyter, G *Same as* DE RUYTER, G. (q.v.)

Ryall, J. fl. 1762 Cart. Brit. M
Executed maps in Bowen and Kitchin's *Royal English atlas.*

Ryland, John 18th cent. Eng. Brit. M
Eng. for J. AND W. KIRBY (q.v.)

Ryther, Augustine fl. 1576–95 Eng. Brit. M X
Expeditionis Hispanorum in Anglia . . . 1588. (Drawn by R. Adams)

Ryûsên *Same as* ISHIKAWA TOSHIYUKI (q.v.)

Sabbadino, Christoforo fl. 1552 Pub. Ital. M

Sabucus, J. 16th cent. Cart. Dut. M
Made map of Transylvania for the Ortelius atlas (1566, 1592).

Saenredam, Jan 1565–1607 Eng. Dut. M

Saevry, S. *See under* HORN, G.; SPEED, J.

Sailmaker, I. *See under* MORTIER, D.

Salamanca, Antonio 1500–62 Eng., Pub. Ital. M
Sometimes signed with the initials 'A. S.'

Salamanca, Franciscus fl. 1550 Cart. Ital. M

Salingen *Same as* VAN SALINGEN (q.v.)

Salt fl. 1814 Cart. Brit.? M

Saltonstall, W. *See under* CARTWRIGHT, S.

Sambucus, Johannes} 1531–84 Physician,
Zsambocky, Janos } Historian Hungarian
} M

Sampier *See under* DE GOUVON SAINT-CYR, L.

Sampson, G. V. 19th cent. Surv. Irish
M
Map of Londonderry, 1813.

Samson *Same as* SANSON (q.v.)

Sances, Antonio and Ciprian 16th/17th
cent. Carts. Port. M

Sánchez, Alexandro 19th cent. Cart.
Span. M
Mapa general de las almas . . . Manila, M.
Sánchez, 1845.

Sánchez, M. *See under* BLANCO, M.;
SÁNCHEZ, A.

Sanctis *Same as* DE SANCTIS (q.v.)

Sanderson, G. fl. 1836 Cart. Brit. M
Maps of Derbyshire, 1836 (eng. J. and C.
Walker); 20 miles round Mansfield,
1835; Nottinghamshire, 1836.

Sanderson, William *c.* 1541–1631 Patron
of Geogs. Brit. G

Sanderus, Antonius 1586–1654 Historian,
Geog. M

Sandford and Maddocks *See under* Morris,
L.

Sandrart, Jacob } 1630–1708 Eng.
Sandrart, Jacobum de} Ger. M
Accuratissima totius africae tabula . . .
Nuremberg, Homann, *c.* 1740.

Sanson, Adrien d. 1708 Cart. Fr. M
Son of Nicolas, Senr.

Sanson, Guillaume d. 1703 Cart. Fr.
M G
Son of Nicolas, Senr.
See also under DE ROSSI, G. G.; HUSSON, P.;
JAILLOT, C. H. A.; JULIEN, R. J.; LE
CLERC, J.; SANSON, N., PÈRE.

Sanson, Nicolas, père 1600–67 Cart.,
Geog. Fr. M
L'Afrique en plusieurs cartes nouvelles . . . Paris,
1656.
Judaea seu terra sancta . . . Amster., Covens
and Mortier, *c.* 1744.

L'Amérique en plusieurs cartes . . . Paris, 1657.
Maps 1 and 8 signed by A. Peyrouin.
L'Amérique en plusieurs cartes nouvelles . . .
Paris, 1662, 1667? Maps 1 and 8 in the
latter signed by A. Peyrouin.
L'Asie en plusieurs cartes nouvelles . . . Paris,
1652–3.
Atlas nouveau . . . Paris, Hubert Jaillot, 1689–
90, 1692–6, 1696. Carts. incl.: N. de Fer,
H. Jaillot, le sieur Vaultier, E. Michalet,
although most are by Sanson himself.
Cartes générales de toutes les parties du monde . . .
Paris, Pierre Mariette, 1658. Publishers'
imprints, in addition to those of Sanson
and Mariette: Melchoir Tavernier. Engs.:
J. Somers, A. Peyrounin, Rivière, R.
Cordier, A. de Plaes or Plaets. Another
edn. by Sanson *père* and *fils*, and Guillaume
Sanson (d. 1703) was pub. by Maliette
in 1670.
Cartes particulières de la France . . . Paris,
Mariette, 1676. Engs. incl.: Johannes
Somer Pruthenus, L. Cordier, R. Cordier.
*Cartes générale de la géographie ancienne et
nouvelle* . . . (with N. Sanson *fils* and
Guillaume Sanson) Paris, Mariette, 1675.
Géographie universelle . . . (with N. Sanson
fils and Guillaume Sanson) Paris, 1675?
Engs. incl.: Jean Somer, R. Cordier,
A. de la Plaetz, A. Peyrounin, Lud.
Cordier, R. Michault.
Description de tout l'univers . . . (with N.
Sanson *fils*) Amster., F. Halma, 1700.
See also under BLAEU, G. DE FER, N.; DE
LA FEUILLE, D.; DE LA FEUILLE, J.; DELISLE,
G.; DESNOS, L. C.; DE WIT, F.; DU
SAUZET, H.; DU VAL D'ABBÉVILLE, P.;
JAILLOT, C. H. A.; LE CLERC, J.; LUYTS, J.;
OTTENS, R.; SCHENK, P.; TASSIN, N.;
TAVERNIER, M.; VISSCHER, N., III.

Sanson, Nicolas *fils* 1626–48 Cart. Fr.
M

Sanson, Pierre Moulard 18th cent. Cart.
Fr.
Grandson of Nicolas Senr.
See also under DU SAUZET, H.

Sanson and Co. *See under* BARBIÉ DU
BOCAGE, J. D.

Santa Cruz, Alonso de 1500–72 Cart.
Span. G (T)

Santarem   *Same as* DE SANTAREM (q.v.)

Santini, P.   fl. 1776   Pub., Cart. Ital.   M
Atlas portatif d'Italie . . .   Venice, 1783.
Atlas universel . . .   Venice, Remondini, 1776–84. Incl. maps by d'Anville.
See also under SCHRAEMBL, F. A.; VON REILLY, F. J. J.

Santucci, Antonio   fl. 1590   Instrument-maker Ital.   O

Sanuto, Giulio   1540–80   Cart. Ital.   G (T)

Sanuto, Livio   *c*. 1530–*c*. 1586   Cart. Ital. M

Sanza   *See under* LÓPEZ, T.

Saunders, Trelawney William   fl. 1847 Pub. Brit.   M

Sauracher, Adalbert   fl. 1584   Cart. Brit.   M

Sauthier, Claude Joseph   18th cent.   Cart. Brit.   M
A map of the inhabited part of Canada . . . Lond., Faden, 1777.
Mappa geographica provinciae Novae Eboraci . . . Nuremberg, Homann's Heirs, 1778.
A topographical map of the northern part of New York island . . . Lond., Faden, 1777.
Sometimes worked with Ratzer, and signed thus: 'Sauthier & Ratzer.'
See also under FADEN, W., SAYER, R.

Savanarola, Raffaelo   18th cent.   Cart. Ital.   M
Pseudonym of Lasor à Varea.
Universus terrarum orbis . . .   2 vols., Patavi, B. Conzatti, 1713.

Saxton, Christopher   *c.* 1542–*c.* 1610   Surv., Cart. Brit.   M
Separate county maps of England and Wales, 1574–9
An atlas of the counties of England and Wales . . . Lond., 1579.
The shires of England and Wales described . . . Lond. 1693, 'Sold by Philip Lea'. County maps indicate roads in accordance with Ogilby. Additions to the atlas incl. 4 maps by Lea and 5 by Seller. 1720, 'Sold by Geo. Willdey'. Willdey's imprint added to the maps. 1749, pub. by Jefferys. *c.* 1770, pub. by Dicey and Co.

An edn. of reproductions was issued about 1883.
The traveller's guide . . .   1687.   A reprint by Philip Lea of Saxton's 20-sheet map of England and Wales, first issued in 1583.
See also under BOWLES, J.; LEA, P.; SCATTER, F., SPEED, J.; TERWOORT, L.; VAN DEN KEERE, P.; WEB, W.; WILLDEY, G.

Sayer, Robert   fl. 1780–1810   Pub., Mapseller Brit.   M
Also traded in partnership with John Bennett as Sayer and Bennett.
Fuller's *Plan of Amelia island* . . .   Lond. 1770.
North America and the West Indies . . .   Lond., 1783.
d'Anville and Robert's *New map of North America* . . .   Lond., 1763.
Chart of Greenland   1783.
The American military pocket atlas   (6 maps) Lond., 1776. Carts.: Samuel Dunn, William Brassier, B. Romans, Capt. Jackson, de Brahm.
The seat of war in New England . . .   Lond., 1775.
An English atlas . . .   Lond., 1787.
Atlas britannique . . .   Lond., 1766.
General atlas . . .   Lond., 1757–94.   Carts. incl.: Thomas Kitchin, Samuel Dunn, Lieut. Campbell, Capt. Clement Lemprière, Robert Mylne, Thomas Jefferys, d'Anville, John Roberts, J. Rennell, Jno. Cary, Claude Joseph Sauthier. Engs. incl.: Wigzeel, W. Palmer, F. Vivares, J. Phillips and W. Harrison, M. Borven, Wm. Faden.
See also under ARMSTRONG, A.; BOWEN, E.

Scalé, Bernard   fl. 1760–87   Surv., Topg. Irish   M
An hiberian atlas   Laurie and Whittle, 1798.

Scaltaglia, Pietro   fl. 1780–84   Cart., Eng. Ital.   G (C and T)

Scapsius, Thomas}   1520–77   Physician,
Schöpf, Th.   }   Cart. Ger.?   M

Scarabelli, Giuseppe   fl. *c.* 1690   Cart. Ital.   G (C and T)

Scatter, Francis   fl. 1579   Eng. Brit.   M
Engraved 'Chestershire' and Staffordshire maps in Saxton's atlas.

Schalbacher, P. J. *See under* SCHRAEMBL, F. A.

Schalekamp, M. *See under* BACHIENE, W. A.

Schalk, Emanuel 18th cent. Eng. Dut. M

Schall von Bell, Johann Adam 1591-1666 Jesuit, Cart. Ger. M

Schanternell, Christoph fl. 1702 Globe-maker Ger. G (T)

Scharrer, Wilhelm fl. 1825-33 Cart., Lithographer Ger. G

Schaudt, Phil. Gottfried fl. 1774 Globe-maker Ger.? G (C)

Schedel, Hartmann 1440-1514 Historian, Geog. Ger. M

Liber cronicarum . . . (with 2 maps) Nuremberg, 1493. The famous 'Nuremberg Chronicle'. *See also under* MÜNZER, H.

Scheibel, Johann Ephraim 1736-1809 Mathematician Ger. G

Scheiner, Christopher b. 1575 Astron. Ger. X

Scheirer, D. *Same as* SCHEYRRER, D. (q.v.)

Schenk, Peter} 1660?-1718/9 Pub., Eng.
Schenck } Dut. M

Americae by A. F. Zürner, Amster., 1709.
L'Amerique Septentrionale, by G. Delisle, Amster., 1708.
Petri Schenkii Hecatompolis . . . Amster., 1702. Reissued at Amster. in 1752 by J. Roman under the title of *Afbeeldinge van eenhondert der voornaamste en sterkte steeden in Europa.*
Atlas contractus . . . Amster., 1705?, 1709? Carts. incl.: Valck, Schenk, Zürner, Ottens, Jaillot, Jean de Lat, Sanson, Sprecht, N. J. Visscher.
Atlas saxonicus novus . . . Amster. and Leipzig, 1752-9. German edn. *Neuer Sachsischer atlas* . . . issued in 1752-8, and incl. maps by Schuchart and Dengelsted.
Le flambeau de la guerre allumée en Rhin . . . Amster., 1735.
Schouwburg van den oorlog . . . Amster., Covens and Mortier, 1730.

Le théâtre de Mars . . . Amster., P. Schenk and A. Braakman, 1706.
Atlantis sylloge compendiosa . . . (with G. Valk) Amster., 1709., Carts. incl.: G. and L. Valk, A. F. Zürner, Sanson and Visscher.
Atlas anglois . . . Lond., D. Mortier, 1715.
See also under BRAAKMAN, A.; DE WIT, F.; HOMANN'S HEIRS; HUSSON, P.; OTTENS, R.; VALCK, G.

Scherer, Heinrich 1628-1704 Jesuit, Mathematician, Cart. Ger. M G

Atlas marinus . . . Monachii, M. M. Rauchin, 1702.
Critica quadripatria . . . Monachii, Mathias Riedl, 1710.
Geographia artificialis . . . Monachii, Mariae Magdalenae Rauchin, 1703.
Geographia hierarchia . . . Monachii, M. M. Rauchin, 1703 (with 21 maps).
Geographia naturalis . . . (with 11 maps) Monachii, M. M. Rauchin, 1703.
Geographia politica . . . (with 25 maps) Monachii, M. M. Rauchin, 1703.
Tabellae geographicae . . . Monachii, M. M. Rauchin, 1703 (with 15 maps).

Scheuchzer, Johann Jakob 1672-1733 Mathematician, Cart. Swiss M
See also under JULIEN R. J.; OTTENS, R.

Scheurleer, H. *See under* D'ANVILLE, J. B.

Scheyrrer, Daniel}
Schyrr } fl. 1624 Cart., Astron.
Scheirer } Ger.? G (C)

Schickhart, Wilhelm 1592-1635 Architect, Cart. Ger. M

Schieble, Erhard *See under* ANDRIVEAU-GOUJON

Schiepp, Christoff fl. 1525 Cart. Ger. G (T)

Schiller, Judas }
Schillerius, Julius} d. 1627 Pub. Ger. X
Ichillerius }

Schimek *See under* SCHRAEMBL, F. A.

Schissler, Christoph} c. 1531-1608 Instru-
Schussler } ment-maker Ger. G

Schissler, Hans Christoph} 1561-1625
Schüssler } Watchmaker Ger. G

Schleich, C. S. *See under* VON SCHLIEBEN, W. E. A.

Schleuen, I. W. 19th cent. Eng. Ger. G (T)

Schlieben *Same as* VON SCHLIEBEN (q.v.)

Schmettau *Same as* VON SCHMETTAU (q.v.)

Schmid, Sebastian fl. 1566 Cart. Swiss M

Schmidburg *Same as* VON SCHMIDBURG (q.v.)

Schmidt, Jodok A. fl. 1828 Cart. Ger. G (T)

Schmidt, O. F. 18th/19th cent. Cart. Ger. M
Post-reise katte durch Deutschland . . . Berlin, Simon Schropp, 1831.
See also under SCHRAEMBL, F. A.; VON REILLY, F. J. J.

Schmidt, R. *See under* WEILAND, C. F.

Schmollinger, W. fl. 1836 Cart. Brit. M
Eng. maps for Moule's *English Counties*.

Schmück, Michael fl. 1593 Cart. Ger. M

Schneewis, Urban 1536-1600 Goldsmith Ger. G

Schneider, Adam Gottlieb and Weigel, J. A. G. *See under* WEIGEL UND SCHNEIDERSCHEN HANDLUNG.

Schneip, Ulrich}
Schneiper } d. 1587 Astron.,
Schneipus } Watchmaker Ger. G

Schniep Christoph fl. 1530 Globe-maker Ger. G

Schoell, F. *See under* VON HUMBOLDT, F. W. H. A.

Schöner, Johann 1477-1547 Geog., Astron. Ger. G (C and T)

Schönfeld, E. fl. 1855 Astron. Ger. X

Schöninger, E. L. fl. 1850 Astron. Ger. G (C and T)

Schoonebeck, Hadrianus (d. 1705), **and Damianus, Peter** Engs. Dut. M

Schöpf, T. *Same as* SCAPSIUS, T. (q.v.)

Schorman, P. fl. 1665 Eng. Dut. M

Schotanus à Sterringa, Bernardus *See under* ALTING, M.; DE WIT, F.; VISSCHER, N., III

Schott, Ernst 19th cent. Globe-maker Ger. G (T)

Schott, Johannes fl. 1520 Printer Ger. M
See also under PTOLEMAEUS, C.

Schouten, Willem Cornelisz 1567-1625 Navigator Dut. M
Iournal ofte beschryvinghe van de wonderlicke reyse ghedaen door Willem Cornelisz Schouten van Hoorn . . . Amster., 1618 (with 3 maps). English edn. (with 2 maps, one woodcut), Lond., 1619.

Schouw, Joakim Frederick 1789-1852 Cart. Ger. M
Pfanzengeographischer atlas . . . Berlin, G. Reimer, 1823.

Schraembl, Franz Anton 1751-1803 Cart. Austrian M
Allgemeiner grosser atlas . . . Vienna, P. J. Schalbacher, 1786-1800. Carts. incl.: d'Anville, Blanco, Bonne, Bowles, Cassini, Crome, Dorret, Djurberg, Endersch, Faden, Kitchin, López, Niebuhr, Pownall, Rizzi-Zannoni, Robert de Vaugondy, Roberts, Santini, Schmid, Schimek, A. von Wenzely, Wussin.

Schreiber, Johann Georg 1676-1745 Cart. Ger. M
Atlas selectus . . . Leipzig, 1749? and other edns.
See also under HOMANN, J. B.

Schrester, Josef fl. 1823 Globe-maker Ger.? G (T)

Schropp, Simon 19th cent. Pub. Ger. M
See also under SCHMIDT, O. F.

Schrotenus, Christian *Same as* SGROOTEN, C. (q.v.)

Schröter, Johann Hieronymus 1745-1816 Astron. Ger. X G

Schuchart *See under* SCHENK, P.

Schultz, B. and J. *Same as* SCULTETO, B. AND J. (q.v.)

Schultz, J. F. 18th/19th cent. Pub. Dan. M
See under DE LOVEHORN, P.

Schüssler *Same as* SCHISSLER (q.v.)

Schut, A. *See under* DANCKERTZ, J.; OTTENS, R.

Schwartzer, Anton Ferdinand 18th cent. Globe-maker Ger. G

Schwinck, G. fl. 1843 Astron. Ger. X

Schwitzky, H. fl. 1810 Cart. Ger. G

Schwytzer, Christopher} fl. 1595 Eng.
Shwytzer } Ger.? M
Eng. a rare map of Sussex for Norden.

Scillius, Joannes } 1538-86 Cart. Dut.
van Schilde, Jan } M
van Schiller, Hans }

Scolari, Stefano 18th cent. Pub. Ital. M

Scoles *See under* PAYNE, J.; WINTERBOTHAM, W.

Scott, B. *See under* BELL, JAS.

Scott, Joseph 18th cent. Cart. Amer. M
An atlas of the United States . . . Philadelphia, F. and R. Bailley, B. Davis, and H. and B. Price, 1796. The earliest atlas of the United States.

Scott, Joseph T. *See under* CAREY, M.; GUTHRIE, W.; PONCE, N.

Scott, R. *See under* MOORE, SIR J.

Scott, Robert b. 1777-1841 Eng. Brit. M

Scott, Thomas 19th cent. Cart. Brit. M
Chart of Van Diemen's Land Edinburgh and Lond., 1824.
Eng. by Charles Thomson of Edinburgh.

Scott, W. *See under* BURNETT, G.

Scottus, Jacobus fl. 1578-92 Cart. ? M
Portolans.

Scull, N., and Heap, G. 18th cent. Carts. Amer. M
A plan . . . of Philadelphia Lond., W. Faden, 1777. There were two distinct issues; the second gave soundings in the Delaware river, etc.
See also under FADEN, W.; JEFFERYS, T.

Scull, William 18th cent. Cart. Brit. M
A map of Pennsylvania . . . Lond., Sayer and Bennett, 1775.

See also under FADEN, W.; JEFFERYS, T.; LE ROUGE, G. L.

Scultetus, Bartolomeus}
(1540-1614) **and Jona** } Mathematicians
Schultz } Ger. M
Sprotta, Jonas Sculte- }
tus }
See also under BLAEU, G; HONDIUS, H.; JANSSON, J.

Seale, Richard William fl. 1732-75 Cart. Eng. Brit. M
See also under FADEN, W.; KELLER, C.; TINDAL, N.

Seaman, James V. 19th cent. Pub. Amer. M
A new general atlas . . . New York, 1820. Map no. 3 eng. by G. B. King. All maps bear imprint of Seaman.

Sebastianus à Regibus Clodiensis *Same as* DI RE (q.v.)

Secco, Fernando Alvero}
Seco } 16th cent. Cart.
Zeccus } Port. M
Portugallia et Algarbia . . . Amster., Blaeu, c. 1660, 1665.
See also under HONDIUS, H.; JANSSON, J.; MERCATOR, G.

Secsnagel, Marcus} fl. 1551 Cart.
Zecsnagel } Austrian M

Sedillau fl. 1682 Cart. Fr. M

Seeley, R. B., and Burnside, W.} *See under*
Seeley, Burnside and Seeley } QUIN, E.

Ségur *Same as* DE SÉGUR (q.v.)

Seibel, J. *See under* THERBU, L.

Seile, Anna fl. 1660-5 Pub. Fr. M

Seile-Trevethen fl. 1652 Cart. ? M

Sekisui *Same as* NAGAKUBO GENSHU (q.v.)

Seligmann, J. M. *See under* LOTTER, T. C.

Seller, Jeremiah fl. 1698-1707? Instrument-maker Brit. M G
Son of JOHN S., I (q.v.)

Seller, John, I d. 1697 Cart., Pub. Brit. M X G
English pilot . . . Lond., J. Darby, 1671; J. Larkin, 1692. Eng. incl.: James Clark.
Atlas maritimus . . . Lond., James Atkinson 1670?, John Darby, 1675 and other edns.

Atlas terrestris . . . Lond., *c.* 1685 and other edns. Most of the maps were by F. de Wit; others were by Joannes de Ram, Hugo Allardt, Philip Lea, and Seller himself.

Atlas contracta . . . 1695.

Maps by J. Seller are reprinted in *The Antiquities of England and Wales* by Francis Grose (Lond., 1777–87).

Atlas minimus . . . Lond., 1679.

The coasting pilot . . . Lond., J. Seller, W. Fisher and J. Wingfield, 1672.

Hydrographia universalis . . . Lond., 1690?

A new systeme: of geography . . . Lond., 1685, 1690.

A mapp of New Jersey in America . . . (with William Fisher) Lond., 1677.

See also under PETTY, SIR, W.; SELLER, JERE-MIAH; SPEED, J.

Seller, John, II d. 1698 Cart. Pub. Brit. M

Selma, F. *See under* DE ANTILLON Y MARZO, I.

Selss, Eduard fl. 1848 Cart. Ger.? G (T)

Seltzlin, David fl. 1575/6 Cart. Ger. M

Selves, Henri 19th cent. Pub. Fr. M

Atlas géographique . . . Paris, 1822–9. All maps bear imprint of Selves.

Semen *See under* ARROWSMITH, A.; DELAMARCHE, F.

Senex, John d. 1740 Cart., Eng., Pub. Brit. M G (C and T) X

Noted, among other things, for road maps. Eng. for R. Budgen (q.v.)

Map of Africa . . . by C. Price, 1711.

Africa 1725.

Ireland Lond., 1712.

A new map of Great Britain Lond., 1714.

Poland . . . Lond., 1725.

A new general atlas . . . Lond., 1721.

Sweden and Norway . . . Lond., *c.* 1725.

Turky in Europe (with J. Maxwell) Lond., 1710.

The VII United provinces (with J. Maxwell) Lond., 1709.

An actual survey of all the principal roads of England and Wales . . . *first perform'd and published by John Ogilby Esq.*, 1719, 1742 and other edns.

Itineraire de toutes les routes d'Angleterre . . . Paris, Desnos, 1766.

Modern geography Lond., T. Bowles and Son, 1708–25.

A map of Virginia Lond., 1785.

See also under NORDEN, J.; WILLIAMS, WILLIAM.

Senior, William 16th cent. Cart. Brit. M

Estate maps.

Sepp, C. and Jan Christian 1739–1811 Pubs., Carts. Dut. M

C. Sepp was the father of J. C. Sepp. Some of their maps are signed: 'C. et I. C. Sepp, pater et filius.'

Nieuwe geographische Nederlandsche reise-en zak-atlas Amster., 1773.

See also under DE LETH, H.

Septala *Same as* SETTALA (q.v.)

Serres, J. T. fl. 1801 Cart. Brit. M

Worked on Bougard's *Little sea torch.*

Servetus, Michael} 16th cent. Cart. Fr.
Villanovanus } M

Woodcut map of parts of N. and S. America, etc., in the Trechsel (Lyons) 1535 edn. of Ptolemy.

Settala, Johann Georg} fl. 1560 Cart. Ital.
Septala } M

Settala, Manfredo} 1600–80 Instrument-
Septala } maker Ital. O

Seutter, Albrecht Carl 18th cent. Cart. Fr. M

Partie orientale de la Nouvelle France, ou du Canada Augsburg, T. Conrad Lotter, *c.* 1740.

See also under SEUTTER, G. M.

Seutter, Georg Mattheus the Older (1678–1757) **and Georg Mattheus the Younger** (1729–60) Cart., Geog., Pub. Ger. M G (C and T)

Africa juxta navigationes et observationes recentissimae Augsburg, *c.* 1740.

Recens elaborata mappa geographica regni Brasiliae Augsburg, *c.* 1720.

Nova designatio insulae Jamaicae Augsburg, *c.* 1740.

Plan von Neu Ebenezer Augsburg, 1740.

Recens edita totius Novi Belgii . . . Augsburg, *c.* 1735.

Atlas minor praecipua orbis terrarum imperia Augsburg, 1744?

Many maps drawn by Albrecht Carl Seutter and eng. by Tobias Conrad Lotter and Andreas Silberstein.

Grosser atlas Augsburg, 1734?

Atlas geographicus . . . Augsburg, 1725. Carts. incl.: Gottfrid Rogg.

Atlas novus indicibus instructus . . . Vienna, Johann Peter van Ghelen, 1730. Carts. incl. G. Rogg. Engs. incl. Melch. Rein.

Atlas novus sive tabulae geographicae totius orbis . . . Augsburg, 1741?; 2 vols. 1745? Carts. incl. G. Rogg. Abr. Drentwett. Engs. incl.; E. Bäck, T. C. Lotter.

See also under HOMANN, J. B.; HOMANN'S HEIRS; LOTTER, T. C.; VALCK, G.

Seutter, M. *same as* GEORG MATTHEUS SEUTTER (q.v.)

Severi Claudius, Johannes} fl. 1514 Pub.
Severinus, Jan } Dut. M
Zepherinus, Jo. }

Seymour, J. H. *See under* CAREY, M.; EDWARD, B.; MARSHALL, J.

Sgrooten, Christopher} c. 1532-1608 Cart.
Schrotenus, Christian } Dut. M

Shallus, Francis *See under* JONES, T. W.; MARSHALL, J.

Sharman, J. fl. 1800 Cart. Brit.? M

Sharp, John fl. 1793 Pub. Brit. M
See also under FOWLER, W.

Sharpe, John 1777-1860 Cart. Brit. M
Sharpe's corresponding atlas . . . Lond., Chapman and Hall, 1849. Eng. by Joseph Wilson Lowry.
Sharpe's students' atlas . . . Lond., Chapman and Hall, 1850. Eng. by Joseph Wilson Lowry.

Sheffield *See under* PALAIRET, J.

Sherman, C. *See under* WILKES, C.

Sherman and Smith *See under* GOODRICH, S. G.

Sherriff, James 19th cent. Cart. Brit. M
Maps of Liverpool district, 1823 (eng. J. H. Franks), Warwickshire, 1798 (eng. B. Baker)

Shiba Kôkan 1747-1818 Cart. Japanese M

Shirley and Hyde *See under* GREENLEAF, M.

Shwytzer *Same as* SCHWYTZER (q.v.)

Sibbald, Sir Robert fl. 1683 Geog. Brit. M

Siborne, William 1797-1849 Historian Brit. M
History of the war in France and Belgium in 1815 . . . Atlas Lond., T. and W. Boone, 1848.

Sibrianus, Leonis fl. 1545 Cart. Dut. M

Sickelmore, Richard, Senr. fl. 1815 Cart. Brit.? M

Sideri, Giorgio *Same as* CALAPODA (q.v.)

Siebert, A. *See under* VON SCHLIEBEN, W. E. A.

Siedentopf, F. R. fl. c. 1825. Globe-maker Ger.? G (T)

Sikkena *See under* VOOGT, C. J.

Silberstein, Andreas *See under* SEUTTER, G. M.

Silva *Same as* SYLVANUS (q.v.)

Sim, W. *See under* D'ANVILLE, J. B. B.

Simmoneau, Carolus 18th cent. Eng. Fr. M G
See also under DELISLE, G.

Simon, Charles *See under* ANDRIVEAU-GOUJON; DUFOUR, A. H.

Simons, Matthew d. 1654 Printer, Pub., Topg. Brit. M
A direction for the English traviller . . . Lond, 1636. Eng. by Jacob van Langeren. Contains triangular distance tables, with a county sketch map eng. with each. A revised edn. was pub. by Thomas Jenner in 1643. In 1668 it was republished by Thomas Jenner, with additions, under the title *A Book of the names of all parishes, market towns, hamlets . . . in England and Wales.*
See also under GARRETT, J.

Simpkin, Marshall and Co. *See under* REID, A.

Simpkins *See under* DE RAPIN-THOYRAS, P.

Simpson, James *See under* CRASKELL, T.

Simpson, Samuel fl. 1746 Geog. Brit. M
The agreeable historian . . . (with maps after
H. Moll) Lond, 1746.

Sims, Valentine *See under* WRIGHT, E.

Sinck *See under* MERCATOR, G.

Singer, Joseph 19th cent. Cart. Brit. M
Map of Cardiganshire, 1803, 2 edns. (surv.
T. Yates).

Singrenius, Joannes *See under* APIANUS,
PETRUS

Sizlin, Johann fl. 1559 Cart. Ger. M
Thought to be author of a map of Württem-
burg, pub. at Tübingen in 1559, and
signed 'J.S.'

Skanke *See under* WESSEL.

Skinner, Andrew fl. 1775-7 Surv. Brit.
M
See also under TAYLOR, G.

Skynner *See under* BEEK, A.

Slater, Isaac fl. 1857 Pub. Brit. M
See also under PIGOT, J., AND CO.

Sloane, O. 18th cent. Cart. Irish M
Map of Queen's County, 1789.

Small, A. *See under* WILKINSON, J.

Smiley, Thomas T. 19th cent. Cart.
Amer. M
An improved atlas . . . Philadelphia, 1824.
A new atlas . . . Philadelphia, 1830, 1832,
1834; Hartford, Belknap and Hamersly,
1838 and many other edns.

Smith, Anthony 18th cent. Pilot, Cart.
Brit. M
*A new and accurate chart of the bay of Chesa-
peake* . . . Lond., Robert Sayer and John
Bennett, 1776. Re-eng. by Le Rouge as *Pilote
American Septentrional* (1778) and *Neptune
Americo-Septentrional* (1780). It is a revision
of a chart by WALTER HOXTON (q.v.)

Smith, B. *See under* PLAYFAIR, J.; REES, A.;
SMITH, CHARLES; WILKINSON, R.

Smith, Caleb 18th cent. Pub. Brit. M
See also under CARRANZA, D. G.

Smith, Charles } fl. 1800-52 Engs.,
Smith, Charles, and Son} Mapsellers Brit.
(thus from 1827 to 1852)} M G (C and T)
Smith's new English atlas . . . Lond., 1804,
1808, 1822 (each map is dated Jan. 6, 1804).

A new map of Yorkshire corrected to 1841 . . .
1841.
*Smith's new map of inland navigation of England
and Wales*, 1803.
Smith's new general atlas Lond., 1808, 1816.
Maps eng. by J. Bye, E. Jones, B. Smith.
Smith's classical atlas . . . Lond., 1835.
Engs. and carts. incl.: Jos. Bye, E. Jones,
Alex. Macpherson.
See also under ANDRIVEAU-GOUJON.

Smith, General D. *See under* CAREY, M.

Smith, Capt. John 1580-1631 Cart.
Brit. M
Virginia 1606 and other edns. Eng. by
WILLIAM HOLE (q.v.)
New England . . . Lond., James Reeue,
1614 (from Smith's *Advertisement for the
unexperienced planters of New England* . . .
Lond., 1631).

Smith, Joseph fl. 1714-29 Printer, Pub.
Brit. M
See also under LE ROUGE, G. L.

Smith, J. (eng.) *See under* THOMSON, J.,
AND CO.

Smith, Roswell C. 19th cent. Cart.
Amer. M
Smith's atlas for schools, academies and families
Philadelphia, W. Marshall and Co.;
Hartford, D. Burgess and Co., 1835;
Hartford, Spalding and Storrs, 1839;
New York, Cady and Burgess, 1839,
1850; and other edns.

Smith, S. fl. 1793 Pub. Brit. M
Successor to MOUNT AND DAVIDSON, (q.v.
under MOUNT W., AND PAGE, T.)

Smith, T. *See under* ELPHINSTONE, J.

Smith, William fl. 1585-98 Cart. Brit.
M
Rouge Dragon Pursuivant.
Various maps, including versions of Saxton.
His map of Cheshire was used by Speed.

Smith, William 1769-1839 Mapseller
Brit. M G (C and T)
*A delineation of the strata of England and
Wales, with part of Scotland* . . . Lond.
W. Cary, 1815.
Author of the 'Anonymous County Maps'.
See British Museum Quarterly, Vol. XXII
No. 3-4, 1960.

Smith and Bowles, J. *See under* Bowles, J.

Smither, James *See under* Carey, M.;
    Guthrie, W.; Jones, T. W.

Smits, Gerard *See under* De Jode, G.

Smyth, Payler *See under* Warburton, J.

Smyth, William Henry, Admiral 1788-
    1865 Astron. Brit. X

Snell, R. 18th cent. Cart. Brit. M
    Map of Monmouthshire, 1785.

Snellius, Rudolph} *c.* 1546-1613 Astron.,
Snel van Roijen } Cosmographer Dut.
             } G

Snodham, Thomas fl. 1603-25 Printer
    Brit. M
    Printer of various edns. of Speed's *Theatre.*

Snowden, George and William 18th cent.
    Pub. Brit. M

Society for the Diffusion of Useful
    Knowledge M G
    Maps of the society . . . Lond., Chapman and
    Hall, 1844, and other edns. Carts. incl.
    W. B. Clarke.
    A series of maps, modern and ancient . . .
    Lond., Baldwin and Cradock, 1829-35.

Society for the Propagation of the Gospel
    in Foreign Parts
    The colonial church atlas . . . Lond.,
    Rivington, 1842. Maps drawn and eng.
    by J. Archer.

Sohr, Karl 19th cent. Geog. Ger. M
    Vollständiger hand-atlas . . . Glogau, C.
    Flemming, 1842-4, 1859. Carts.: F.
    Handtke, J. G. B. Klose, S. Theinert, A.
    Tschierschky. Reissued in 1865 as *Voll-
    ständiger universal-handatlas* . . . same
    place and pub.

Soimonov, Feodor 1682-1780 Sailor,
    Cart. Russ. M

Solinus, Caius Solinus 16th cent. Geog.
    Swiss? M
    Rerum toto orbe memorabilium . . . Basel, 1538
    (with 2 woodcut maps).
    See also under Apianus, Petrus.

Solomeïn *See under* Nakovalnik

Soly, Michel 17th cent. Pub. Fr. M
    See also under De Herrera, Antoine.

Somers, Jean   } 17th cent.
Sommer   } Pub. Eng.

Somer Pruthenus, Johannes} Fr. M
    See also under Blaeu, G.; Sanson, N., père.

Sonetti *Same as* di li Sonetti (q.v.)

Sonis, I. *See under* Ponce, N.

Sophianos, Nikolaos 16th cent. Cart.
    Greek M

Sörensen, Jens 1646-1723 Hydrographer
    Dan. M

Sorte, Christoforo fl. *c.* 1510 Cart. Ital.
    M

Sotzmann, Daniel Friedrich 1754-1840
    Cart. Ger. M G (T)
    Series of small maps of Prussia, Berlin, *c.*
    1814.
    Karte von Polen . . . Berlin, 1793, 1796.
    See also under Bossi, L.; Gaspari, A. C.;
    von Reilly, F. J. J.; Weigel und
    Schneiderschen Handlung.

Soulier, E. *See under* Andriveau-Goujon,
    J.

Spafarieff, Sir Leonty } 1765-1845 Soldier,
Spafar'ev, Leontiï Vas-} Cart. Russ. M
         il'evich}
    Atlas of the gulf of Finland . . . St. Petersburg,
    1821, 1823. Text in English and Russ.

Spalding and Storrs *See under* Smith, R. C.

Spano, Antonio d. 1615 Cart. Span.
    G (T)

Sparke, Michael *See under* Cartwright, S.;
    Mercator, G.

Sparrman fl. 1785 Cart. Ger.? M

Sparrow, Thomas 18th cent. Cart.
    Brit. M
    Plan of Colchester 1767.

Specht *See under* Ottens, R.; Visscher,
    N., III

Speckel, Daniel fl. 1576 Architect, Cart.
    Ger. M

Speed, John   } 1552-1629 Cart. Brit. M
Spédo, Iohanne}
    Africae described Lond., Basset and Chiswell,
    1676. Eng. by Abraham Goos.
    A newe map of Germany . . . 1626; Basset
    and Chiswell, 1676.
    Greece Basset and Chiswell, Lond., 1676.
    A new mape of ye XVII United Provinces . . .
    Lond., 1676.

The mape of Hungari . . . Lond., 1676.

Italia . . . Lond., Basset and Chiswell, 1676.

A new mappe of the Romane empire Lond., Basset and Chiswell, 1676.

A new mape of Tartary Lond., Geo. Humble, 1626; Basset and Chiswell, 1676.

The Turkish empire Lond., 1626; Basset and Chiswell, 1676.

A new description of Carolina Lond., Basset and Chiswell, 1676. Eng. by Francis Lamb.

A map of Virginia and Maryland (unsigned) Lond., Basset and Chiswell, 1676. Eng. by F. Lamb.

A prospect of the most famous parts of the world . . . 1646; 'Printed by John Legatt for William Humble' . . . 1650.

The Theatre of the empire of Great Britaine: presenting an exact geography of the kingdomes of England, Scotland, Ireland, and the isles adioyning: with the shires, hundreds, cities and shire-townes, within ye kingdome of England, divided and described by John Speed. Imprinted at London. Anno, cum privilegio, 1611 [-1612]. And are to be solde by John Sudbury & George Humble, in Popes-head alley at ye signe of ye white horse. fol. 9½″ × 15″. In 4 parts, comprising books 1 to 4 of Speed's *History of England.* 67 maps. Isle of Man by Thomas Durham; Isle of Wight by William White. Remainder by John Speed from surveys of Norden and Saxton. Eng. mostly by Jodocus Hondius. They have English text printed on their backs. It is a shortened version of Camden's *Britannia*, except in the case of Norfolk; this is by Sir William Spelman. Printed by W. Hall. Most of the maps bear the imprint of Sudbury and Humble, some of Humble alone.

Before the issue of the above atlas the maps had, between 1605 and 1610, been issued separately, without text on their backs. A collection of these is in the British Museum, and in many cases they were printed before the engraver had added his name and dates. But note that the 1713, 1743 and 1770 edns. are also without text on their backs.

SOME OTHER EDITIONS
with brief details of some of their variations

1614 (Sudbury and Humble). Date on title page altered to 1614. With a sheet following title page, listing the parts of the 'empire', and 'The Achievement of our Soveraigne King James . . .'. This is followed by a dedication to the King, the Royal Arms, four leaves addressed to the reader, verses of J. Sanderson and index to the maps.

1616 (Sudbury and Humble). *Theatrum Imperii Magnae Britanniae . . .* The only edn. with Latin text. Trs. by Philemon Holland. This edn. is of great rarity.

1627-31 (George Humble). Includes, facing the title page, a portrait of Speed, eng. by S. Saevry. Text, incl. preliminary pages, reset in different type, with double-line border.

1650-62 (Roger Rea the elder and younger). The whole of the text reset in smaller type. New ornamental band at top of 'The British Empire' and following pages. Circular space instead of shield in middle of band. In the preliminary pages the author's address and the verses each occupy a whole sheet. In part I imprint on maps altered from 'J. Sudbury and G. Humble' to 'Roger Rea'.

1676 (Thomas Bassett and Richard Chiswell). Title page re-engraved, with many alterations, by R. White. Arms of Charles II replace those of James II in preliminary pages. Text completely reset. New initials on the text at back of maps. Imprints now thus: 'Sold by Thomas Bassett in Fleetstreet, and by Richard Chiswell in St. Pauls Church yard.'

1713 and 1743 (Henry Overton). Title page is printed from the eng. plate of 1611 edn., but original title replaced by this: *England fully described in a compleat sett of mapps of ye county's of England and Wales, with their islands, containing, in all, 58 mapps.* In the 1743 edn. the words 'by John Speed' follow '58 mapps'. In both edns. maps are reprints of the 1676 edn. above, but with main roads eng. on all plates. Also the Bassett and Chiswell imprint is replaced by this: 'Henry Overton at the white horse without Newgate, London.' Maps have plain backs.

1770 (C. Dicey and Co.). $11\frac{1}{2}'' \times 17\frac{1}{4}''$. Title now *The English atlas, or a complete set of maps of the counties in England and Wales.* Of the maps, 47 are from the Overton edn. of 1743 (see above); 5 are copied from maps by Jansson, 4 of which have the imprint 'Sutton Nicholls sculp.'; 2 others are by the Overtons, 1 by J. the other by H., and there are 2 road maps. All but 4 have the imprint of C. Dicey and Co. Maps have plain backs.

Complete details of the foregoing edns. of Speed's *Theatre* will be found in Chubb *The Printed maps and atlases of Great Britain* (see *Bibliography*).

Collections, i.e. not complete atlases, were issued in 1680 (Seller), c. 1690 (Christopher Browne) 1696 (Browne, John and Henry Overton), 1710 (ditto).

Miniature copies of *Saxton's* maps, wrongly attributed to Speed, and reduced and eng. by Pieter van den Keere, were issued in *England Wales and Scotland described and abridged with ye historic relation of things worthy memory from a farr larger voulume done by John Speed.* 63 maps approx. $5'' \times 3\frac{5}{8}''$. Signed 'Petrus Kaerius caelavit'. They had first been issued in 1599, and, with a Latin title page, in 1617 (Amster., Jansson). Other edns. were issued in 1627, 1630, 1646, 1662, 1666, 1676.
See also under Blome, R.; Durham, T.; Hollar, W.; Humble, G. and W.

Speer, Capt. Joseph Smith fl. 1766 Cart. Brit. M
The West India pilot (unsigned) Lond., W. Griffin, 1766; S. Hooper, 1771.

Spelman, Sir W. *See under* Speed, J.

Spiegel, Joh. Baptist fl. 1722 Globe-maker Ger.? G (T)

Spiesshaimer, J. *Same as* Cuspinianus, J. (q.v.)

Spilbergh, J. *See under* Mortier, D.

Spilsbury, T. *See under* de Brahm, J. G. W.

Spilsbury, W. and C. *See under* Gaultier, A. E. C.

Spirinx, Nicolas fl. 1606-43 Cart. Fr. M G (C and T)

Spitzer, A. fl. 1764 Instrument-maker Austrian O

Spofforth, Robert fl. 1701 Eng. Brit. M

Sporer, E. *See under* Meyer, J.

Sporer, Hanns 15th/16th cent. Cart. Ger. M
Das ist die mapa mūdi . . . Augsburg, early 16th cent. Woodcut.

Spotswood, W.}
Spottiswood } *See under* Malham, J.

Sprange, J. fl. 1779 Surv. Brit. M
Surveyed for R. Budgen (q.v.)

Sprecher à Berneck, } *See under* Blaeu,
            Fortunato} G.; Hondius,
Sprecher von Bernegg,} H.; Jansson, J.;
            Fortunatus} Mercator, G.

Sprecht *See under* Beek, A.; Schenk, P.

Sprint, John fl. 1698-1727 Bookseller Brit. M

Sprotta, Jonas Scultetus *Same as* Sculteto, B. (q.v.)

Spurner von Merz, Karl 1803-92 Cart. Ger. M
Atlas antiquus . . . Gotha, I. Perthes, 1850.
Atlas zur geschichte von Bayern Gotha, J. Perthes, 1838.
Dr Karl von Spurner's historisch-geographischer hand-atlas 3 vols., Gotha, J. Perthes, 1846-51 and later edns.
Historisch-geographischer hand-atlas zur geschichte der staaten Europa's . . . Gotha, J. Perthes, 1846.

Stabius, Johann fl. 1515 Cart. Austrian M
See also under Dürer, A.

Stackhouse, Thomas 1706-84 Pub., Cart. Brit. M
An universal atlas Lond., S. J. Neele, 1798 and other edns.

Stampfer, Jacob c. 1505-79 Cart. Ger. G (T) O

Stampioen, Jan fl. 1651-3 Cart. Dut. M

Stanford, E. 19th cent. Pub. Brit. M
Stanford's library map of Australasia Lond., 1859.
See also under Andriveau-Goujon.

Starck-man, P. *See under* Blaeu, G.; de Fer, N.; Desnos, L. C.

Stark, John 19th cent. Pub. Brit. M
See under THOMPSON, J.

Starkenburg See under OTTENS, R.

Starling, Thomas fl. 1819-31 Eng.
Brit. M
Geographical annual . . . Lond., 1830, 1833
and other edns. Printed by Perkins and
Bacon.
See also under CREIGHTON, R.

Staunton, Sir George Leonard 1737-1801
Statesman Brit. M
An authentic account of an embassy from the
king of Great Britain to the emperor of
China . . . Atlas Lond., G. Nicol, 1798.
Carts.: H. W. Parish, J. Barrow, T.
Barrow.

Stead, J. W. 18th cent. Cart. Brit. M
Map of Jersey, 1799.

Stedman, C. 18th cent. Historian Brit. M
The history of . . . the American war 2 vols.,
Lond., 1794. Contains many fine battle
plans.

Steel, David fl. 1800 Pub. Brit. M

Steel, Penelope }
Steel and Goddard} See under HORSBURGH, J.

Steen, C. V. D. Same as GEMMA FRISIUS
(q.v.)

Stefano Same as DI STEFANO (q.v.)

Stein, Christian Gottfried Daniel 1771-
1830 Geog. Ger. M
Neuer atlas der ganzen erde . . . Leipzig, J. C.
Hinrichs, 1833–4 and later edns. Maps by
M. Riedig, F. W. Striet.

Stella, Tileman}
Stoltz   } 1525–89 Cart. Dut.
Stolz   }
See under BLAEU, G.; HONDIUS, H.; HORN,
G.; JANSSON, J.

Stelliola, N. A. Same as STIGLIOLA, C. (q.v.)

Stempel, Georg-Gerard fl. 1587-98 Cart.
Ger. M

Stender, Gotthard Friedrich 1714-96
Theologian Ger. G

Stent (or **Stint**), **Peter** fl. 1659 Pub.
Mapseller Brit. M
Published an edn. of Symonson's map of
Kent.

Stephanius, Sigurd fl. 1570 Cart.
Icelandic M

Stephenson, John fl. 1786 Cart. Brit.
M
Laurie and Whittle's channel pilot (2 charts)
with the collaboration of G. Burn.
Lond., 1794-1803. Includes also charts
by James Grosvenor, J. Ross, Christopher
Collins, Capt. Wm. Price, Francis Owen,
Capt. Dobree, Capt. Clement Lemprière,
Lieut. James Cook, John Thomas and
William Denys, A. Tovey and N.
Ginver, Capt. John Hammond, Capt.
Geo. Watson.
See also under LAURIE, R.

Sterne, Thomas fl. 1619-31 Globe-
maker Brit. G

Stetter, Joh. Jac. See under DE WIT, F.

Stevens, Henry 18th cent. Cart. Brit.
M

Stevinkhof, I. See under VOOGT, C. J.

Stieler, Adolf and August 1775-1836
Carts. Ger. M
Schul-atlas . . . Gotha, I. Perthes, 1841.
See also under GASPARI, A. C.

Stier, Martin 1630-69 Engineer, Cart.
Austrian M

Stigliola, Collantonius } 1547-1623
Stelliola, Nicolò Antonio} Astron., Cart.
                              } Ital. M

Stiles, S. See under GOODRICH, S. G.

Stimmer, Christoph} fl. 1544 Cart. Ger.
Meister Christoph } M
Sometimes signed with the initials 'CHS'.

Stobie, James and Matthew fl. 1770-83
Surv. Brit. M
Various Scottish county maps.

Stobnicza, Jan d. c. 1530 Cart. Ger. M

Stockdale, John 1739-1814 Bookseller,
Pub. Brit. M
New British atlas . . . Lond., 1805. Made
up of maps drawn by E. Noble and eng.
by J. Cary for the first edn. of Gough's
Camden's Britannia 1789, 1806.
Map of England and Wales 1809.
See also under ANDREWS, J.; CARY, J.;
CHAUCHARD, CAPT.; FILSON, J.; LÓPEZ,
T.; MENTELLE, E.; MORSE, J.

Stöffler, Johannes 1452-1531 Astron.
   Ger. G (C)

Stoke, J. 18th cent. Pub. Brit. M

Stokes, Gabriel 18th cent. Cart. Irish
   M
   Map of Dublin, 1750 (eng. S. Wheatley).

Stolz Same as Stella (q.v.)

Stone, D. S. See under Burr, D. H.

Stone, Edward James 19th cent. Astron.
   S. African X

Stopius, Nicolaus 16th cent. Cart. Dut.
   M
   Sometimes signed with the initials 'N. St.'

Stout, J. D. See under Willetts, J.

Strabo c. 63 B.C. A.D. 25 Geog. Roman
   M
   *Strabonis noblissimi et doctissimi philosophi ac
   geographi rerum geographicarum commentary
   libris xvii . . .* Basle, Henricpetrina, 1571.

Strachey, ? 18th cent. Cart. Brit. M
   Map of Somersetshire, 1736.

Strahan, W., and Cadell, T. See under
   Cook, Capt. J.; Hawkesworth, J.

Strahlenberg Same as von Strahlenberg
   (q.v.)

Streit, Friedrich Wilhelm See under
   Fischer, W.; Gaspari, A. C.; Stein,
   C. G. D.

Stridbeck, Johann, Junr. 1640-1716
   Printer Ger. M

Strong, C. D. See under Bradford, T. G.;
   Goodrich, S. G.

Strubitz, Maciej c. 1520-89 Cart. Polish
   M

Strunz fl. 1810 Cart. Ger. G

Stuart, James 18th cent. Eng. Brit. M
   Map of Cheshire, 1794; 2 edns.

Stucchi, Stanislao 19th cent. Cart. Ital.
   M
   Grande atlante universale . . . Milan, 1826.

Stuers Same as de Stuers (q.v.)

Stukeley, William 18th cent. Cart.
   Brit. M
   Map of Holland (i.e. in Lincolnshire), 1723.

Stumpffius, Johannes} 1500-77 Theologian,
Stumpff   } Historian Swiss M

Sturt, John 1658-1730 Eng. Brit. M
   Maps in Camden's *Britannia* (1695 edn. of
   R. Morden).

Stuurlieden, Ervaren See under Voogt,
   C. J.

Subias, Joaquin 18th cent. Cart. Span.
   M
   *Carte nueva corografica del virreynado de S.
   Fé de Bogota . . .* Bogota, c. 1760.

Suchet, Louis Gabriel, Duc d'Albufera
   1770-1826 Soldier Fr. M
   Mémoires du maréchal Suchet . . . Atlas Paris
   and Lond., Colburn, 1828-9.

Sucholdez See under von Reilly, F. J. J.

Sudbury, John fl. 1610-15 Bookseller
   Brit. M
   Was joint publisher with George Humble of
   two edns. of Speed's *Theatre*. Sometimes
   they are designated by their initials, thus:
   J. S. and G. H.

Sudlow, E. fl. 1784-90 Eng. Brit. M
   See also under de Rapin-Thoyras, P.

Surhon, Jean and Jacques}
Surhonius   } fl. 1555 Cart.
Surchon   } Fr. M
   See also under Jansson, J.; Mercator, G.;
   Tassin, N.; Tavernier, M.; van den
   Keere, P.

Surhonio, Joanne 17th cent. Cart. Fr.
   M
   Picardia, Regis Belgica Amster., Blaeu,
   c. 1660.
   See also under Blaeu, G.; Ortelius, A.

Sutorium, C. See under Botero, G.

Svart, Olof Hansson }
Oernehufvuds, Olaus} 1606-44 Cart.
Johannis Gothus} Swedish M

Swale, Abel} fl. 1665-99 Bookseller, Pub.
Swalle   } Brit. M
   Pub. the 1698 edn. of Ogilby's *Britannia*.

Swart, Stephen See under de la Feuille, J.;
   Visscher, N., III.

Sweerts See under de Wit, F.

Swelinc, Ian 17th cent. Cart. Dut. M
   Maps in *Histoire de la nouvelle France . . .*
   Paris, I. Millot, 1609.
   See also I. Millot.

234

Swire, William, and Hutchings, W. F.
fl. 1830 Surv. Brit. M
Map of Cheshire 1830; ditto, with railways added, 1831.

Switzer, Christopher fl. 1593-1611 Eng.
Swiss M

Sydow *Same as* VON SYDOW (q.v.)

Sylvanus, Bernardo} 16th cent. Cart.
Silva } Ital. M

Symeone, Gabriel} 1509-75 Cart. Ital.
Symeoneus } resident in France M

Symonds, H. D., and Ridgway, J. *See under* RUSSELL, JOHN

Symons *See under* TAVERNIER, M.

Symonson, Philip fl. 1592-8 Surv., Cart.
Brit. M
Map of Kent, 1596 (eng. Charles Whitwell) and subs. edns.
See also under STENT, PETER.

Sÿmonsz, E. *See under* BLAEU, G.

Tabbert, Stralsunder *Same as* VON STRAHLENBERG, P. J. (q.v.)

Tabourot, Stephan} 1547-90 Cart.,
Tabourotius } Barrister Fr. M

Tadataka Ino fl. 1800-16 Cart. Japanese M

Taitbout de Marigny, E. 19th cent.
Cart. Fr. M
Atlas de la mer Noire et de la mer d'Azov Odessa, Nitzsche, 1850.
Plans de golfes, baies, ports, et rades de la mer Noire et de la mer d'Azov Odessa, A. Braun, 1830.

Takahashi, Kageyasu fl. 1810 Astron.,
Cart. Japanese M

Takebara, Sjun fl. 1769 Cart. Japanese M

Tallis, L. fl. 1850 Pub. Brit. M
See also under DUGDALE, T.

Tanner, B. *See under* GUTHRIE, W.;
MARSHALL, J.; WINTERBOTHAM, W.

Tanner, Henry Schenk}
Tanner and Marshall } 1786-1858 Eng.,
} Pub., Cart. Amer.
Tanner, Vallance, } M
Kearny and Co.}
Atlas classica . . . Philadelphia, 1840.

Atlas of the United States . . . Philadelphia, 1835.
A new American atlas . . . Philadelphia, 1818-23, 1819-21, 1823, 1825, 1825-33, 1839 (some maps by Thomas R. Tanner, Stephen T. Austen).
A new pocket atlas of the United States . . . Philadelphia, 1828 (some eng. by J. Knight).
A new universal atlas . . . Philadelphia, 1833-4, 1836; Carey and Hart, 1842-3, 1844. Engs.: W. Brose, G. W. Boynton, F. Dankworth, E. B. Dawson, E. Gillingham, W. Haviland, J. Knight, J. and W. W. Warr.
A new general atlas . . . Philadelphia, 1828.
See also under EDWARD, B.; LAURIE, R.; LUCAS, F.; MAYO, R.; MELISH, J.; PAYNE, J.; PINKERTON, J.

Tanner, Thomas R. *See under* TANNER, H. S.

Tannstetter, Georg} 1482-1535 Cart.
Collimitius } Austrian M

Tardieu, Ambroise 1788-1841 Cart.
Fr. M G
Atlas universel . . . Paris, Furne et Cie, 1842. Incl. maps by J. T. Thunot-Duvotenay.
See also under ARROWSMITH, A.; CREVIER, J. B. L.; DUMAS, G. M.

Tardieu, Antoine François *See under* MENTELLE, E.

Tardieu, H. *See under* MALTE-BRUN, C.

Tardieu, Jean Baptiste Pierre 1746-1816 Eng. Fr. G

Tardieu, Pierre François *See under* BALD, W.; LORRAIN, A.; MENTELLE, E.

Tardo, Joannes *See under* TAVERNIER, M.

Tassin, Nicolas fl. 1634 Cart. Fr. M
Les plans et profils de toutes les principales villes et lieux considérables de France . . . Paris, S. Cramoisy, 1634, 1636; M. Tavernier, 1638.
Les cartes générales de toutes les provinces de France . . . Paris, N. Berey, 1640-3. Carts. incl.; Bertius, Hondius, Boisseau, Almada, Surhon, Sanson, Jansson, Baudoin, Blaeu, Danckertz, in addition to Tassin himself.
See also under JEFFERYS, T.; TAVERNIER, M.

Tatsu, J. *See under* Dumont d'Urville, J.S.C.

Tatton, Gabriel 16th cent. Hydrographer, Cart. Brit. M

Maris pacifici . . . 1600. Eng. by Benjamin Wright.

Noua et rece terrarum et regnorum Californiae . . . 1616. Eng. by Wright.

Tatton, M. fl. 1616 Cart. Brit. M

Tavernier, Melchoir (1544–1641), **Gabriel, Jean Baptiste** (1605–89), **Daniel, Melchoir, Junr.** (1594–1665). Pubs., Carts. Fr. M

Carte d'Allemagne . . . Paris, 1635.

Théâtre géographique du royaume de France . . . Paris, 1634. Carts. incl.: P. Bertius, Cornelis Danckertz, I. Hondius, H. Hondius, Jan Jansonius, Sanson, Petit Bourbon, Hardy, Evert, Symons, Beins, Tassin, I. van Damme, d'Armendale, Jean Jubrien, J. Le Clerc, Joannes Tardo, de Classun, I. Surhonius, I. Fayanus, J. C. Visscher, Gaspar Baudoin, de Chattillon.

See also under Bertius, P.; Sanson, N., père; Tassin, N.

Taylor, Lieut. Alexander fl. 1783 Cart. Irish M

Map of Kildare, 1783 (eng. Downes).

See also under Faden, W.

Taylor, George, and Skinner, Andrew fl. 1805–13 Cart. Brit. M

Map of Co. Louth, 1778 (eng. G. Terry).

Taylor, Isaac fl. 1765–95 Cart. Brit. M Prolific maker of English county maps.

Taylor, Janet fl. 1846 Cart. Brit. X

Taylor, John 19th cent. Cart., Eng. Brit. M

Map of 10 to 14 miles round Dublin, 1816.

Taylor, Thomas fl. 1670–1724 Bookseller, Mapseller, Pub. Brit. M

England exactly described . . . Lond., 1715, 1716 (road maps). Eng. by W. Hollar, Richard Blome, Richard Palmer.

The gentleman's pocket companion . . . Lond., 1722.

See also under Blome, R.

Taylor, William fl. 1700–23 Pub., Bookseller Brit. M

Pub. Gibson's edn. of Camden's *Britannia* (1722).

Teesdale, Henry, and Co. fl. 1828–45 Pub. Brit. M

New British atlas . . . Lond., 1829, 1831, 1832, 1835, c. 1848. The maps are those of Robert Rowe in his *English atlas* (1816) with the imprints erased.

Map of the county of York . . . 1826. A reprint of Greenwood's map of 1806.

A new general atlas of the world 1835.

See also under Dower, J.; Hennet, G.

Teixeira Albernas, João 1627–66 Cart. Port. M

Teixeira, Father Ludovico}
Tesseira } fl. 1595 Cart.
Tiexiera, Luiz } Ital. M
Georgius, Ludovicus }
Made the first printed European map of Japan; it was pub. by Ortelius.

Insulae Açores Amster., 1665.

See also under Ortelius, A.

Templeux *Same as* de Templeux, D. (q.v.)

Temporal, Jean } fl. 1592 Cart.
du Temps } Fr. M
Temporarius, Johann}

Ten-Have (ten-Have) *See under* de Wit, F.; Visscher, N., III

Terborch, Gerard}
Ter Borch } c. 1616–81 Cart.,
Borch } Painter Dut. G

Terbuggen, Hendrik 16th cent. Cart Dut. M

Terry, Garnet fl. 1777–98 Eng. Brit. M

See also under de Rapin-Thoyras, P.; Taylor, G.

Terwood, Leonard} fl. 1575–6 Eng.
Tervoort } Flemish M
Eng. Saxton's Cornwall, Somerset, Southampton and Warwick.

Terzuolo *See under* Duval, H. L. N.

Tesseira *Same as* Teixeira (q.v.)

Teunissen *Same as* Jacobsz, T. (q.v.)

Thackara, J. } *See under* Guthrie,
Thackara and Vallance} W.; Mayo, R.

Theinert, A. *See under* Sohr, K.

Theodorus, Peter d. 1596? Astron. Dut.? G (C)

Therbu, L. 18th cent. Cart. Fr. M
Les plans de la guerre de sept ans ... Frankfurt-am-Main, 1789-91. Eng. by G. T. Cöntgen. Maps 10-17 are signed: 'Dessiné par J. Seibel, Bombardier.' The remainder are by Therbu.

Theunisz, Jacob}
Theunis }　*Same as* JACOBSZ, T. (q.v.)

Thevenot, Melchisedech 17th cent. Explorer Fr. M
Recueil de voyages de M. Thevenot ... Paris, Estienne Michallet, 1681. With 3 maps eng. by Liebaux.

Thevet, André 1502-90 Cart., Cosmographer Fr. M
'Cosmographe du Roy.'
Le novveau monde . . . (woodcut) Paris, Guillaume Chaudiere, 1581.

Thierry, D. *See under* MALLET, A. M.

Thomas, John *See under* STEPHENSON, J.

Thomas and Andrews *See* ARROWSMITH, AARON

Thomassino, Filippo ? Pub ? M

Thompson, J. (Eng.) *See under* WHITE, F.

Thompson, John 19th cent. Cart. Brit. M
A new general atlas Edinburgh, John Stark, 1830.

Thomson, Charles 19th cent. Cart. Brit. M
Eng. Thomas Scott's *Chart of Van Diemen's Land*, 1824.

Thomson, John, and Co. fl. 1814-69 Pubs. Brit. M
A new general atlas . . . Edinburgh, 1817, 1827; Edinburgh and Lond. (Baldwin, Cradock and Joy) 1821. Carts. incl.: G. Buchanan, James Wyld. Engs.: N. R. Hewitt, S. I. Neele, Wm. Dasauville, Kirkwood and Son, J. and G. Menzies, T. Clerk, J. Smith, J. Moffat, Sidney Hall.
See also under EDGEWORTH, W.; WYLD, J.

Thomson, M. *See under* WALKER, J.

Thonawer, J. *Same as* DONAUER, J. (q.v.)

Thorlaksson } 1542-1627
Gudbrandus Thorlacius} Bishop, Cart.
　　　　　　　　　} Icelandic M

Thorne, Robert 16th cent. Cart. Brit. M
'Orbis universalis descriptio', woodcut map in Richard Hakluyt's *Divers voyages touching the discouerie of America* Lond., Thomas Woodcocke, 1582.

Thornhill, Sir James 1675-1734 Artist Brit. X
Drew the figures for the constellations in Flamsteed's *Atlas Cælestis*.

Thornton *See under* MOUNT.

Thornton, John 1641-1708 Hydrographer, Pub. Brit. M
Atlas maritimus . . . Lond., 1700? With a world map by Edward Wright after Mercator. Engs. incl.: F. Lamb, Ia. Clerk.
See also under VAN KEULEN, J. I

Thorp, Joshua 19th cent. Cart. Brit. M
Map of Yorkshire, 1822 (eng. S. J. Neele) and subs. edns.

Thorpe, Thomas 18th cent. Cart. Brit. M
An actual survey of the city of Bath 1742.

Thuilier *See under* DE WIT, F.

Thuillier, Col. (Later General Sir) H. C. fl. 1860 Cart. Brit. M
His map of Bengal and N.W. India was issued several times.

Thunot-Duvotenay, J. T. *See under* TARDIEU, A.

Thurneysser, Leonard 1530-96 Physician, Astron., Cart. Ger. M

Thym, Moses fl. 1609 Cart. Ger. M

Ticknor, W. D. *See under* BRADFORD, T. G.

Tiddeman, Mark 18th cent. Cart. Brit. M
A draught of New York . . . Lond., W. Mount and T. Page, 1737.

Tideman, Ph. *See under* DE WIT, F.

Tiebout, C. *See under* CAREY, M.; COLLES, C.; GUTHRIE, W.; PAYNE, J.

Tilleman, Stella *Same as* STELLA, T. (q.v.)

Tillemont, Jean Nicolas}
　　　　de Tralage} d. 1699 Geog.
Tillemon }　Fr. M G
See also under DESNOS, L. C.; HOMANN, J. B.; HOMANN'S HEIRS; JULIEN, R. J.; NOLIN, J. B.

Tilman, Giorgio fl. 1570 Eng. Ital. M

Tilt and Bogue fl. 1842 Pubs. Brit. M

Tindal, Nicolas 1687-1714 Historian Brit. M
Maps and plans of Tindal's Continuation of Rapin's History of England Lond., Harrison, 1785-9? Incl. maps by R. W. Seale.

Tinney, John d. 1761 Bookseller, Eng., Pub. Brit. M
Joint pub. of many of Emanuel Bowen's maps.
See also under Pɪɴᴇ, J.

Tirion, Isaac d. 1769 Cart., Pub. Dut. M
Atlas van Zeeland . . . Amster., 1760.
Nieuwe en beknopte hand-atlas . . . Amster., 1730-69, 1744-69.
Nieuwe en keurige reis-atlas door de XVII Nederlanden . . . Amster., J. de Groot and G. Warnars, 1793.

Tisley, S. 19th cent. Globe-maker Brit. G

Titon du Tillet, Everard 1677-1762 Poet Fr. G

Tjarde *See under* Oᴛᴛᴇɴs, R.

Todesco, Nicolo fl. 1478 Pub. Ital. M

Tofiño de San Miguel, Vicente 1732?-1795 Cart. Span. M
Cartas maritimes de las costas de España 1786-88
Plan of the harbour of Cadiz Lond., Wm. Faden, 1805.
New military map of Spain and Portugal (with Thomas López) Lond., Arrowsmith, c. 1817.
Atlas maritimo de España Madrid, 1787, 1789. Maps by Joseph Varela y Ulloa and Principe de la Paz, in addition to Tofiño.

Tokunai early 19th cent. Cart. Japanese M

Tollmann *See under* Hᴏᴍᴀɴɴ's Hᴇɪʀs.

Tolomeo, Claudio *Same as* Pᴛᴏʟᴇᴍʏ, C. (q.v.)

Tompion, Thomas 1638-1713 Clock-maker Brit. G

Toms, William Henry fl. 1723-58 Eng., Printseller M
Eng. H. Popple's *A map of the British empire in America* Lond., 1732.
See also under Bᴀᴅᴇsʟᴀᴅᴇ, T.; Kᴇʟʟᴇʀ, C.

Tonaver, J. *Same as* Dᴏɴᴀᴜᴇʀ, J. (q.v.)

Torlacius *Same as* Tʜᴏʀʟᴀᴋssᴏɴ (q.v.)

Torreño, Nuño Garcia 16th cent. Pilot, Cart. Span. M

Torricelli, Joseph fl. 1730 Instrument-maker Ital. O

Tosi, F. *See under* Rᴏsᴀᴄᴄɪᴏ, Gɪᴜsᴇᴘᴘᴇ

Tourneisen *See under* ᴅᴇ Bᴏᴜʀɢᴏɪɴɢ, J. F.

Tovey, A. *See under* Sᴛᴇᴘʜᴇɴsᴏɴ, J.

Townson, R. *See under* Kᴏʀᴀʙɪɴsᴋʏ, J. M.

Tramezini, Michaelis} 16th cent. Pub.
Tramezinus, Michael} Ital. M
Pub. a map of the world in two hemispheres, 1554.

Transilvanus, Maximilian} fl. 1520 Cart.
Transsylvanus } Austrian? G (T)

Treffler, Christopher fl. 1680 Cart., Instrument-maker Ger. G (T) O

Treschel, Gaspar and Melchoir fl. 1535-41 Pub., Cart. Fr. M
Produced the first edn. of Ptolemy printed in France. Pub. Servetus's American map.
See also under Pᴛᴏʟᴇᴍᴀᴇᴜs, C.

Trevethen fl. 1652 Cart.? M

Triere, Ph. *See under* ᴅᴇ Lᴀᴘᴇ́ʀᴏᴜsᴇ, J. F. ᴅᴇ G.

Truchet, Olivier 16th cent. Cart. Fr. M

Truschet et Hoyau fl. 1551 Carts. Fr. M

Tryon *See under* Lᴇ Rᴏᴜɢᴇ

Tschierschky, A. *See under* Sᴏʜʀ, K.

Tschudi, Egidius } 1505-72 Historian
Tschudo, Aegidio} Ger. M
See also under Oʀᴛᴇʟɪᴜs, A.

Tuke, John fl. 1787 Cart. Brit. M
Map of Yorkshire, 1787 and subs. edns.

Tunnicliff, William fl. 1787-91 Topg. Brit. M
A topographical survey of the counties of Hants, Wilts, Dorset, Somerset, Devon and Cornwall (with 7 maps), Salisbury, B. C. Collins, 1791.

Turellus, Petrus fl. 1575 Astrologer, Philosopher Cart. Fr. M

Turmair, Johannes *Same as* AVENTINUS (q.v.)

Turner, James *See under* ASTON, J.; EVANS, L.; FISHER, J.

Turpin, H. fl. 1750 Pub. Brit. M
See also under MORDEN, R.

Türst, Konrad *c.* 1450–1503 Mathematician, Cart. Ger. M

Tyler, C. S. 19th cent. Pub. Brit. M

Tymms, Samuel fl. 1832–43 Topg. Brit. M
See also under CAMDEN, W.

Tyrer, James *See under* FADEN, W.

Tyson, J. Washington 19th cent. Cart. Amer. M
An atlas of ancient and modern history . . . Philadelphia, S. A. Mitchell, 1845.

Ulpius *Same as* VULPIUS, E. (q.v.)

Univertzagt, G. I. *See under* DELISLE, J. N.

Unzer, August Wilhelm *See under* BOBRIK, H.

Vacani, Camillo 1784–1862 Historian Ital. M
Atlante topografico–militare per servire alla storia delle campagne e degl' Italiani in Ispagna al MDCCCVIII al MDCCCXIII Milan, 1823.

Vadagnino *Same as* DI VAVASSORE (q.v.)

Vadianus   } 1484–1551 Geog.
Von Watt, Joachim} Swiss M
Epitome trium terrae partium . . . Zürich, 1534 (with woodcut world map).

Valck, Gerhard} 1651/2–1726 Cart., Pub.
Valk   } Dut. M G (C and T) X
Father of Leonhard V.
Nova totius geographica telluris projectio Amster., 1720?, 1748?. In addition to Valck, carts. incl., according to edn.: G. Sanson, P. Schenck, M. Seutter, Homann, N. Visscher III, de Wit, Schuchart, Leonhard Valk, Flandro, Covens, Mortier, Casimirus, Abraham and Carlo Allard, Nolin, Delisle, Joan Blaeu, Petty, de Fer, Coronelli, Jaillot, Müller, Kraus, Gerard van Keulen.

See also under BRAAKMAN, A.; CELLARIUS, A.; DE WIT, F.; HOMANN'S HEIRS; HUSSON, P.; OTTENS, R.; SCHENK, P.

Valck, Leonhard} 1675–1755 Cart. Dut.
Valk   } M G (C and T)
Son of Gerard V.
See also under DE WIT, F.; HOMANN'S HEIRS; SCHENK, P.; VALCK, G.

Valet, Jean *See under* MENTELLE, E.

Valgrisi, Vincenzo fl. 1561–2 Pub., Cart. Ital. M
Made a copy of Gastaldi's oval world map.
See also under PTOLEMAEUS, C.

Vallance, J. *See under* CAREY, M.; GUTHRIE, W.; LAURIE, R.; MAYO, R.; MELISH, J.; PONCE, N.

Vallardi, P. and G. *See under* BOSSI, L.

Vallemont *Same as* DE VALLEMONT (q.v.)

Vallet, J. E. J. fl. 1673 Eng., Pub. Fr. X
See also under DESNOS, L. C.; PHILIPPE DE PRÉTOT, E. A.

Valvasor} *Same as* DI VAVASSORE (q.v.)
Valassor }

van Aelst, Nicolas 16th cent. Eng. Dut. M
Eng. for LAFRÉRY (q.v.)

van Alphen, Pieter fl. 1660 Cart. Dut. M
Nieuwe zee atlas ofte water werelt Rotterdam, 1660, 1682.

van Bos, Jacob Belga} fl. 1551–63 Eng.
van Bossius   } Dut. M
Busius   }
Worked for LAFRÉRY (q.v.)

van Cleef *See under* VAN DEN BOSCH, J.

Vancouver, George 1758–98 Explorer Brit. M
Voyage of discovery to the North Pacific Ocean ... *Atlas* Lond., G. G. and J. Robinson and J. Edwards, 1798. A Fr. edn. was pub. at Paris in 1799.

van Crickenbourg, Jehan fl. 1506 Cart. Dut. M

van Damme, Jean 17th cent. Cart. Dut. M
Les environs de l'estang de longpendu . . . Amster., Blaeu, *c.* 1660.
See also under JANSSON, J.; HONDIUS, H.; TAVERNIER, M.

van den Bosch, Johannus 1780-1844
Cart. Dut. M
Atlas der overzeeche . . . Nederlanden
's Gravenhage en Amsterdam, van Cleef,
1818.

van den Broeck, A. *See under* OTTENS, R.;
VISSCHER, N., II.

van den Ende, Josua 17th cent. Eng.
Dut. M
Eng. for G. AND J. BLAEU, G. BLAEU, H.
HONDIUS, G. HORN (qq.v.).

van den Keere, Pieter}
Keer, Peter   } Eng., Bookseller
Kaerius, Petrus   } Dut. M G (T)
Kerius   }
Brother-in-law of Hondius.
Miniature English county maps after Saxton
(listed under SPEED, J. [q.v.])
La Germanie inferieure . . . Amster., 1621,
1622. Incl. maps by C. J. [N. J.] Visscher,
Martini, Surhon and Emmius.
Petri Kaerii Germania inferior . . . Amster.,
1622.
See also under DU SAUZET, H.; HONDIUS, H.;
HONDIUS, J.; HORN, G.; JANSSON, J.;
LANGENES, B.; MERCATOR, G.; SPEED, J.

van der Aa, Pieter } 1659-1733 Pub. Dut.
Vander Aa   } M
Atlas nouveau et curieux . . . 2 vols. Leyden,
1714, 1728.
La galerie agréable du monde . . . 66 vols.
(with 6-48 maps apiece) Leyden, 1729.
Nouvel atlas . . . 1714.
*Représentation, où l'on voit un grand nombre des
isles, côtes, rivières et ports de mer . . .*
Leyden, 1730?.
Cartes des itineraires et voïages modernes . . .
Leyden, 1707.
See also under COVENS, J.; MORTIER, P.;
OTTENS, R.

van der Beke, Pieter ? Cart. Dut. M

van der Berg, Petrus 17th cent. Geog.,
Pub. Dut. Dut. M
Brother-in-law of JOD. HONDIUS (q.v.)

van der Borcht, Petrus} 16th cent. Eng.,
Verborcht   } Cart. Dut. M
Worked with Plantin (q.v.)

van der Burght, Willebordus *See under*
HONDIUS, H.; JANSSON, J.

van der Ende, Josua *See* VAN DEN ENDE, J.

van der Heyden, Gaspard and Peter *Same
as* À MYRICA, G. AND P. (q.v.)

van der Loss 17th cent. Cart. Dut. M

Vandermaelen, Philippe Marie Guillaume
1795-1869 Cart. Belgian M
Atlas universel . . . Brussels, 1827 and later
edns. Lithographed maps by H. Ode.

van der Putte, Bernard 16th cent. Cart.
Dut. M
Worked with Plantin (q.v.)

van Deutecum *Same as* DOETECUM (q.v.)

van Deventer, Jacob 16th cent. Cart.
Dut. M
Worked with PLANTIN (q.v.)

van Doet, J. 17th cent. Cart. Dut. M

van Gouwen, G. *See under* COVENS, J.; DE
LA FEUILLE, J.; DE WIT, F.; OTTENS, R.

van Herrenvelt, E. *See under* RAYNAL,
G. T. F.

van Hoirne, Jan *Same as* À HORN, J. (q.v.)

van Keulen, Gerard d. 1726 Cart.
Dut. M
The new sea map of the Spanish Zee . . .
Amster., c. 1690.
*De groote nieuwe vermeerderde zee-atlas,
ofter water-waereld . . .* Amster., 1720.
See also under VALCK, G.; VAN KEULEN,
JOANNES, II; VOOGT, C. J.

van Keulen, Gerard Hulst *See under*
VOOGT, C. J.

van Keulen, Joannes, I 1654-1711 Cart.
Dut. M
*The great and newly enlarged sea atlas or
water-world . . .* 3 vols. Amster., 1682-6.
Carts. incl.: C. J. Vooght, John Thornton,
Ioanne Blaeu, George Harwar, Anthony
Williams. Engs. incl.: Ia. Clark, F.
Lamb.

van Keulen, Joannes, II d. 1755 Cart.
Dut. M G(T) X
*Atlas of charts for navigation of the British
isles* Amster., c. 1730.
Le grand nouvel atlas de la mer . . . Amster.,
1696. Carts. incl.: de Ram, de Vries, C.
Voogt; there is a *Stellatum planisphaerium*
by Ludovico Vlasblom.

De groote nieuwe vermeerderde zee-atlas . . .
Amster., 1682-4 (5 vols.), 1695, 1734.
Carts. incl.: N. J. Visscher, Teunis Ys, C.
Vooght, Gerard van Keulen.
See also under LAURIE, R.; VISSCHER, N.,
III; VOOGT, C. J.

van Keulen, Joannes, III 1676-1763 Cart.
Dut. M
De nieuwe groote lichtende zee fakkel . . .
Amster., 1753 (with J. de Marre).
See also under MACKENZIE, M.; OTTENS, R.;
VAN LOON, J.

van Keulen, Robijn fl. 1682-98 Cart.
Dut. M
Several atlases. Editions appeared as late as
1736.
*The new sea map of the channel betwixt England
and France* Amster., c. 1700.
Nouvelle carte des côstes de Guinée, d'Or . . .
Amster., c. 1695.

van Krevelt, A. *See under* RAYNAL, G. T. F.

van Langren, Arnold Florentius fl. 1600
Cart. Dut. M G (C and T)
See also under HONDIUS, H.; JANSSON, J.

van Langren, Hendrik Florentius b. c. 1574
Cart. Dut. M G

van Langren, Jacob }
        Floris} 17th cent. Eng.
van Langeren, Jacob} Dut. M G (C and
      Florentius} T)
Father of Arnold, Hendrik and Michaele
van L.
A direction for the English traviller . . . Lond.,
T. Jenner, 1643.
See also under SIMONS, M.

van Langren, Michaele Florentius 17th
cent. Cart. Dut. M G
See also under BLAEU, G.

van Lochom, Pierre and Michael fl. 1660
Cart. Dut. M
Made maps for B. J. BRIOT (q.v.).

van Loon, Gillis *See under* VAN LOON,
JOANNES

van Loon, H. *See under* DE FER, N.; DESNOS,
L. C.; NOLIN, J. B.

van Loon, Joannes or Jan fl. 1661-68
Cart., Pub. Dut. M

Noort-ster ofte zee atlas 1661 (with Gillis van
Loon), 1666, 1668 (with W. Goeree,
Middleburg).
Die nieuwe groote lichtende zee-fackel . . .
(with C. J. Voogt), Amster., J. van Keulen,
1699-1702.
See also under BRAAKMAN, A.; DE WIT, F.;
JANSSON, J.; OTTENS, R.; VISSCHER, N.,
II; VISSCHER, N., III.

van Luchtenburg, A. *See under* DANCKERTZ,
J.; LE CLERC, J.

van Medtman, M. *See under* HUSSON, P.

Vannelli, Guiliano d. 1517 Engineer
Ital. G

van Os, Pieter *See under* BROUCKNER, I.

van Roijan, Snell fl. 1618 Astron. Dut.
X

van Salingen, Simon fl. 1601 Cart.
Dut. M
Made a map of Scandinavia.

van Schilde}
van Schiller} *Same as* SCILLIUS, J.

van Schley, J. *See under* RAYNAL, G. T. F.

Vante, Gabriello 1452-84 Globe-maker
Ital. G

van Verden, Karl fl. 1720 Sailor, Cart.
Russ. M

van Wijk Roelandszoon, Jacobus 1781-
1847 Cart. Dut. M
*Verhandeling over de nederlandsche ontdekkingen,
in Amerika, Australië, et Indiën en de
Poolanden . . . Atlas* (with R. G. Bennet)
Utrecht, J. Altheer, 1827.

Vapovsky, B. *Same as* WAPOWSKI, B. (q.v.)

Varela y Ulloa, Josef 18th cent. Cart.
Span. M
Plano del rio Uraguay c. 1784.
See also under TOFIÑO DE SAN MIGUEL, V.

Varen, Bernhard fl. 1622-50 Geog.,
Physician Dut. G

Varlé, P. C. *See under* PONCE, N.

Vauban *Same as* DE VAUBAN (q.v.)

Vaughan, Robert fl. 1656 Cart. Brit.
M
The Mapp of Warwick-shire . . . 1656
See also under DUGDALE, SIR W.

Vaughan, William *See under* MASON, J.

Vaugondy *Same as* DE VAUGONDY (q.v.)

Vaultier, Le Sieur} *See under* DE FER, N.;
Vaulthier } SANSON, N., PÈRE

Vavassore *Same as* DI VAVASSORE (q.v.)

Vaz Durado, F. *Same as* DOURADO, F. (q.v.)

Vazquez, Francisco 18th cent. Cart. Span. M
Atlas elementar Madrid, Pantaleon Aznar, 1786.

Vedel, A. *Same as* VELLIUS, A. (q.v.)

Veen, Adrian 1572–1613 Cart. Dut. M

Veldico, Willem fl. 1507–10 Cart. Dut. G

Velleius, Andreas } 1542–1616 Cart.
Vedel, Anders Sörensen} Dan. M

Velserus, Marcus} fl. 1598 Humanist,
Welser } Pub. Dut. M

Vendosme, L. *See under* BOISSEAU, J.

Venetus de Vitalibus, ⎫
       Bernardus} fl. 1507–8 Pub.
Venetum de Vitalibus,} Ital. M
       Bernadum}

Verardus, Carolus (15th cent. Geog.) **and Colombus, Christopher** (*c.* 1436–1506 Sailor) Span. M
In laudem serenissimi Ferdinandi Hispaniarum regis Bethicae & regni Granatae obsidio victoria & triūphus . . . (with woodcut map of W. Indies). Basle, Johann Bergman, 1494.

Verbiest, Père Ferdinand 1623–88 Astron. Fr. G (C) O

Verborcht, P. *Same as* VAN DER BORCHT, P. (q.v.)

Verden *Same as* VAN VERDEN (q.v.)

Verdussen, H. and C. *See under* DE AFFERDEN, F.

Vergerius, Ludwig fl. 1550 Cart. Ger. M

Vernor and Hood *See under* WALKER, J.

Verrazano, Giovanni da *See under* DA VERRAZANO

Verrazano, Girolamo da *See under* DA VERRAZANO

Verreyken fl. 1552 Cart. Dut. M

Vertue, George 1684–1756 Eng. Brit. M
Eng. for H. MOLL (q.v.).

Vespucci, Giovanni Juan 16th cent. Cart. Ital. M
Totivs.orbis.descriptio . . . 1524. With copperplate map of the world.

Viani, Mattio di Venezia fl. 1780–84 Cart. Ital. G (T)

Vico, Enea fl. 1542, Pub. Ital. M
Sometimes signed with the initials 'E.V.'

Viegas, Caspar Luiz fl. 1534 Cart. Port. M

Vietch, J. *See under* DICKSON, J.

Vieth, Gerhard Ulrich Anton 1763–1836 Cart. Ger. M
Atlas der alten welt . . . (with Karl Philipp Funke) Weimar, 1819.

Vigliarolo, Domenico } fl. 1530–80 Cart.
Vigliarolus, Dominicus} Ital. M

Villanovanus *Same as* SERVETUS, M. (q.v.)

Villaroel *Same as* DE VILLAROEL (q.v.)

Vinchio, Pietro Maria da fl. 1745 Cart. Ital. G (C and T)

Vinckeboons, Jan} 17th cent. Cart., Eng.
Vingboons } Dut. M G

Vindelicor, A. } *See under*
Vindelicorum, Augusta} HOMANN'S HEIRS

Virtue, George *See under* BARCLAY, J.; MOULE, T.

Visconti fl. 1699 Soldier, Cart. Ital.? M

Vischer, Georg Mathias 1628–96 Cart., Topg., Cosmographer Austrian M

Visscher, J. C. *See under* TAVERNIER, M.

Visscher, Nikolaus Joannis} 1587–1637
      or Claes Jansz} Cart., Pub.
Piscator } Dut. M
Novi Belgii . . . Amster., *c.* 1651
See also under BEEK, A.; BRAAKMAN, A.; JULIEN, R. J.; MERCATOR, G.; OTTENS, R.; SCHENK, P.; VAN DEN KEERE, P.; VAN KEULEN, J., II.

Visscher, Nikolaus, II} 1618–79 Cart., Pub.
Piscator } Dut. M
Son of Nikolaus J. V.
Atlas contractus orbis terrarum . . . Amster., 1666?. Carts. incl.: A. Goos, F. de Wit, A. van den Broeck, R. de Hooge, Joh. van Loon.

Visscher, Nikolaus, III⎫ 1639–1709　Cart.,
　　or Nikolaas⎬ Pub.　Dut.　M
Piscator⎭
Son of Nikolaus V. II.
Indiae orientalis . . . Amster., *c.* 1680.
*A new mapp of the kingdome of England and
　Wales . . .* Lond., Lea and Overton, and
　Amster., 1686.
Novi Belgii . . . Amster., *c.* 1685. (Also
　re-engraved and issued by J. Danckertz,
　though unacknowledged, *c.* 1700.)
Atlas minor . . . Amster., *c.* 1660–91, 1684?,
　1690? 1692?, 1710?, 1712?, 1717?. Carts.
　incl., according to edn.: N. P. Berchem,
　Ioannem Ianssonium, F. de Wit, Isaaco
　Massa, Moses Pitt, R. de Hooge, Stephani
　Swart, Wolfgango Lazio, Carolus Allard,
　Hardy, Joh. van Loon, G. de l'Isle,
　Christopher Browne, N. ten-Have,
　Damien de Templeux, Sr. du Frestoy,
　Samson (i.e. Sanson), Abraham Ortelius,
　G. Blaeu, J. Blaeu, J. van Keulen.
N. Visscheri Germania inferior . . . Amster.,
　1680?. Carts.: Justus Danckers, F. de
　Wit, J. Janssonius, J. Laurenberg, B.
　Schotanus. Some maps have vignettes
　eng. by Rom. de Hooge.
Variae tabulae geographicae . . . Amster.,
　1700?, 1709?. Covens and Mortier, 1730.
　Carts. incl.: Delisle, Allard, de Wit,
　Ottens, Danckertz, Halma, Specht, Lake-
　man, Sanson, Jaillot.
See also under COVENS, J.; DE LA FEUILLE, J.;
　DE WIT, F.; HOMANN, J. B.; HOMANN'S
　HEIRS; HUSSON P,; JANSSON, J.; VALCK,
　G.; WOLFGANG, A.

Vitruvio fl. 1550　Artist, Cart.　Ital.　M

Vivares, F. *See under* SAYER, R.

Vivien de Saint-Martin, Louis 1802–97
　Cart., Geog.　Fr.　M
Atlas universel . . . Paris, Ménard et Desenne,
　1825, 1827, 1834 and subs. edns.
Historie de la géographie . . . Atlas Paris,
　Hachette et Cie, 1873–4.

Vlasblom, Ludovico *See under* VAN KEULEN,
　JOANNES, II; VOOGT, C. J.

Voigt, Johann fl. 1681　Astron.　Ger.　X

Volckmar, Johann M. fl. 1642　Cart.
　Dut.?　G (T)

Volkamer, Johann Christoph 1644–1720
　Botanist, Cart.　Ger.　M

Volkamer, Johann Magnus d. 1752　Art
　Collector　Ger.　G

Vollier, Pierre 1649–1715　Geog.　Fr.　G

Volpaja, Girolamo Camillo fl. 1560
　Instrument-maker　Ital.　G O

Volpi, Giuseppe Antonio fl. 1680
　Instrument-maker　Ital.　G O

Völter, Daniel 1814–65　Cart.　Ger.　M
Schul-atlas . . . Esslingen, J. M. Dannheimer,
　1840. Maps drawn and lithographed by
　Ed. Winckelmann.

Voltius, Vincentius⎫
　　　Demetrius⎬ fl. 1593–1607　Cart.
Volcius⎭ M
Portolans.

von Aitzing, Michael⎫
Aitzinger⎪ 1530–*c.* 1593
Eitzing⎬ Historian, Diplomat
Eyzinger⎭ Austrian　M
Belgici leonis chorographia, 1587.

von Anse, L. *See under* DE WIT, F.

von Baarle *Same as* BARLÄEUS, C. (q.v.)

von Breydenbach, Bernhard d. 1497
　Geog.　Ger.　M
Peregrinatio in terram sanctam . . . (with map
　of Palestine) Mainz, Erhard Reuwick,
　1486.

von Elekes, Franz fl. 1831　Cart.　Ger.
　G (C and T)

**von Herberstein, Sigismund or Siegmund,
　Freiherr** 1486–1566　Statesman, Cart.
　Austrian　M
Muscovia . . . Vienna, 1549.

**von Humboldt, Friedrich Wilhelm
　Heinrich Alexander** 1769–1859
　Explorer　Ger.　M
Atlas géographique . . . de la Nouvelle-Espagne . . .
　Paris, F. Schoell, 1811; Paris, G. Dufour et
　Cie, 1812.
Voyage de Humboldt et Bonpland . . . Atlas
　Paris, F. Scheoll, 1805–39, 1814–20.

von Kausler, Franz Georg Friedrich
　1794–1848　Cart.　Ger.　M
Atlas des plus mémorables batailles . . .
　Carlsruhe and Freiburg, B. Herder,
　1831–7.

von Kotzebue, Otto 1787-1846 Sailor
Russ. M
*Atlas to the voyage of Lieutenant Kotzebue in
the ship Rurik . . . Atlas* St. Petersburg,
1823 (title in Russ.).

von Krusenstern, Adam Johann 1770-
1846 Cart. Ger. M
атласЬ южнато моря (*Atlas de l'Ocean
Pacifique*) St. Petersburg, 1826-7, 1827-38.

von Lauchen, J. *Same as* RHETICUS, J. G.
(q.v.)

von Littrow, J. fl. 1839 Astron. Ger. X

von Mädler, Johann Heinrich fl. 1834
Astron. Ger. X

Von Marsigli, Aloysius Ferdinand, Graf
See under MARSIGLI, L. F. (q.v.)

von Metzburg, Georg Ignaz Frieherr
1735-98 Cart., Mathematician
Austrian M

von Mollis *Same as* GLAREANUS, H. (q.v.)

von Parijs, Sylvester *c.* 1500-*c.* 1576 Eng.
Dut. M

von Reilly, Franz Johann Joseph 18th
cent. Pub. Austrian M
Atlas universale . . . Vienna, 1799.
Schauplaz der welt . . . Atlas . . . Vienna,
1789-91. Maps eng. by Ign. Albrecht,
Andreas Withalm.
Grosser Deutscher atlas Vienna, 1796. Carts
incl.: d'Anville, Has, de Vaugondy,
Djurberg and Roberts, López, Kitchin,
Dorret, Jefferys, Cassini and Julien, Faden,
Sotzmann, Müller, Schmidt and Santini,
Sucholdez and Endersch, O. A.
Wangensteen and I. N. Wilse.

von Roesch, Jakob Friedrich, Ritter
1743-1841 Historian Ger. M
*Collection de quarante deux plans de batailles,
sièges et affaires les plus mémorables de la
guerre de sept ans . . .* Frankfurt-am-Main,
J. L. Jaeger, 1796.

von Rothenburg, R. *See under* MEYER,
JOSEPH

von Schlieben, Wilhelm Ernst August
1781-1839 Cart. Ger. M
Atlas von . Amerika . . . Leipzig, G. J.
Göschen, 1830 (lithographed maps).
Atlas von Europa . . . Leipzig, G. J. Göschen,
1829-30.

*Lehrgebäude der geographie mit historischen,
statistischen und geschichtlichen andeutungen . . .*
Leipzig, Göeschen, 1828-30. Most maps
signed by Schlieben. Engs. incl.: J. C.
Ausfeld, R. Dreykorn, C. Grünewald,
C. Martin, A. G. and J. M. Mossner, C. S.
Schleich, Junr., A. Siebert.

von Schmettau, Samuel, Graf 1684-1751
Soldier Ger. M

von Schmidburg, G. R. *See under* GASPARI,
A. C.

von Strahlenberg, Philipp } 18th cent.
Johann } Cart. Ger.
Tabbert, Stralsunder } M

von Sydow, Emil 1812-73 Cart. Ger.
M
E. von Sydow's gradnetz-atlas . . . Gotha, J.
Perthes, 1847.
E. von Sydow's hydrographischer atlas . . .
Gotha, J. Perthes, 1847 and later edns.

von Tschudi *Same as* TSCHUDI, E. (q.v.)

von Watt, Joachim *Same as* VADIANUS
(q.v.)

von Wedell, Rudolph 19th cent. Cart.
Ger. M
Historisch-geographischer hand-atlas . . .
Glogau, C. Flemming, 1843.

von Wenzeley, A. *See under* SCHRAEMBL,
F. A.

Voogt, Claes Jansz d. 1696 Cart. Dut.
M
Die nieuwe groote lichtende zee-fakkel Amster.,
G. H. van Keulen, 1682. Maps are by
Gerard and Johannes van Keulen, pre-
decessors of Gerard Hulst van Keulen; they
include maps by the navigators Bertrand,
Frezier, Kraay, Lynslager, and Sikkena.
*Le nueva, y grande relumbrante antorcha de la
mer . . .* Amster., J. van Keulen, 1700?.
Carts., in addition to Voogt: Nikolaus
de Vries, P. Pickart, I. Stevinkhof,
Ervaren Stuurlieden, Gerard van Keulen,
Ludovicus Vlasblom.
See also under LOOTS, J.; OTTENS, R.; VAN
KEULEN, JOHANNES, I AND II; VAN LOON, J.

Vopel, Caspar }
Vopellius, Kaspar } 1511-1564 Cart. Ger.
Vopell } G (C and T) O
Medebach }

Vries *Same as* DE VRIES (q.v.)

Vrints, Vrientius or Vrients, Johannes Baptista b. 1552 Pub. Dut. M
Pub. edns. of Ortelius's atlas.

Vuillemin, Alexandre A. b. 1812 Cart. Fr. M
Atlas universel de géographie . . . Paris, J. Langlumé, 1847. Maps eng. by Lale.
See also under ANDRIVEAU-GOUJON.

Vulpius, Euphrosinus} 16th cent. Instru-
Ulpius } ment-maker Ital.
G (T) O

Wächtler, Ferdinand Friedrich 16th cent. Silversmith Ger. G

Waesburg *See under* JANSSON, J.

Waghenaer, Lucas Jansz} d. *c.* 1593 Cart.
** or Luc Janszoon**} Dut. M
Aurigarius }
Speculum nauticum . . . Amster., Cornelius Nicolai (i.e. Cornelis Claesz) 1591.
The mariners mirrour . . . Lond., 1588. Engs. incl.: J. Hondius, Johannes Rutlinger, Theodore de Bry.
Spieghel der zeevaerdt . . . Leyden, Plantin, 1583-5.
Dv miroir de la navigation . . . Amster., Corneille Nicolaus, 1590. Maps eng. and signed by Ioannes à Doeticum.

Walch, Jean fl. 1790 Cart. Fr. M
Charte de l'Afrique Augsburg, Martin Will, *c.* 1790.

Walckenaer, Charles Athanase, Baron 1771-1852 Geog. Fr. M
Atlas de la géographie ancienne historique et comparée des Gaules . . . Paris, P. Dufart, 1839.

Waldseemüller, Martin}
Hilacomilus } 1470-1518 Cart.
Hyacomilus } Ger. M G (T)
Ilacomilus }
Made the first map to bear the name America.
See also under APIANUS, PETRUS.

Walker, John (1759-1830), **Alex. and Charles** Carts. Brit. M
A general map of India Lond., J. Horsburgh, 1825.
Map of the countries on the N.W. frontier . . . *from the surveys of Lieut. J. Wood* . . . *Major R. Leech* Lond., E. India House, 1841, 1846.

Map of Sikh territory . . . Lond., E. India Co., 1846.
Map of the United States . . . (John and Alex.) Lond., and Liverpool, 1827.
Royal atlas (John and Charles) Lond., 1837, 1873, Longman, Rees and Co.
Hobson's fox-hunting atlas . . . (John and Charles with W. C. Hobson) Lond., 1850, and various other edns. from *c.* 1866 to *c.* 1880.
An atlas to Walker's geography . . . Dublin, T. M. Bates, 1797; Lond., Vernor and Hood; Darton and Harvey, 1802.
Walker's universal atlas . . . Lond., F. C. Rivington, G. Wilkie. Maps drawn and eng. by M. Thomson. Various English county maps.
See also under DRINKWATER, J.; FADEN, W.; GREENWOOD, C. AND J.; JONES, F.; LOCKWOOD, A.

Wall, G. A. *See under* PATTESON, E.

Wallace, J. 17th cent. Cart. Brit. M
Map of the Orkneys, 1693.

Wallis, James fl. 1810 Eng., Pub. Brit. M
Wallis's new pocket edition of the English counties . . . Lond., 1810, *c.* 1814.
Maps were reissued in 1819 in *Martin's sportsman's almanack* with altered imprints; in the same year they were used in *Lewis's New traveller's guide*. In the *c.* 1814 edn. plate numbers have been added to the maps at the top right-hand side.
Wallis's new British atlas . . . 1812; Lond., S. A. Oddy. Maps are close copies of those in Cole and Roper's *British atlas* (1810).
The Panorama . . . Lond., W. H. Reid, 1820. The maps closely resemble those of R. Miller of 1810, but mail-coach routes are more prominently indicated.
See also ELLIS, G.; REID, W. H.

Wallis, John 18th cent. Cart. Brit. M
Circular map of 22 miles round London, 1783; map of environs of London (with J. Cary), 1783.

Walpoole, George Augustus fl. 1784 Topg. Brit. M
The new British traveller . . . Lond., 1784, Alexander Hogg.

Walser, Gabriel} 1695-1776 Cart. Swiss
Walsero } M
Walserum }
Schweitzer-geographie . . . Zurich, Orell, Gessner et Cie, 1770.
See also under HOMANN'S HEIRS; LOTTER, T. C.

Walsh fl. 1802 Cart. Brit. M

Walter, H. 19th cent. Surv. Brit. M
Map of Windsor Forest, 1823.

Walthoe, J. *See under* WELLS, E.

Walton, Robert 1618-88 Cart. Brit. M

Walton and Maberly fl. 1858 Astron. Brit. X

Wangensteen, O. A. *See under* VON REILLY, F. J. J.

Wapowski, Bernard} d. 1535 Cart.
Vapovsky } Polish M

Warburton, John, Bland, J., Smyth, P. fl. 1749 Survs., Carts. Brit. M
Maps of Hertfordshire, *c.* 1749 (eng. N. Hill), and Middlesex, 1749 (ditto).

Ward, T. *See under* WELLS, E.

Ward and Lock *See under* PETERMANN, A. H.

Wardlow, W. 18th cent. Pub. Brit. M

Warin *See under* DE GOUVION SAINT CYR, L.

Warner, B. *See under* GUTHRIE, W.

Warner, John 18th cent. Eng., Cart. Brit. M
The courses of the rivers Rappahannock and Potowmack in Virginia . . . 1737 and other edns.
See also under CUBITT, T.; GARDNER, W.

Warr, J. and W. W. *See under* TANNER, H. S.

Warren, William 1806-79 Geog. Amer. M
Atlas to Warren's system of geography Portland, W. Hyde, 1843.

Warthabeth } fl. 1665-95
Wuscan Theodorus Wertha-} Printer
beth von Herouanni } Armenian
M
Worked in Amster.

Watson, A. 18th cent. Pub. Brit. M

Watson, Capt. Geo. *See under* STEPHENSON, G.

Watson, J. F. *See under* MAYO, R.

Watt *Same as* VADIANUS (q.v.)

Watté, John 18th cent. Cart. Brit. M
Map of Bedford Level, 1777.

Wayne, C. P. *See under* MARSHALL, J.

W.B. fl. 1590 Cart. Brit.? M
Playing-card maps signed: 'W.B. inven. 1590.'

Web, William} fl. 1645 Bookseller, Pub.
Webb } Brit. M
Pub. an edn. of Saxton's county maps.

Webb, Rev. T. W. fl. 1859 Astron. Brit. X

Wechel, C. fl. 1546 Pub. Fr. M
Pub. at Paris an edn. of Ptolemy with a Greek text.

Wedell *Same as* VON WEDELL (q.v.)

Weidner, I. G. L. *See under* GASPARI, A. C.

Weigel, B. J. C. *See under* GÖTZ, A.

Weigel, Christoph, the} 1634-1725 Eng.,
Elder, or Johann } Pub. Ger. M
Christoph } G
Weigelio }
Weigeln }
Descriptio orbis antiqui . . . Nuremberg, 1720-60. Same as J. D. Köhler's atlas of the same title.
Atlas portatilis . . . Nuremberg, 1720, 1724 (with one map by Adam Friedrich Zürner), 1745.
See also under FABER, S.; KÖHLER, J. D.

Weigel, Christoph, the Younger d. 1746 Cart., Pub. Ger. M

Weigel, Erhard 1625-99 Astron. Ger. G (C and T) X O
Signed: 'Erhardi Weigelii Mathem. Prof. P.'

Weigel, H.} fl. 1559 Cart. Ger. M
Wurm, H. }
Sometimes signed with the initials 'H.W.'

Weigel und Schneider-}
schen Handlung } 18th cent. Pub.
Schneider, A. G. and } Ger. M
Weigel }
Schneideri-Weigeliana }
Karte von Australien oder Polynesien Nuremberg, *c.* 1796.

Atlas der geographie von der bekannten ganzen welt . . . Nuremberg, 1794–1805. Carts. incl.: C. Mannert and D. F. Sotzmann.
See also under D'ANVILLE, J. B. B.

Weiland, Carl Ferdinand d. 1847 Cart. Ger. M
Atlas von Amerika . . . Weimar, 1824–8. Maps copied from those of H. C. Carey and I. Lea.
Allgemeiner hand-atlas . . . Weimar, 1848. All maps bear Weiland's name, except no. 69 (Mexico) which bears that of H. Kiepert. Engs.: Carl Jungmann, W. Kratz, Senr., Anson, Karl Jos. Maedel, Senr., L. Beyer, G. Hengsen, Senr., Mädel med., J. Maedel, Junr., C. Poppey, R. Schmidt, L. V. Kleinknecht, J. C. Gerrich, L. Bernhardt.
See also under GASPARI, A. C.; REICHARD, H. A. O.

Weiner, Peter fl. 1568 Cart. Ger. M

Weiss, J. H. 18th cent. Cart. Swiss M
Atlas suisse Aarau, J. R. Meyer, 1786–1802.

Welland, J. *See under* GODFRAY, H.

Weller, E. *See under* McLEOD, W.

Wellington, Lieut. fl. 1710 Cart. Brit. G (T)

Wells, Edward 1667–1727 Cart. Brit. M
A new sett of maps . . . Lond., J. and J. Bonwicke, S. Birt, T. Osborne, E. Wicksteed and T. Cooper, c. 1700; Oxford 'at the Theater', 1700; London, R. Bonwicke, J. Walthoe, R. Wilkin, T. Ward, c. 1705; and other edns.

Welser, M. *Same as* VELSERUS, M. (q.v.)

Wenzeley *Same as* VON WENZELEY (q.v.)

Werner, F. B. *See under* HOMANN'S HEIRS.

Werner, Johannes 1468–1522 Cart., Astron. Ger. M G

Werner, S. W. 18th cent. Cart. Brit.? M
See also under FADEN, W.

Wesenaer, N. fl. 1661 Cart. Dut.? M

Wessel and Skanke 18th cent. Cart. Swedish? M

West and Blake *See under* ADAMS, DANIEL.

Westenberg, Joanne *See under* BLAEU, G.; HONDIUS, H.; JANSSON, J.

Westermaÿr, C. *See under* GASPARI, A. C.

Westphalia *Same as* DE WESTPHALIA (q.v.)

Wheatley, S. *See under* STOKES, GABRIEL

Whitaker, C. and W. B. fl. 1823 Pub. Brit. M X
The travellers' pocket atlas . . . Lond., 1823. Eng. by Neele and Son.
See also under PHILIPPS, SIR R.

Whitchurch, Guilielmus *See under* COOK, CAPT. J.

White, Frances 18th cent. Cart. Brit. M
Map of York and Ainsty, 1785 (eng. J. Thompson).

White, Gallaher and White *See under* BRIGHAM, J. C.

White, John fl. 1585–93 Cart. Brit. M
Map, 'Americae pars . . .' in Thomas Hariot's *Merveilleux et estrange rapport . . . des commoditez qui se trouvent en Virginia . . .* Frankfurt, Theodore de Bry, 1590.

White, Robert fl. 1770–93 Astron. Brit. X

White, William 16th cent. Cart. Brit. M
Constructed Speed's map of the Isle of Wight (*see* SPEED, J.).

Whitehead, A. G. *See under* BROWN, T.

Whittle, James *See* LAURIE, R.

Whitwell, Charles fl. 1593–1606 Eng. Brit. M
Eng. Norden's map of Surrey, Symonson's map of Kent.

Whyman, J. 18th cent. Surv. Brit. M
Surveyed J. Prior's map of Leicestershire, 1779.

Wicheringe, Bartholdo *See under* BLAEU, G.; HONDIUS, H.; JANSSON, J.; MERCATOR, G.

Wicksteed, E. *See under* WELLS, E.

Wied, Anton 1500–58 Cart. Danzig M

Wieland fl. 1723–50 Soldier, Cart. Ger. M

Wightman, T. *See under* CUMMINGS, J. A.

Wigzeel}
Wigzel } *See under* SAYER, R.

Wijk *Same as* VAN WIJK ROELANDSZOON (q.v.)

Wild, James 19th cent. Pub., Cart. Brit. M
Tasmania or Van Diemen's Land., c. 1846.

Wild, Joseph fl. 1697–1701 Bookseller, Pub. Brit. M
Camden's *Britannia abridg'd* 1701.

Wiley and Long *See under* BRADFORD, T. G.

Wilkes, Charles 1798–1877 Explorer Amer. M
Narrative of the United States exploring expedition during the years 1838, 1839, 1840, 1841, 1842 . . . Atlas Philadelphia, C. Sherman, 1850–8. 2 vols.

Wilkes, John fl. 1810–28 Topg. Brit. M

Wilkie, G. *See under* WALKER, J.

Wilkin, R. *See under* WELLS, E.

Wilkinson, James 1757–1825 Author Amer. M
Diagrams and plans illustrative of the principal battles and military affairs treated of in Memoirs of my own times . . . Philadelphia, A. Small, 1816.

Wilkinson, Robert fl. 1785 Pub. Brit. M
Successor to John Bowles.
Atlas classica . . . Lond., 1797, 1797–1805. Engs.: B. Baker, E. Bourne, T. Conder, W. T. Davis, W. Palmer, J. Roper.
A general atlas . . . Lond., 1794, 1800–2, 1800–3, 1800–8?. Engs. incl.: E. Bourne, T. Conder, W. Harrison, I. Puke, B. Smith, according to edn.
See also under ARROWSMITH, A.; KITCHIN, T.

Wilkinson, W. C. 18th cent. Cart. Brit. M
See also under FADEN, W.

Will, Martin fl. 1790 Pub. Ger. M
Charte de l'Afrique by Jean Walch (q.v.)

Willard, Emma Hart 1787–1870 Geog. Amer. M
Willard's atlas . . . Hartford, O. D. Cooke and Co., 1826.

Willdey (or **Wildey**), **George** d. 1737 Mapseller and Toymaker Brit. M
An atlas of the world 1717.

Map of 30 miles round London 1720.
Pub. Lea's reissue of Saxton's map of Shropshire, 1690.
See also under AUSTEN, S.; SAXTON, C.

Willemsen, G. fl. 1588 Cart. Dut. M
Die caerte van de oost ende est zee.

Willemsz, Govert 16th cent. Cart. Dut. M
Sea-charts.

Willett, Mark fl. 1822 Cart. Brit. M

Willetts, Jacob 18th/19th cent. Educator Amer. M
Atlas of the world Poughkeepsie, P. Potter, 1814, 1820. Engs.: J. D. Stout, J. Lewis.
Atlas designed to illustrate Willett's Geography . . . Poughkeepsie, P. Potter and Co., 1818.

Williams, Anthony *See under* VAN KEULEN, J. I.

Williams, Calvin S. 19th cent. Pub. Amer. M
A new general atlas . . . New Haven, 1832.

Williams, Richard 18th cent. Cart. Amer. M
A plan of Boston . . . (i.e. U. S. A.) Lond., Andrew Dury, 1776.

Williams, William 18th cent. Cart. Brit. M
Map of Denbigshire, c. 1770 (eng. J. Senex).

Williams, W. *See under* MITCHELL, S. A.

Williamson, James fl. 1810 Surv. Irish M
Map of Co. Down.

Williamson, W. A. 19th cent. Pub. Brit. M

Willis, John 18th cent. Cart. Brit. M
Map of 10 miles round Newbury, 1768.

Willius, P. fl. 1686 Cart. Ger. M

Wilmers, Wilhelm b. 1815 Theologian Ger. G

Wilse, I. N. *See under* VON REILLY, F. J. J.

Wilson, James }
Wilson & Sons } 1763–1855 Carts. Amer.
Wilson's & Co. } G (C and T)
The first Amer. globe-maker, W. was a farmer of Bradford, Vermont. His headquarters as a cart. were at Albany.

Wilson, John *See under* CAREY, H. C.;
HEATHER, W.

Wilson, the Rev. John Marius fl. 1854-7
Topg. Brit. M

Wilson, William 18th cent. Pub. Brit.
M

Wiltsch, Johann Elieser Theodor 19th
cent. Historian Ger. M
Kirchenhistorischer atlas . . . Gotha, J.
Perthes, 1843.

Winckelmann, Ed. *See under* VÖLTER, D.

Wing, John 18th cent. Brit. M
Map of north level of fens, 1749 (eng. N. Hill).

Wingfield, J. *See under* SELLER, J.

Winstanley, H. *See under* MORTIER, D.

Winter, A. *Same as* DE WINTER, A. (q.v.)

Winter, D. *See under* MOLL, H.

Winterbotham, William 1763-1829
Cart. Amer. M
The American atlas . . . New York, J.
Reid, 1796. Engs. incl.: Scoles, B. Tanner,
A. Anderson, D. Martin, Roberts.

Winzelberger, David} fl. 1577-97 Cart.
Winzenberger } Ger. M

Wissenburg, Wolfgang} 1496-1575
Wyssenburger } Theologian, Cart.
Swiss M

Wit } *Same as* DE WIT (q.v.)
Witt}

Withalm, Andreas *See under* VON REILLY,
F. J. J.

Witsen, Nicolas 1641-1717 Cart. Dut.
M
See also under DANCKERTZ, J.; RENARD, L.

Witte, Wilhelmine 19th cent. Cart.
Ger. M

Woerl, J. E. 19th cent. Cart. Ger. M
Carte de la France . . . Fribourg, B. Herder,
1833?
Atlas über alle theile der erde . . . Carlsruhe
and Freiburg, 1842.

Wolfe, John fl. 1598 Printer Brit. M
Printed an English edn. of Linschóten.

Wolff, Jeremias 1663-1720 Pub., Eng.
Ger. M
Pub. Delisle's *America Septentrionalis* (Augsburg, 1735).

Wolfgang, Abraham 17th cent. Pub.
Dut. M
Atlas minor . . . Amster., 1689. Carts.
incl.: A. Barreo, W. Blaeu, F. Fer, I.
de Ram, J. B. Labanna, I. Massa, Ortelius,
de Witt, N. Visscher III.

Wolters, J. *See under* CLUVER, P.

Wood, Basil 18th cent. Cart. Brit. M
Map of Shropshire, c. 1710. Eng. B. Cole.

Wood, Lieut. J. *See under* WALKER, J. AND
C.

Wood, William c. 1580-1639 Geog:
Brit. M
New England prospect . . . Lond., Thomas
Cotes for Iohn Bellamie, 1634. Contains
a woodcut map of 'The south part of
New-England.'

Woodbridge, William Channing 1794-
1854 Geog. Amer. M
Woodbridge's larger atlas . . . Hartford, S. G.
Goodrich, 1822.
Modern atlas on a new plan . . . Hartford,
O. D. Cooke and Co., 1831; Belknap and
Hamersley, 1843, and other edns.
School atlas . . . Hartford, O. D. Cooke and
Sons, Ltd., 1821; Beach and Beckwith,
1835, and many other edns.

Woodcocke, Thomas *See under* THORNE,
R.

Woodman, James *See under* NORDEN, J.

Woodthorpe, V. *See under* LAURIE, R.

Woodville, William *See under* LAURIE, R.;
NORRIS, R.

Woolaston, Francis fl. 1811 Astron.
Brit. X

Woolsey, R. fl. 1802 Astron. Brit. X

Worcester, Joseph Emerson 1784-1865
Cart. Amer. M
An historical atlas . . . Boston, Hilliard,
Gray, Little and Wilkins, 1827, and other
edns.
Worcester's outline maps . . . Boston,
Hilliard, Gray, Little and Wilkins, 1829.

Wortels *Same as* ORTEL,

Woutneel, Hans or John fl. 1603 Eng.
Dut. M

Wren, Matthew fl. 1766 Cart. Irish M
Map of Co. Louth, 1766.

Wright, A. *See under* REID, A.

Wright, Benjamin fl. 1596–1613 Eng.
Brit. M
Eng. for G. and M. Tatton, and world map
for JOHN BLAGRAVE (q.v.) Celestial chart
in Blagrave's *Astrolabrium Uranicum
Generale,* 1596.
See also under LANGENES, B.

Wright, D. *See under* FADEN, W.

Wright, Edward *c.* 1558–1615 Cart.
Brit. M G
Made a world map for Hakluyt's *Voyages.*
It is rare and much sought after. Also a
chart of the W. coast of Europe and the
Azores in his own *Certaine errors of
navigation* . . . By E. W., Lond., Valentine
Sims, 1599.
See also under THORNTON, J.

Wright, G. 1740–83 Instrument-maker
Brit. G

Wright, Thomas fl. 1782 Globe-maker
Brit. G

Wurm, Hans *Same as* WEIGEL, H. (q.v.)

Wussin *See under* SCHRAEMBL, F. A.

Wyld, James, the Elder (1790–1836) **and
James, the Younger** 1812–87 Pubs.
Brit. M G
A general atlas . . . Edinburgh, J. Thomson
and Co., 1819.
Map of South Australia etc. Lond., *c.* 1850.
Chart of New Zealand . . . Lond., 1851.
The island of New Zealand Lond., *c.* 1851.
Cape district. Cape of Good Hope . . . Lond.,
1838.
Map of Syria . . . Lond., 1840.
*Map of the countries lying between Turkey
and Birmah* . . . Lond., 1839.
An emigrant's atlas . . . Lond., 1848.
*Maps and plans, showing the principal move-
ments . . . in which the British army were
engaged* . . . *1808 to 1814* . . . Lond., 1840.
A new general atlas . . . Lond., 1840?, 1852.
Carts, incl.: Louis Stanislas d'Arcy
de la Rochette, d'Anville, Jasper Nantiat.
See also under BANSEMER, J. M.; DE LA
ROCHETTE; THOMSON, J. AND CO.

Wynkyn de Worde d. 1534? Printer
Alsatian M
Worked in England.

*Almanack with charts of the coasting parts of
England* 1520.

Wyss, Johann Rudolf 1781–1830 Cart.
Swiss M
Hand atlas für reisende in das Berner oberland . . .
Bern, J. J. Bürgdorfer, 1816.

Wyssenburger *Same as* WISSENBERG.

Wytfleet, Cornelius} fl. 1597 Cart. Brit.
Wytfliet, Corneille} M
Descriptionis Ptolemaicae augumentum
Lovanii, Johannis Bogardi, 1597; Gerard
Riuj, 1598; Draci Franciscum Fabri,
1603. The first atlas to deal exclusively
with America.
*Historie universelle des Indes Orientales et
Ocidentales* . . . (with Anthoine Magin)
Douay, F. Fabri, 1605, 1611.

Yamasita Sigemasa fl. 1742–9 Cart.
Japanese M

Yates, George 18th cent. Surv. Brit.
M
Map of Glamorgan, 1799.

Yates, T. *See under* SINGER, J.

Yates, William and Billings, T. fl. 1786–
93 Survs., Carts. Brit. M
Compiled maps of Lincolnshire and (Yates
only), Staffordshire, Warwickshire (as
W. Yates and Sons), Lancashire, Liverpool
district (with G. Perry).

Yeager, J. *See under* CAREY, H. C.;
MARSHALL, J.

Yeakell, Thomas, and Gardner, W.
fl. 1778–83 Cart. Brit. M
Maps of Sussex, 1778–83. Completed by
T. Gream, 1795.

Ygel, Wahrmund d. 1611 Cart.
Austrian M

Young, J. H. *See under* MITCHELL, S. A.

Young and Delleker *See under* MACPHERSON,
D., AND A.

Ys, Teunis *See under* VAN KEULEN,
JOANNES, I

Zainer, Gunther *See under* ISIDORE OF
SEVILLE

Zaleski Falkenhagen, Piotr *See under*
Bansemer, J. M.

250

Zallieri, Bolognino} fl. 1566–70 Pub.,
Zaltierius } Ital. M
See also under BALLINO, G.; LAFRÉRY, A.

Zannoni, Rizzi *See under* RIZZI-ZANNONI,
G. A.

Zatta, Antonio 1757–97 Cart., Pub. Ital.
M
Atlante novissimo 4 vols. Venice, 1779–85.
Nuovo atlante (with Giacomo Zatta)
Venice, 1799, 1800 and other edns.

Zatta, Giacomo *See under* ZATTA, ANTONIO

Zeccus *Same as* SECCO (q.v.)

Zecsnagel *Same as* SECSNAGEL (q.v.)

Zeiller, Martin} 1589–1661 Geog. Ger.
Zeiler } M
*Topographia electorate Brandenburgici et ducatus
Pomeraniae* . . . Frankfurt-am-Main, M.
Merian, 1652?.
Topographia Helvetiae . . . Frankfurt-am-
Main, Merian, 1654.
Topographia Italiae . . . Frankfurt-am-Main,
Merian, 1688.

Zell, Christoph d. 1590 Printer Dut. M

Zell, Heinrich d. 1564 Cart. Ger. M

Zeno, Nicolo}
Zeni } 16th cent. Cart. Ital. M
Cenus }
'Carta de navegar . . .' Venice, 1558 (wood-
cut). In *De I commentarii del viaggio in
Persia di M. Calterino Zeno* . . . Venice,
Francesco Marcolini, 1558.

Zenoi, Dominico} fl. 1552–69 Eng. Ital.
Cenoi } M

Zepherinus, J. *Same as* SEVERI, C. (q.v.)

Zeune, Johann August 1778–1853 Geog.
Ger. G

Zibermayer, Mathias fl. 1822 Cart. Ger.
G

Ziegler, Jacobus 1470–1549 Cart. Ger.
M
Terrae sanctae . . . (8 woodcut maps)
Strasbourg, Wendelin Rihel, 1536.

Ziletti, Giordano fl. 1564–74 Cart. Ital.
M
Atlas with 64/65 copper-plate maps.

Zimmermann, Johann Jakob 1644–93
Mathematician, Cart. Ger. G

Zipter, J. *See under* MEYER, J.

Zoll 16th cent. Cart. Ger. M
Map of Europe.

Zollmann, Fridericus fl. 1717 Cart.
Dut. M
See also under HOMANN, J. B.; HOMANN'S
HEIRS.

Zsam Bocky, J. *Same as* SAMBUCUS, J. (q.v.)

Zubov, Alexej 1682–1743? Eng. Russ.
M

Zumbach von Koesfeld, Lothar 1661–
1727 Mathematician, Astron. Ger. G

Zündt, Mathes} 16th cent. Eng., Cart.
Zyndt } Ger. M
Town plans.
Maps of Corsica, Cyprus, Hungary, Malta.

Zürner, Adam Friedrich 1680–1742 Cart.
Ger. M
Americae tam septentrionalis quam meridionalis
Amster., Peter Schenk, 1709.
See also under HOMANN, J. B.; HOMANN'S
HEIRS; OTTENS, R.; SCHENK, P.; WEIGEL,
C.

Zwick, Johann fl. 1540 Cart. Ger. M

Zyndt, M. *Same as* ZÜNDT, M. (q.v.)

Nominal Index to the Chapters and Appendix